Cyclical Catastrophe

PLANETARY WARS IN THE SOLAR AND SATURN SYSTEM

Georgia Balbin

TET Publishing

TET Publishing
4727 North Broadway
Boulder, CO 80304
www.tetpub.com

Printed in the United States of America
Book design by Sara Patton

ISBN 0-9763452-0-x

Contents

Acknowledgments

Fred Templeton, who taught at the University of Colorado at Boulder for many years, was the first to read the material on Olmec Crossed Bands, making the valuable suggestion that it be clear on what the X means to us as well.

Michael Coe felt that I had a good grasp of the Olmec culture, although he wholeheartedly rejected Velikovsky.

Vine Deloria was the first to read *Cyclical Catastrophe* in its entirety and we discussed it over several pleasant lunches. He was very encouraging.

My husband, Paul Balbin, provided the kind of support that enables serious research to go forward. That he has never been particularly interested in the work itself probably says more about his character as a husband than otherwise. Chauffeuring me to various ruins in Mexico, particularly down on the Gulf Coast, was not his idea of a good time. The book wouldn't have been the same without him.

Every one of my children has grilled me about various aspects of *Cyclical Catastrophe,* most often in an effort to grasp precessional motion, both regular and catastrophic: Amy Willey, Lynn Thiel, Robert Cull, Mary Ellen Hopkins, and Jane Barone. Criticism was always helpful, tending to clear things up on both sides. And hopefully they were satisfied.

Intelligent questions and real excitement came from my older grandchildren: Rob, Allan, and Eric Willey. Matt Stewart read and enjoyed several chapters while he was still in high school, very encouraging at the time.

My son-in-law, Dan Willey, was always up for discussion, bringing the Zeeman effect into focus at one point. I never did bring it into the book, but it probably belongs there.

My friend and neighbor, Lars Wahlin, was always trying to tell me something about the behavior of a bicycle wheel when we talked about precessional motion. One day he brought down the bicycle wheel and when I held the axle of the spinning wheel in my hands and tried to tilt it, finally grasped what he was trying to tell me, the gyroscopic force pair. He is a good and patient friend.

Illustrations go far beyond mere artwork since the originals were created by the ancients to carry information to those who came after, to us. The thanks we owe them can hardly be expressed. My dear friend and best gardening buddy, Bruce Migdal, had agreed to do the illustrations, but died before he could begin work. The loss was great, but in the process of doing most of the artwork myself, more information became clear which required some rewriting. You don't really see something until you try to draw it.

Astronomical maps were generated on a Macintosh iBook with the Voyager III program of Carina Software who kindly allowed reproduction of astronomical maps. All subsequent drawing on the maps is, of course, my own. Carina Software is located at P.O. Box 2108, Danville, CA 94526, www.carinasoft.com.

The map projections are usually centered on points in the sky other than the conventional polar one. A projection centered on the pole of the ecliptic allows us to see the tilts of Earth in proper perspective, as straight lines.

And many thanks to all the supportive friends who very kindly inquired "how the book was going" over long years.

Author's Preface

It has always seemed, to me at least, that life is a conspiracy to teach you more than you want to know. By the time this book began to come about, I had learned simply to yield to it. The necessary reading of some of the world's great inspirational literature was a great pleasure and support to the work. I was often conscious of working through the gunas described in the Bhagavad Gita, knowing that whatever was happening now—illumination, ignorance, or activity—it would change.

Although encouraged to write by many of my teachers during formal education, I always postponed this pleasurable activity to a later period of life when, presumably, there would be something to say. After reaching middle age, with plenty of living under my belt in what now appears to be a cyclical pattern of catastrophe, I sat down to write a novel. By then it was a lifelong habit to jump in and figure something out rather than initiate study or practice. The wisdom of the ages was always available at the library.

It was at the birth of my second grandchild that my son-in-law, Dan Willey, gave me a copy of *Worlds in Collision*, introducing the provoking idea that some of the planets in our solar system had interacted violently within historical time. Science fiction paled beside the carefully researched, well-referenced *Worlds in Collision.* I read it through several times and began catching up with the events of the 20-odd years since it had been published.

The excitement was over. Scientific orthodoxy had managed to make the name of Immanuel Velikovsky unrespectable, but his ideas still seemed sound to me, his arguments sensible. That a cometary Venus had interacted with the Earth sometime in the 2nd millennium B.C. was indicated by the historical and mythological record. That the orbit of Mars became upset by Venus so that it, too, threatened Earth in the 8th and 7th centuries B.C. was sufficiently well supported to encourage further research. In the years since the book had been published, many of Velikovsky's predictions, the natural fallout from his ideas, had come to pass. Surely this man should be taken more seriously.

The mythological aspects of the narrative caught at me: that Venus had twice bested Mars in battle (as in the 5th and 21st books of *The Iliad*), that Mars had a love affair with Aphrodite the Moon, that Mars had fathered Romulus and Remus, the twin kings of early Rome. The possibility of fanciful stories reflecting reality had a strange allure. As they say in dream work, the idea resonated with my bones. That Velikovsky expected dialog and received contumely prompted anger on my part and a strong desire to help.

At the time it hardly seemed possible, but without thinking about it too hard, I was soon embarked on my own study of historical catastrophism. Led along at first by Velikovsky's references, I soon branched out on my own. Any knowledge I could bring to bear was more in the nature of the interests of a lifetime: natural history in general with a particular fondness for picking out the constellations of stars in the night sky, wide reading in mythology since childhood, and decades of serious dream work. This last was in part a preparation for my projected writing career. Robert Louis Stevenson says somewhere that a group of brownies in his dreams assisted him at critical points in his writing, and I wasn't about to pass up any help.

The novel proceeded in fits and starts as I learned how to let it happen. A reading of Robert Graves' *White Goddess* convinced me to invoke my muse and that seemed to help. It was terribly exciting when the characters actually began talking and interacting with each other. Soon I couldn't be sure what would come out of their mouths. It must have been about that time that I noticed historical catastrophism creeping into the work. By the time Part III was underway, a college student, Ed Winson, was in hot pursuit of interplanetary contacts that had been witnessed by the ancient Olmecs of the Gulf Coast of Mexico. This was probably because the extensive literature on the Maya was overwhelming to me. The Olmecs were less forbidding because less known. Taking a page from Velikovsky, I imagined them as real people trying to communicate.

Necessary research began to slow writing on the novel, but I just got on with it, beginning with standard works in the field. The Olmecs have no writing as we know it and speak only through their symbols and massive statuary, so I began looking at these, wondering about the handful of symbols they were beginning to use. Where did symbols come from, and how did they turn into writing? Many early people claimed that a god had taught writing to them; some were clear on it having come from the sky. The crossed band symbol was the first to yield its secret. Other paths were opening. I began to read psychology and the history of religion. Most important, I began to dream about my work.

Needless to say, my hero Ed found what he was looking for while I had acquired a body of evidence indicating catastrophic events during Olmec times that had burst the frame of the novel and now required its own setting. Not only did the material fit Velikovsky's scenario for the Earth-Mars Wars of the 8th and 7th centuries B.C., but it located the end of the Jaguar Age of the ancient Mexicans. Ed and I (he will always seem like a valuable colleague) also located a stela at Yaxchilan that indicated yet another previous age.

But what to do with the information? The ghosts of the past had already been laid, principally by orthodox astronomy. In 1950, for whatever reason, the collective psyche was not ready to hear the testimony of the ancients. Not perceiving myself as a scholar, I hesitated at the thought of writing a serious paper, yet felt responsible for the material. After wrestling with the problem for a while, I finally sat down and wrote "Olmec Crossed Bands: An Analog of Time." My angst was relieved and I went on to complete my novel, *Family Tree.*

Then the bears began to appear in both waking and dreaming reality. After receiving help on the Olmec material, I had a good idea what was coming down. There would be work on the tilting of the terrestrial pole as discussed by Velikovsky, involving the movement of the pole from one bear to another, Ursa Major to Ursa Minor in the 8th and 7th centuries B.C. It had been clearly indicated in "Crossed Bands," but now I would examine the phenomenon in the Old World. Although it has never been taken seriously by the establishment, the tilt is potential dynamite, one of the most powerful elements in Velikovsky's reconstruction of history. Once show that the Earth had tilted in the 8th century B.C. and the game was up. I went to work cheerfully, though advising my muse that we should be getting back to more creative work soon. The research and writing flowed as smooth as paint and the days slid by like beads on a string.

The bears led me back to Greek and Roman mythology, but now there would be another level of meaning. Myth, like dream, is endlessly fecund. There is always another level to be explored. Obscure books fell into my hands and I began spending more time in the art and architecture section of the library, making copies of drawings and photographs that seemed significant. In *Mankind in Amnesia,* Velikovsky remarks that the mystery religions are based on knowledge of historical catastrophe, so these subjects were added to the list as well.

The plan was to sort out events of the 8th and 7th centuries B.C., but it immediately became apparent that much of the material had an existence outside our linear time and referred to events in the more distant past as well as in the period under discussion. I opened files on other ages, other periods of destruction, and these too began to fill up. Models of other ages came into focus, each one attracting information like a magnet. I was slipping, sliding into the past, seeing different skies. The concepts were strange and more than a little disorienting.

The overall pattern of previous world ages gradually acquired cohesion. The mighty hero Hercules came to my aid, tracing out a new celestial equator. It was like a frog or snake dropping into my lap, so startling that it was weeks before I thought to check the angle with the ecliptic, to help locate it in time. At some point Hesiod's Age of Heroes (the Jaguar Age) snapped in place around me like a solution crystallizing, with "a new Heaven and a new Earth." The pole points to the Great Bear, Ursa Major. The celestial equator, one previous to that of Hercules, forms a higher angle with the ecliptic; the sun goes farther north and south than we could have imagined. Bull and Scorpion are firmly stationed at the equinoxes, but they will be there for only 700-odd years. How lightly we have used the term New Age, and how much more it comprehends!

This time I did not resist writing a paper, realizing the importance of organizing the material. "Great Bear at the Pole" composed itself quickly, followed by an analysis of a persistent folk tale, Bearson, which supported the movement of the pole as well as hinting that each world age might have its own psychological set. There was no letup in the flow of material. As it happened, Hercules and his patron Athena had a great deal to say about the two visitations of the planet Venus in the 15th century B.C. More Greek mythology, Mithraic Mysteries, and the Bible led back to the Golden Age and the Saturn Configuration, all against a frame of space and time that Santillana and Dechend make almost palpable in *Hamlet's Mill.*

Velikovsky breached the walls of uniformitarianism in two places: at the terrestrial encounters with Mars in the 8th and 7th centuries B.C., and at the catastrophes caused by Venus in the 15th century B.C., as well as opening a crack indicating interaction with Jupiter earlier. It was left to David Talbott to illuminate the Golden Age of Kronos, the First Time of the Egyptians, in his groundbreaking *Saturn Myth.* Like Velikovsky, he took the ancients seriously, deducing that they were living under an alien sky ruled by the planet Saturn.

Talbott had already penetrated much of the confusion surrounding our oldest polar god in his development of the Saturn Configuration, an order of the heavens which preceded the present one. With this firm anchor to tie up to, there seemed little point in projecting another paper. There was plenty of material for a book which would also lend space for other subjects that had surfaced: the fate of the goddess, the curtain of forgetfulness that rings down between ages, and that most engrossing mythological creature, the werewolf. After working with epics and sagas, it also seemed possible to line up some history that has been adrift in the ages: the Mahabharata War and the end of the Shang dynasty.

It was an easy project to get behind, vindicating Velikovsky and the ancient astronomers who have been telling us about past ages and getting short shrift. In retrospect, the quote with which Velikovsky opened *Worlds in Collision* seems like an invitation: "Which part of the work belongs to us?" Mine had begun.

No new terminology is necessary. What we need is already in place, including the five world ages named by the Greek mythographer Hesiod who wrote in the century following the Earth-Mars Wars. Gold, Silver and Bronze Age, Age of Heroes, and Iron Age are embedded in the culture, familiar to at least some part of the population. To them need only be added the names of the constellations at the pole during those ages and the solstice and equinox positions of the constellations of the zodiac.

Work in depth is left to the future, since it is important at present to have an integral, if skeletal, work in place treating the entire series of world ages. Only thus can we see the coherent pattern of Earth motions indicated by myth and legend, indicating a new kind of order in the universe. If it is time for the paradigm to shift, the ideas will enable it to happen rather than the weight of the book or number of references. *Cyclical Catastrophe* is deliberately open, leaving plenty of room for anyone to stick in an oar and begin to pull. We will find the ship of state, of spaceship Earth, and the ship of the heavens to be the most durable of metaphors.

I have recounted major dreams from which I drew inspiration. If Einstein, Neils Bohr, and Descartes (examples are plentiful in the literature) used their dreams as jumping-off places in their thought, surely these elegant messages should be used to unlock myth, art, and legend, which appear to be spun from the same filaments. Life is not so easy that we can afford to overlook a source of help. The thinkers mentioned above did not make their mark by being close-minded. As William James has suggested, we should not close the books on reality prematurely.

The best introduction to this work is a fresh reading of *Worlds in Collision.* Velikovsky was often accused of making intuitive leaps, but so many of the holes have been filled in by accepted work in many fields that the book reads more easily today. The textbooks have been rewritten, and Jupiter is no longer a frozen waste but something closer to the failed star that Velikovsky said it was. Venus is no longer a watery sister planet, its great heat now well understood but difficult to account for, its atmosphere massive enough to crush the first space probe. The surface of the planet Mars clearly shows it to be a victim of catastrophe.

Worlds in Collision has enjoyed a long and successful publishing history despite its deliberately never having been revised. There are those who flatter themselves that science has caught up with Velikovsky's far-reaching ideas as one after another of his predictions have been validated. This assumption is unwise. Although this seminal thinker's foot has occasionally slipped, even as yours or mine, he is still ahead of most of us.

Velikovsky drew on myth, legend, and historical sources to create his historical reconstruction. Having begun with the preliterate culture of the Olmecs, I learned to make use of art and symbol as well, while ritual, too, speaks of alien skies and historical catastrophe. Later, architecture would reveal its secrets. Modern man has come to think of cultures without writing as primitive indeed. But when myth, legend, ritual and symbol, art and architecture are considered together, we will find that they constitute an ancient media which speaks with one voice on cyclical catastrophe.

In his preface to *Worlds in Collision,* Velikovsky issued a caveat: "If these two men of science are sacrosanct [Darwin and Newton], then this book is a heresy." I would rest a warning to my readers on other grounds. If you cannot think metaphorically, if you cannot perceive layers of meaning in life, if you are content with an endless array of ad hoc explanations for conditions in the natural world, then don't read this book. It will only upset you. It is only for those who are ready to see pattern, ready to move into the next paradigm, who can grasp the reality of the cosmos perceived by the ancients and by Velikovsky and a few of his heirs.

On the other hand, if you sometimes look at the night sky and wonder how the constellations got their names, if you enjoyed fairy tales as a child or presently read myth and poetry, you might enjoy this trip into the cyclical past to see what our ancestors survived to give us life and culture. We will travel over a strong but fragile-looking bridge made of unlikely materials. And where we are going, no government or corporate entity stands by to help. There is only the changeful matrix of Mother Earth and Father Sky, and the God we have chosen to believe in. When we finally connect with our courageous and resourceful ancestors, we will have a better idea of who we are and what we are capable of.

In the following chapters, I will strengthen the mythological model begun by Velikovsky and extend it into the past to the Saturn Configuration of David Talbott. The theory of what causes the Precession of the Equinoxes grows directly out of the ancients' perception of two different orders of the universe. We can probably expect it to cohere with the theories of the electric plasma universe now being developed. This can only play havoc with the current views of orthodox science, but be that as it may, it is time to put a few more bricks on the wall.

An important element in chaos theory is the discovery of so-called "attractor fields." These reveal that behind what appear to be inexplicable, chaotic, and random events is actually a hidden energy field whose pattern influences the appearance of "stochastic," or random, data. These patterns are discernible behind what appear to be senseless or incomprehensible occurrences in nature and include explanations of global and environmental changes and weather patterns as well as the actual beating of the human heart.

– David R. Hawkins, *The Eye of the I*

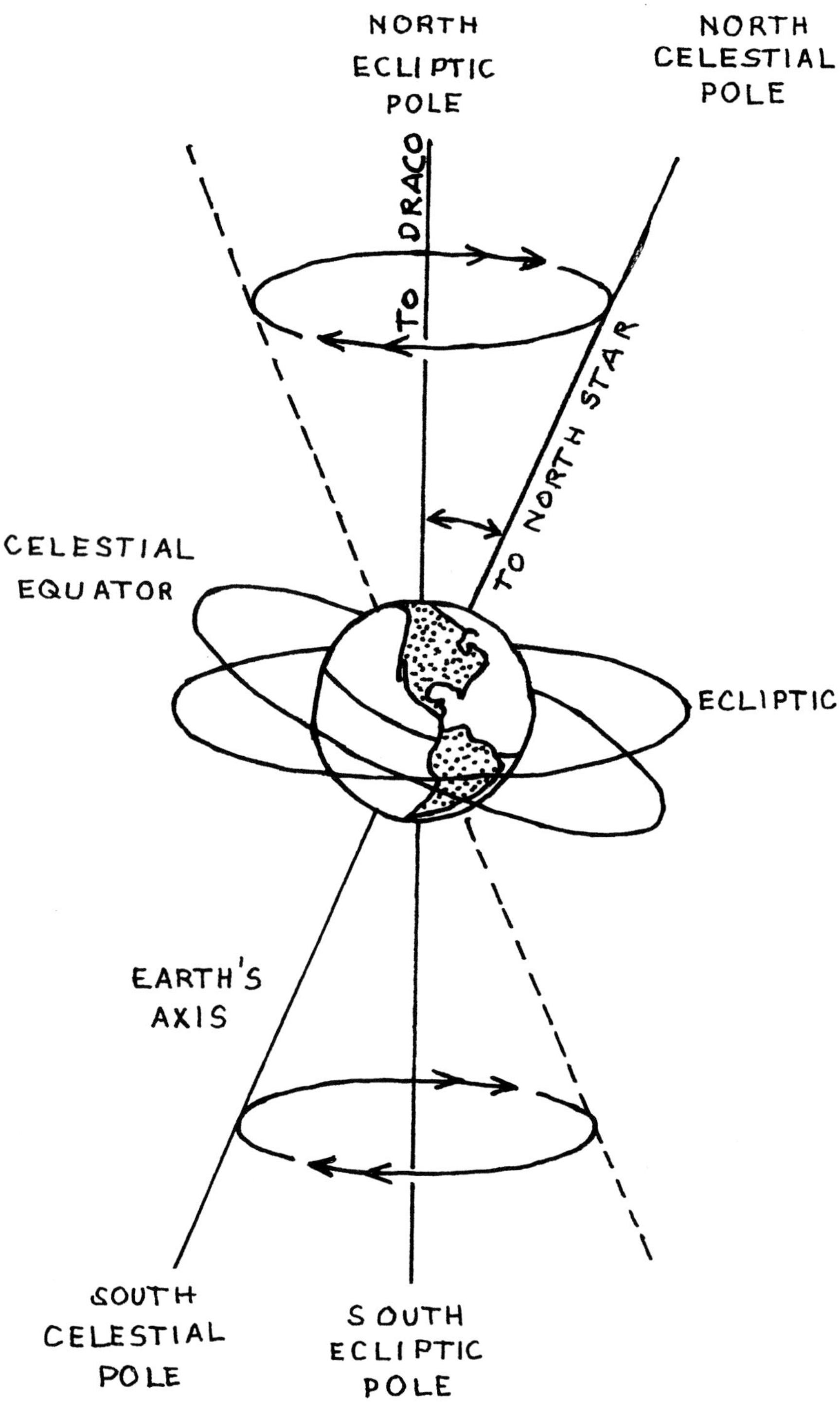

Earth's present precessional motion. Each top-like spin would theoretically take about 26,000 years.

1. Differing Views of the Past

. . . you Greeks are always children . . . young in your minds, which hold no store of old belief based on long tradition, no knowledge hoary with age. The reason is this. There have been, and will be hereafter, many and divers destructions of mankind, the greatest by fire and water, though other lesser ones are due to countless other causes . . . with you and other peoples again and again life has only lately been enriched with letters and all the necessaries of civilization when once more, after the usual period of years, the torrents from heaven sweep down like a pestilence leaving only the rude and unlettered among you. And so you start again like children, knowing nothing of what existed in ancient times here or in your own country.

– Plato, *Timaeus*

On 4-Ix, sky and Earth were tilted.

– Napuctun, "Book of Chilam Balam of Chumayel"

When we look over our shoulder as children of Western Civilization, we see a view of the past generally agreed upon, a rise in the development of culture, the origins of which reach into gray antiquity, eventually into the caves of a more primitive man. Anomalies are conveniently swept under the rug. A few thinkers have attempted to trace the rise and fall of different civilizations, doubtless out of some concern for our own. We are occasionally reminded that those who refuse to study history will be forced to repeat it, though it has also been noticed that what we call history is actually created in the present and written and rewritten under some compulsion. We may perceive a remarkable record of achievement or a stumbling through events that does not bear repeating, but there is something comforting about the agreed-upon underlying structure of our history.

It is our history, and we are here, more or less on top of it. There are rumblings beneath us that not even Lloyds of London can overlook,[1] but our tight, figure-to-ground consciousness keeps us focused on daily events, and there are myriad ways to distract ourselves in leisure time. Despite what is going on in our nation's capital or in the street outside our home, the narrative of the soap opera will likely continue. We tend to talk a great deal or listen to others talk. Whether we read extensively or not, we are adrift on a sea of paper and at least peripherally connected to an information highway. The words spin reality about us, giving us a sense of our existence. If we are unsure of where we are going, at least we think we know where we have been.

Just behind the recent decades of glittering technology, we can look back on the steady development of an industrial society, the frontier that we wrested from those who were already here, the rise of the civilization of the Renaissance in Europe. Continuing into the past, there were the Middle Ages when the beautiful cathedrals to the Virgin were built,

[1] *The Economist* il V327, May 29, 1993, p. 77. Also Nicolas Bray, *Wall Street Journal,* Feb. 13, 1996.

the Dark Ages when the torch of knowledge was held by the Muslim world, then the orderly Romans with their carefully thought out ideas about government, the Greeks who laid the foundations of our learning.

The vision blurs somewhat on looking further into the past. Egyptians and Babylonians are shadowy, mysterious figures whose languages, written in alien characters, seem strange and impenetrable. Bronze, copper, and stone age appear to be more mixed than we might wish. Go back far enough and there are only primitive nomads and cave men with their pitiful tools. How lucky we are to live in the present enlightened age with millennia of slow development behind us.

Our schools teach ancient history, and textbooks speak confidently of the Code of Hammurabi and how Champollion translated the Rosetta Stone so that we can read ancient texts. Unfortunately, when we go to the library to examine the science of old civilizations, we run into difficulty. The ancients were apparently careless about recordkeeping and observations of physical phenomena. Their calendars are a scandal, only 360 days. Although they watched the sky relentlessly, particularly the planet Venus, they were unable to come up with correct figures to describe her cycles. And this despite the fact that the planet figured largely in most ancient religions. What altered perception of our morning and evening star could result in the use of a name like "Queen of Heaven?"[2] Sumerians and Babylonians addressed prayers:

> To the pure flame that fills the heavens,
> To the light of Heaven, Ishtar, who shines
> like the sun.[3]

The 8th and 7th centuries B.C., seemed to be a particularly difficult time for calendar makers. Although any school child can count the days from one full Moon to another, and any reader of this book could track equinox and solstice by setting a post in the ground and observing its shadow,[4] the records of the ancients contain what are considered to be outrageous errors. In the days of Hezekiah, the Jewish New Year, an equinoctial feast, was once moved backward nine days from the 10th to the 1st day of the first month. Curiously, the same time period is reflected in Babylonian tablets where the equinox is recorded on one tablet on the 15th, on another on the 6th of Nisan.[5] In other Babylonian records the equinox appears to move by a full month.[6] The Jews are on record with three different calendars, one giving only numbers rather than names to the months.[7] The early Romans apparently couldn't count the days in the months or the months in the year.[8]

---

[2] Jeremiah 7:18, 44:17-25.

[3] S. Langdon, "A Hymn to Ishtar as the Planet Venus and to Idin-Dagan as Tammuz," *Journal of the Royal Asiatic Society*, 1926, p. 8.

[4] "Nothing is easier to determine than a solstice or equinox," remarks Norman Lockyer in *Dawn of Astronomy* (Boston, M.I.T. Press, 1964) p. 62. Native Americans used the natural features of the land as markers. See "Sun Dagger," an excellent video of the calendric marker in Chaco Canyon. Also "Sun Dagger at Fajada Butte," *National Geographic*, Nov. 1982, Vol. 162, No. 5. Also A. L. Preston, "The Discovery of 18 Calendric Petroglyph Sites in Arizona," *The Bulletin of the Center for Archaeoastronomy* (College Park, Maryland, 1982) Vol. V, No. 4, Oct.–Dec.

[5] Immanuel Velikovsky, *Worlds in Collision*, pp. 348–349.

[6] B. L. Van der Waerden, "Babylonian Astronomy III, The Earliest Astronomical Compilations," *Journal of Near Eastern Studies*, Vol. X, 1951, p. 20.

[7] Julian Morgenstern, "The Three Calendars of Ancient Israel" (Hebrew Union College Annual, 1924) Vol. 1, p. 13.

[8] Plutarch, Numa, XVIII.3.

You would almost have to agree that these ancestors of ours weren't quite as bright as we are. There is a word for it in German: urdummheit, old dumbness. Thus, astronomical observations made before the 7th century B.C. are not trusted by orthodoxy. But if one goes back further in time, before the 8th and 7th centuries B.C., there is substantial agreement among the calendars of many cultures. Babylonians, Egyptians, Assyrians, the Maya, Chinese, and Peruvians all seemed content with a year of 360 days. Both Babylonians and Egyptians divided their zodiac into 36 decans of 10 days each.[9] When the "missing" days were later added to the various calendars, they were often considered to be unlucky.

Because we do not take the ancients at their word, their civilizations exist under a cloud. Despite rigorous work by good men, and all the aid that modern science can bring to bear, the history of Egypt has resisted penetration. Perhaps under the spell of the exotic past, early workers in the field were led to push the first dynasties further into the past than they belonged, errors of millennia rather than centuries. The dates have been in a kind of free fall ever since, always being revised downward. Manetho's list of dynasties is often used for sequence, but it agrees with little else. Sothic dating, by the heliacal (at the same time as the sun) rise of Sirius, is found in many texts.

There is wide disagreement among scholars. Errors made early on have not been dealt with, and our vision of Egyptian history has grown by default. Sir Alan Gardiner, in *Egypt of the Pharaohs*, is blunt: "It must never be forgotten that we are dealing with a civilization thousands of years old and one of which only tiny fragments have survived. What is proudly advertised as Egyptian history is merely a collection of rags and tatters."[10]

But Egyptian history is the measure of Middle Eastern and Mediterranean civilization. The histories of Babylon and Sumer, Assyria and Greece are cut to its Procrustean bed, raising problems of dating and sequence that cannot be resolved in the accepted chronology. The Dark Ages of Greece, culminating in the appearance of Homer, are still unaccounted for.

Oddly, as we delve into the remote past, the caricature of the cave man gives way to the image of a more thoughtful, cooperative, creative early man. Agricultural problems were solved and commerce carried on. A grasp of numbers and notation which are thought to have existed over 8,000 years ago has recently been uncovered by Marshack.[11] Art of a high order was created. The history of mining is particularly instructive. Though salt, ocher, and flints were mined long before metals, Rickard notes that the mines of Assyria, Egypt, Greece, and Rome, created for mining metal, show little improvement over early flint mines.[12] Intelligent activity by homo sapiens continues to be pushed farther and farther into the past, leaving us with an impossibly long time before an accurate calendar was devised. Eventually ancient man accomplished remarkable feats of geodesy and architecture, measured the diameter of the Earth, and built pyramids and other marvelous structures, all apparently without determining that there were 365¼ days in the year.

But Western Civilization is not the only one with a comprehensive view of the past. Other civilizations, some ancient, some still extant, with different world views, different styles of

[9] Velikovsky, "The Year of 360 Days," *Worlds in Collision*, pp. 326–355.

[10] Sir Alan Gardiner, *Egypt of the Pharaohs* (London, Oxford University Press, 1961) p. 25.

[11] Alexander Marshack, *The Roots of Civilization* (New York, McGraw-Hill, 1972) p. 23.

[12] T. A. Rickard, *Man and Metals* (New York and London, McGraw-Hill, 1932) Vol. I, p. 83.

consciousness, look back on a different past. Rather than a gradual accretion of knowledge and culture, they see a succession of world ages separated by violent cataclysm, the present civilization comprehending only the most recent age. In some of these worldwide destructions, human life was all but eliminated. Afterward civilization would begin again, sometimes rising to great spiritual and technological heights before subsequent catastrophe.

The ancient Mexicans record five (some sources say seven) different suns that were ages. Five are represented on the Aztec calendar stone: Atl, Quiauitl, Ehecatl, Oceotl, and Ollin. Accounts differ as to the order of these suns, but the names are understood to describe the destructions in which they were, or will be, brought to an end: water, rain of fire, wind, jaguar, and earthquake. The present age is that of Ollin: earthquake, movement or rubber, the root of the name of the Olmecs, the rubber people of the Gulf Coast who were the first to play the ball game, tlachtli.

The culture of India calls their past ages "yugas," of which there were four: Krita, representing the Golden Age; then Treta, Dvapari, and Kali. A yuga is preceded and followed by a dawn and a twilight that connect the ages together.[13] The word yuga comes from the idiom of dicing, and the last cast, Kali, is considered a bad one indeed. It puts one in mind of the five ages of Hesiod, gradually degenerating: golden, silver, bronze, Age of Heroes, and the present Iron Age.

The Maya are well known to have a long cultural memory. They record repeated destructions in the *Popol Vuh*, where the gods try making human beings with various materials and subsequently destroy them, finally achieving success making men of corn.[14] Hopi legends recede into a past which may antedate the cycles to be covered in these pages. Following Vedic culture, both Buddhism and Jainism employ cyclical concepts of time, an eternal repetition of a fundamental rhythm of creation and destruction.[15]

From the above it can be seen that some men perceive a different reality than we do. Deeper treatment of their art, ritual, and religion reveals a remarkably connected view. The past is something they do not leave behind as we are accustomed to, but access directly in seasonal and synodic commemoration of events that have great power for them. The Hopi priests turn the sun back in ritual at the Winter Solstice because they have an awareness that it once traveled further south. Although the impulse to celebrate an important event is similar to that in Western Civilization, it is on a different level than, say, the Fourth of July or Guy Fawkes Night.

Western Civilization does have similar roots, but they have been obscured by the Judeo-Christian tradition that founds our culture. The English festal system is a ghost of older, more powerful connections to Earth and Sky. And certain aspects of older systems shine through our culture. It has been noticed that the chief appeal of the Catholic religion is not the masculine Trinity, but the same mother and son configuration that is reflected in many old Mediterranean and Middle Eastern religions.[16] And while today's polar god may drive a sleigh and reindeer rather than horse and chariot, he commands the Winter Solstice as Saturn did in Roman times.

[13] Mircea Eliade, *The Myth of the Eternal Return*, p. 113.

[14] Adrian Recinos, trans., *Popol Vuh*, Part I.

[15] Eliade, *Eternal Return*, p. 115.

[16] Robert Graves, *White Goddess* (New York, Farrar, Straus and Giroux, 1966) p. 61.

The two views of the past, that of Western Civilization and that of more Earth-connected cultures, do not square up to each other. The differing world views can hardly dialog, the former having a profane view of the universe as raw material to be manipulated, the latter seeing the world as sacred and alive. They would be as difficult to integrate as science and art, as hard to fit together as, say, the Bible into our version of history. For the Bible also contains stories of terrible destruction: the Fall, the Flood, the overturning of the cities of the plain in the time of Abraham, the events surrounding the Exodus, and the time of Isaiah and Jeremiah.

It is a problem that has troubled many thoughtful Christians and Jews, that the events of the Exodus make no appearance in Egyptian history. It was during speculation on this apparent paradox that Immanuel Velikovsky came upon the Papyrus Ipuwer which appeared to be written during a period of great destruction in Egypt. He connected it with the Exodus of the Jews from Egypt, the plagues, the parting of the sea of passage, and eventually rewrote the history of the period.[17] While doing so, he addressed many problems that historians have wrestled with for decades, notably the Dark Ages of Greece, which have as little existence in fact as they do in archaeological digs.

Most remarkably, Velikovsky's researches led him into a rewriting of the history of the cosmos as well. In studying myth, legend, and records of the past, he began to realize that the present order of the solar system does not go further back in time than the 7th century B.C., that before that time Earth and other planets were on different orbits, that the pole of the Earth had been tilted and great destruction visited on the planet on a number of occasions.

Velikovsky identified a cometary Venus as the agent which twice approached the Earth in the 15th century B.C., causing worldwide cataclysm: a rain of fire, tidal waves, earthquakes, and vulcanism of epic proportions. Once disturbed, the solar system fell awry again in the 8th century B.C., when interactions between Mars and Venus resulted in Mars threatening the Earth in another series of close approaches. He told this amazing story in *Worlds in Collision,* which became a bestseller in 1950 even though its pages bristled with notes and references. It was one of the most controversial books in publishing history.

Most striking in the relation of these planetary wars within our own solar system is the weight given to the testimony of ancient witnesses. Velikovsky credits these almost forgotten men with sense and both the desire and ability to record what they saw. It is the Christian ethic extended across the centuries. These men were like ourselves: built homes, raised their families, engaged in commerce, and complained about taxes. Unlike us, however, they watched the skies—and with good reason. It is only recently that Western Civilization has made some progress in integrating groups into our culture who are different, who have been thought of as "other," notably women and blacks. It is time to consider extending full humanity to our own ancestors, those who have gone before.

Velikovsky's work can be seen as synthesizing the history of two radically different world views. His history of Earth and cosmos has demonstrated high predictive value for our science. One could call it a narrative hypothesis rather than a scientific one. The point is that it is valuable. One story will make this clear.

Amidst the hoopla surrounding our projected landing on the Moon, Velikovsky suggested in several letters to the authorities

[17] *Ages In Chaos, Peoples of the Sea, Ramses II and His Time.*

that any rocks brought back from the Moon be marked as to their orientation in situ. Because of magnetic interaction between Mars and the Moon in the 7th century B.C., he was expecting strong remanent magnetism to be found on the Moon. His suggestion was not followed and U.S. taxpayers had to fund yet another mission to the Moon to learn what they might have discovered the first time around. Velikovsky's prediction was correct and the Moon rocks exhibited strong remanent magnetism.[18]

Well aware of his intuitive ability, Velikovsky preferred to call his predictions "advance claims." Unfortunately, this obscures comparison between narrative and scientific hypotheses, both of which are attempts to impose order on chaos and depend on predictability to establish their usefulness.[19]

The concept of world ages separated by periods of destruction is not entirely foreign to scientific orthodoxy. Although geology has generally supported a theory of gradual change, there is implicit acceptance of more violent change in the way it distinguishes eras, periods, and epochs that vary greatly, one to another, displaying entirely different flora and fauna. These changes, of course, are thought to be in the remote past rather than the recent cyclic past with which we are concerned here. Science generally accepts the view that the dinosaurs perished at the end of the Cretaceous period in an intense upheaval, and that the Earth has repeatedly been struck with large objects from outer space. Considered outrageous in 1950, the idea of a comet striking the Earth has become commonplace.

Another body of evidence supports widespread periodic destruction in the 1st and 2nd millenniums B.C. Claude Schaeffer's extensive excavations throughout the Middle East are examined in his *Stratigraphie comparee.*[20] The charts of all the locations indicate destruction levels at the same points in time. Schaeffer's dating is problematical since it must fit into the agreed-upon Egyptian sequence.

Schaeffer's work is difficult to dismiss. The various destruction levels (usually described in the literature as hiatus) represent thick layers of dust and ashes containing no sign of human habitation. It is at points like these that we run into a different order of time, for the chain of linear events can only be traced so far into the past. In order to penetrate further, we must switch over to cyclical time and world ages separated by cataclysm.

The present paradigm of our culture, with its linear, cause-and-effect world view, enjoys all the support of a linear language. Yet a cyclical vision has been creeping in. At the middle of the 20th century, Eliade traced it in *The Myth of the Eternal Return,* citing Spengler, Toynbee, and Sorokin. What he saw as ". . . a certain revival in the theory of cycles . . ."[21] is now bursting upon us full bore. Originating in history, the concept is now widely applied in many fields, providing structure for the soft sciences and, since Kuhn's *Structure of Scientific Revolutions,* for the hard sciences as well. Business is highly attentive to cycles, and they are the daily bread of technicians who deal with commodities.

The popularity of Eastern ideas provided much of the ambiance of the 1960s, when intimations of a New Age were revealed to us.

[18] Robert Treash, "Magnetic Remanence in Lunar Rocks," *Pensee,* 1972, Vol. 2, #2, May, pp. 21–22.

[19] Since Heisenberg, physics has been content with probability.

[20] Claude Schaeffer, *Stratigraphie comparee,* charts on end papers.

[21] Eliade, *Eternal Return,* p. 146.

Significantly, this was linked with a particularly long cycle, the 26,000-year Precession of the Equinoxes, during which the Earth completes a slow spin similar to that of a top. The general public learned that neither science nor astrology has kept the precessional clock running. What we think of as the First Point of Aries, from which both astronomical time and astrological charts are calculated, is now in the constellation of Pisces, and we will enter the Age of Aquarius, as sung in the musical *Hair*, within the next century. This new awareness of real sky time seems to have brought both equinox and solstice into greater prominence and wider celebration.

Eliade sheds light on the change from cyclic, cosmic time to linear, historical time. "Judaism represents an innovation of the first importance. For Judaism, time has a beginning and will have an end. The idea of cyclic time is left behind. Yahweh no longer manifests himself in cosmic time (like the gods of other religions), but in historical time, which is irreversible . . . His gestures are personal interventions in history . . ."[22] Christianity goes further since its faith hangs upon an historical revelation. Time has changed. The historical event becomes decisive. The terrible compulsion to write history and rewrite history, continually changing it according to present views, begins at this point.

By its own admission, science is deeply antihistorical, leading the pack in rewriting history from the present. Kuhn describes tradition-bound periods of "normal science" punctuated by revolutionary periods of breakthrough when the paradigm changes. But the rewriting of textbooks afterwards reminds us more of history assimilated to myth. By the time it appears in print, the chain of events (usually involving something akin to the burning of Bruno) is barely recognizable. But let Kuhn speak.

> Why dignify what science's best and most persistent efforts have made it possible to discard. The depreciation of historical fact is deeply and probably functionally ingrained in the ideology of the scientific profession, the same profession that places the highest of all values on details of other sorts.[23]

"Whoops! We goofed again," is transformed into an orderly search for the truth. So much for the historical record.

It is not possible to get away with this sleight of hand scot free, however. "The theory of the Big Bang and the modern doctrine of universal evolution bear a striking resemblance to this fundamental myth of our culture."[24] Rupert Sheldrake refers to the Biblical Genesis, but the resemblance fits other creation myths as well. With more poetic wording, the Big Bang could stand next to any of them.[25] As it struggles to deny its own past, Western science loses conscious connection with the natural roots of the civilization and is thereby doomed to create its own mythology, unconsciously patterned after the real one.

The current popularity of Eastern thought suggests that Western man may be weary of "making history," ready to settle into a more transcendental style of time. Cyclical vision is coming into focus, and when the final connections are made, the present age, the Iron Age of

[22] Eliade, *The Sacred and the Profane*, p. 110.

[23] Thomas Kuhn, *The Structure of Scientific Revolutions* (Chicago, Univ. of Chicago Press, 1962) p. 137.

[24] Rupert Sheldrake, *The Presence of the Past* (New York, Random House, 1988) p. 258.

[25] But see Eric Lerner, *The Big Bang Never Happened* (New York, Random House, 1992).

Hesiod, will drop smoothly into the larger conception of Velikovsky. The complimentarity of reason and intuition will be realized as we move into a new paradigm.

Under the spell of urdummheit, we have grown accustomed to an image of ancient man as a rather dull fellow, unable to mark equinox or solstice correctly or count the days in the year. In the succeeding chapters we will learn to the contrary that sky events and cycles were primary to our ancestors, recorded carefully over millennia.

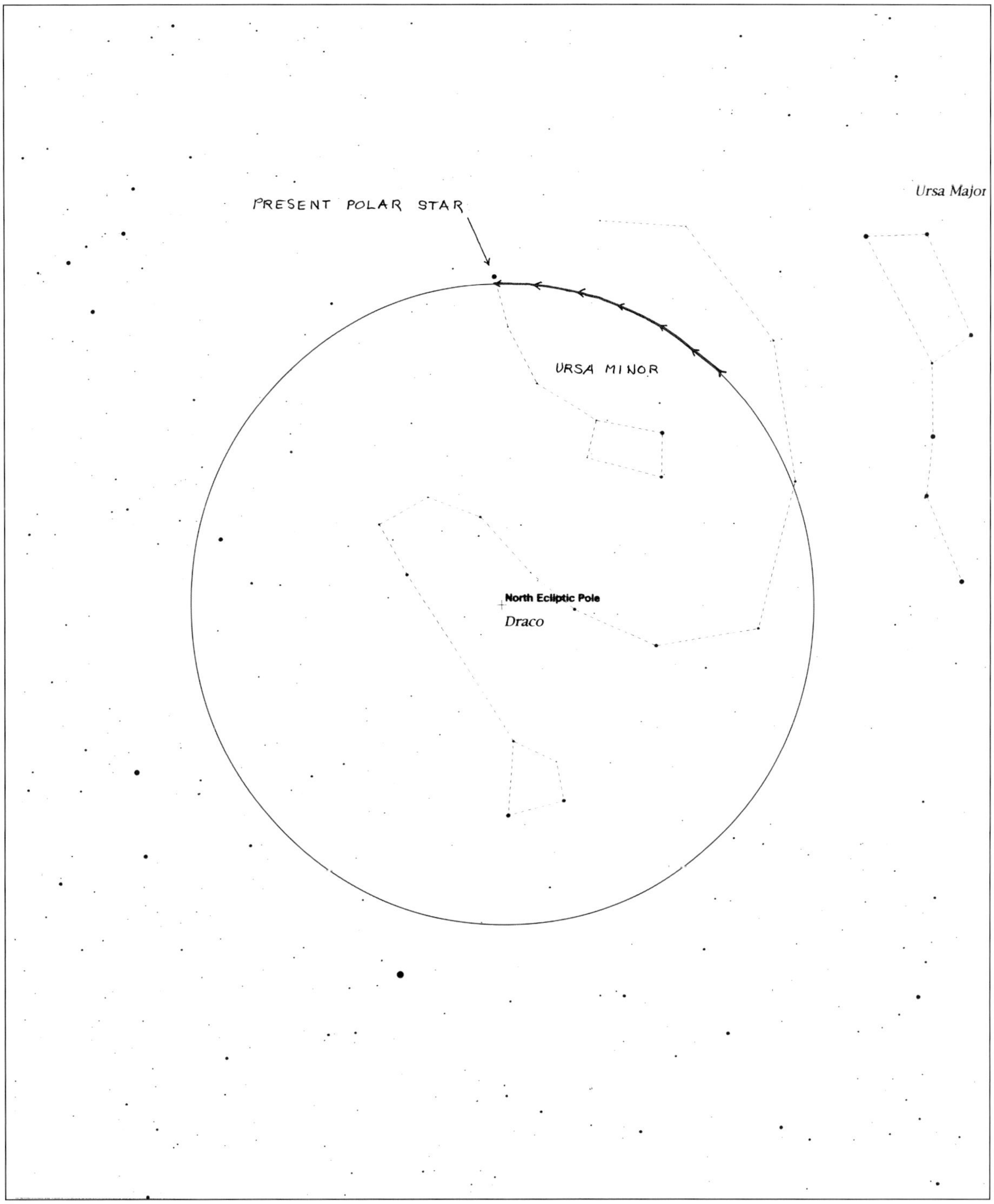

The precessional path of Earth's north pole in the sky as retrocalculated by orthodox astronomy. But ancient media would agree with this model only from 687 B.C. to the present, the heavily-marked segment of the line.

"Sun and Moon Riding in the Same Chariot."

2. Recent Visible Gods: Sun and Moon

. . . for archaic man, reality is a function of the imitation of a celestial archetype.

– Mircea Eliade, *The Myth of the Eternal Return*

Thy will be done on earth as it is in heaven.

– The Lord's Prayer

Most men who have walked the Earth over the recent series of world ages have acknowledged the existence of a god or gods. As Eliade remarks, ". . . the completely profane world, the wholly de-sacralized cosmos is a recent discovery in the history of the human spirit."[1] But there is an order among gods: manifest or immanent, visible or invisible. The oldest gods are invisible or immanent, exerting their power from behind things.[2] Since they are usually understood to be at a distance, they are not called upon except in the most desperate of circumstances.[3]

The most beloved gods, those to whom we feel closest, are those who have appeared to us: Jesus Christ, Jina, the Buddha, and Krishna, and that extended family of planetary gods who combined comforting visibility with a record of unusual powers credited to them by the ancients.

It is a simple truth that the sky itself, particularly at night, provokes feelings of awe, infinity, and transcendence in the human breast, so that it is not surprising that man has found significant images and most of his gods in the heavens. Many cultures speak of a Golden Age or Garden of Eden. In this paradise, Heaven and Earth were still connected and man lived in an ideal relationship with the gods, all his needs provided for. After a "Fall" or "descent of lordship," memories of this happy time initiated what we call religion. The word comes from the Latin verb ligio, meaning to tie or bind, "re" suggesting a desire to retie or reconnect Heaven and Earth, to return to the ideal, original time. This desire, in turn, is expressed in man's continuing efforts to take patterns from the sky and repeat them, to the best of his ability, on Earth.[4]

Down through the ages, man has struggled to bring Heaven to Earth, recording sky events in myth and art, in legend and ritual. The constant repetition of these patterns provides one way for us to unlock the past. The desire of archaic man to reconnect with original time has created a kind of history of the cosmos, a record written in an unfamiliar style, but still accessible to us.

From giant Jupiter to vastly smaller Mercury, the gods and goddesses were visible in the sky, giving us the dualistic ideas that cluster around above and below, spirit higher than flesh, etc. Man having gained position in space may change this, but the old metaphors still have power. For most of us, if there is a god, he is above. The mythological record was

[1] Eliade, *Sacred and Profane*, p. 13.

[2] Frank Waters, *Book of the Hopi* (New York, Viking Press, 1963) p. 9. Also Nuttall, *Fundamental Principles*, p. 32.

[3] Eliade, "The Remote God," *Sacred and Profane*, pp. 121–128.

[4] Eliade, "Archetypes and Repetition," *Eternal Return*, pp. 3–48.

first written in the heavens, alien skies and catastrophic events visible to all who raised their eyes, and charged with religious meaning.

The beginning that we perceive through myth and legend, the Golden Age or Garden of Eden, Tep Zepi or the First Time of the Egyptians, was the beginning for humankind; the family of visible planetary gods is the first that we can now remember. Whether we need to push further into the past under what has been the burden of history remains to be seen, but at least we can find our place in the most recent set of cycles. Perhaps setting the past in order, the great work which Velikovsky began, will provide the balance and self-knowledge we need to proceed.

Sun, Moon, and even planets have often been connected with time. Since the Babylonians, they have been represented by the days of the week: Saturn day, sun day, Moon day, Tiw's day (Tiw-pater, Jupiter), etc.[5] The names vary from one culture to another, but we recognize the members of our solar system, former gods, most firmly time-connected in the case of the two who still dominate our skies. The sun marks off days, seasons, and year, but the Moon is out of sync, no longer marking off the months perfectly.[6] Because they have such a powerful dualistic connection, we will consider them together.

The "golden rope" of the sun[7] is the base line of the solar system, connected to the Earth in imagination, expressing the force that holds everything together. For us, this base line is the ecliptic, defined by the yearly path of the sun through the sky; for the ancients, it described a level plane in space, the surface of the watery abyss.[8] At the edge of this space ocean they saw the constellations of the zodiac, familiar to some of us through either astronomy or astrology. Directly above is the north pole of the ecliptic, the "empty place in the sky" unmarked by any star, from which God hung the Earth according to the book of Job.[9] Below is the south pole of the ecliptic near the first-magnitude star Canopus.[10]

Seen from Earth, the sun appears to move through the 12 constellations of the zodiac in the course of a year. Even though constellations and signs do not agree perfectly, 30 degrees are allotted to each sign for a total of 360. The planets are also seen to move through the zodiac, but on a broader path on each side of that of the sun. Their movement is contained, almost as if they were bouncing slowly back and forth between walls. It is this motion which inspired the ball game tlachtli,[11] of Mesoamerica, played by Olmec, Maya, and Mexica, a game brought from Heaven to Earth.

Complicating this orderly arrangement for us is the 23.4-degree tilt of the Earth, which we must love, however, because it gives us our beautiful changing seasons. All the planets are tilted at different angles; Mars at slightly over 25 degrees is closest to that of Earth. The pole of Uranus actually lies almost in the plane of the ecliptic, the planet on its side as it were.

5 Wednesday and Thursday follow Woden and Thor, while Friday is named after the twin deities of the Vanir, Freyr, and Freyja.

6 In Aristophanes' *The Clouds* (line 615), the Moon complains that the days are not being kept according to her reckoning.

7 Plato, *Theaetetus*, 153 C.

8 The surface of the watery abyss is sometimes confused with the celestial equator, but Santillana correctly equates the "whirlpool" of myth with the ecliptical world. *Hamlet's Mill,* p. 239.

9 Job 26:7.

10 The pole of the ecliptic and pole of precession are identical.

11 Livio Stecchini, appendix to *Secrets of the Great Pyramid,* p. 295.

As we shall see, throughout the recent series of planetary catastrophes, it is the attitude of the terrestrial pole that changes, the ecliptic and its poles that endure.

At the center of our solar system, the sun offers stability to the planets. Ruling the present sky, recognized as a powerful force, its heat and light the most basic support for life on Earth, it is small wonder that the sun would be perceived as a god by primitive man, or even by the early civilizations of Mexico and Egypt. The sun of the present age, the Iron Age of Hesiod, was rather early equated with intelligence. In the 1st century B.C., Cicero speaks of the sun as "the lord, chief, and ruler of the other lights, the mind and guiding principal of the universe."[12]

The sun equates with intellect and consciousness, shedding light on the subject, illuminating ideas. Our language is rich in metaphors connecting light and intelligence, with only lingering glimmers of a religious sense. Our conception of the sun implies power and sovereignty; Louis XIV of France was the Sun King. On the other hand, the Moon is the queen of night, its reflected light and dark setting more expressive of feeling and intuition, of the unconscious. Any powers the Moon may have are dark ones and suspect, her connection with human sexuality for one. Our last goddess has grown weak. Even Juliet chides Romeo, "O swear not by the Moon, the inconstant Moon . . ."[13]

At one time the Moon was the goddess of love, Aphrodite of the Greeks, Diana of the Romans. Curiously, if we go back far enough, we find that the Moon is given great consideration by the ancients. The New Moon feasts of the Jews are mentioned in the Bible in the same breath as the sabbath.[14] To understand the New Moon feasts, we must move, not just into the past, but into cyclical time, into Hesiod's Age of Heroes, the period be tween the visits of Venus and those of Mars, 1450 B.C. to 747 B.C.[15]

Throughout the Age of Heroes, sun and Moon enjoyed an entirely different situation vis-a-vis one another from that at present. Helius the Sun had not yet been supplanted by the intellectual Apollo; he was not even in the direct descent of the gods of Olympus, merely the offspring of two Titans, Hyperion and Theia.[16] Helius the Sun was married to Aphrodite the Moon and they shared a temple at Corinth.[17] In a red-figured krater in the Louvre, they ride in the same chariot. Following Velikovsky, there were 360 days in the year and 30 in each month.[18] Thus a Moon exactly equaled a month in the Age of Heroes. The ease of keeping a calendar can hardly be overestimated, and it is the Moon who would have ruled the heavens as practical timekeeper, accounting for much of her greater prominence.

Today, weak as it is, the light of the Moon is what we cherish. New Age people frequently honor the full Moon. Whether others are interested in a moonlit walk, cross country skiing, or visiting friends in the country, it is the light of the Moon that is valued. Not so

[12] Cicero, *Republic,* trans. C. W. Keyes (Cambridge, Harvard Univ. Press, 1966) p. 269 (6.17) ". . . the sun is the soul, or, more intelligibly, the mind of the universe, the ruling principal and divinity of Nature." See also Pliny, *Natural History*, Book II, 13.

[13] Shakespeare, *Romeo and Juliet,* act 2, scene 2.

[14] Amos 8:5; I Samuel 20:5, 6; II Kings: 4–23; Isaiah 1:13; Hosea 2:11; Ezekiel 46:1, 3.

[15] We use Velikovsky's dating.

[16] Hesiod, *Theogony,* pp. 372–375.

[17] Robert Graves, *Greek Myths* 67.2.

[18] Velikovsky, *Worlds in Collision,* "Year of 360 Days," pp. 326–341.

with the Jews of the Age of Heroes. They held a feast at the new Moon, during the darkest part of the cycle. These New Moon feasts occurred on the first of each month[19] while full Moon was on the 15th, a situation that cannot obtain in the present order of the universe.[20] We must either abuse our ancestors for stupidity or recognize that they were observing very real 30-day months. Plutarch echoes the cycle. "The moon . . . causes solar eclipses, always on the 30th of the month."[21] Across the Atlantic, ". . . the months were counted from one neomenia to another, that is to say from one appearance of the new moon to another."[22]

It was not until Mars began its series of approaches to Earth in the middle of the 8th century B.C., throwing the Moon into turmoil, that the Jewish god suddenly began to disapprove of New Moon feasts, the sabbath, and other appointed feasts.[23] Isaiah naturally credited his god as being behind Mars in the shakeup in the heavens, and quoted him. "Your new moons and your appointed feasts my soul hateth; they are a trouble to me."[24] God spoke on the same subject to Hosea. "I will also cause all her mirth to cease, her feast days, her new moons, and her sabbaths, and all her solemn feasts."[25] Strange sentiments, but his intent was clearly registered in the heavens. With the new irregularity in the Moon's cycle, who could know when the New Moon, sabbath, or other appointed feasts would fall?

The Earth-Mars Wars lasted 60 years, from 747 to 687 B.C.,[26] a period during which Earth's year gained 5¼ days.[27] But it was the Moon, so much smaller and weaker than Earth, that suffered the most grief. The Martian visitations stole half a day from its cycle, reducing it from 30 days to 29½, but this does not give a clear picture of the wide variation in the Moon's cycle within the 60-year period.

After the first visit of Mars in the days of Romulus, the year consisted of ten months, the first one named after Mars, the second for Venus.[28] "For during the reign of Romulus, they had been irrational and irregular in their fixing of the months, reckoning some at less than 20 days, some at 35, and some at more; they had no idea of the inequality in the annual motions of the sun and moon, but held to this principle only, that the year should consist of 360 days."[29] Plutarch obviously agrees with modern orthodoxy, but if we take the ancients at their word, the calendar, following the heavens, was in an uproar.

Dates like "the 33rd day of the month" are cited in Babylonian tablets of the period,[30] while the Romans at that time had a nundina or market every nine days that reflected the

[19] Velikovsky, *Worlds in Collision,* p. 330. Reference is to Roscher, *Lexikon der griech. und Rom. Mythologie,* Col. 892. The importance of the new Moon in the Age of Heroes will be discussed in greater depth in the chapter on Olympic Games.

[20] The days of new Moon and full Moon were apparently in addition to four 7-day weeks in a 30-day month.

[21] Plutarch, Isis and Osiris, 368:44. Also 372:52.

[22] Serna, *Manual de los Ministros de las Indias*, Annales of Nat'l Museum of Mexico, 1899.

[23] I Samuel 20:24, 27.

[24] Isaiah 1:14.

[25] Hosea 2:11.

[26] The Venus-Mars wars, which Velikovsky connected with the founding of the Olympic Games in 776 B.C., will be addressed in the following chapter and elsewhere.

[27] Velikovsky, *Worlds in Collision,* pp. 332–337, 354.

[28] Ibid., "Years of Ten Months," pp. 341–344.

[29] Plutarch, "Numa" XVIII.

[30] Velikovsky, *Worlds in Collision,* p. 341.

longer weeks.[31] Later, after another catastrophe in the series, two months were added to the beginning of the calendar under Numa,[32] destroying the sense of our September, October, November, and December, which actually carry the numerical meaning of seven, eight, nine, and ten. Again, either our ancestors never looked at the sky or couldn't count, or they were describing remarkable behavior of the Moon.

Characteristically, Greek myth records the interaction between Mars and the Moon as a love affair between Ares and Aphrodite. The master storyteller, Homer, relates the tale in Book VIII of the *Odyssey*.

The myth has a fascinating element, the fine bronze net made by the smith Hephaestus to catch his wife Aphrodite[33] and Ares in bed together. The smith was successful and invited the other gods to witness their shame. The event must have taken place early in the morning, for "... Helius as he rose, saw them at their sport ..."[34] The Moon and Mars must have been "caught" in a magnetic interchange, certainly a celestial event to be recorded. Aphrodite is credited with being the mother of Deimos and Phobos, names of Ares' mythological children[35] as well as Mars' planetary satellites.

The sun was involved in at least four unusual motions during this period of the Earth-Mars Wars. We must be attentive to the apparent movement of the sun in order to sort out two different kinds of Earth motions.

"According to Latin historians, on the very day of Rome's foundation, the sun was disrupted in its movement and the world was darkened."[36] This event took place in the same year as the "commotion of Uzziah," as well as the beginning of an astronomical era in the Middle East, the year 747 B.C.[37] Unfortunately, in this instance, there is no indication of which way the sun moved.

The second unusual motion was at the "commotion" on the day King Ahaz was buried, when the sunset was hastened. "This cosmic disturbance is described in the Talmud, in the Midrashim, and referred to by the Fathers of the Church."[38] At the burial of Ahaz, then, since the day was shortened, the sun moved forward.

A third unusual motion of the sun is described in the Thyestes myth, the subject of a play by Seneca. Velikovsky quoted Seneca at length in *Worlds in Collision*. This well-educated Roman of distinguished family, born in Spain (Hispania), a senator under Caligula and Claudius, and the teacher of Nero, was a student of myth and a playwright fond of using mythological themes and allusions. His *Thyestes* certainly contains descriptions of the serial catastrophes of the Earth-Mars Wars, but muddled together as if they all happened at the same time.

The myth of Thyestes tells of the struggle with his brother Atreus for the throne of Mycenae. Apollodorus puts it clearly, stating the premise twice. "But Zeus sent Hermes to

[31] O'Neill, *Night of the Gods*, p. 711.

[32] Plutarch, "Numa" XVIII.

[33] Aphrodite is here married to Hephaestus, serial marriage and affairs being popular among Olympians.

[34] Graves, *Greek Myths*, 18.a.

[35] Ibid., 18.a.

[36] Velikovsky, *Worlds in Collision*, p. 237. "The fires lighted on the Parilia are said to derive from an episode during the founding [of Rome]." George Dumezil, *Archaic Roman Religion* (Chicago 1970) p. 57. Parilia was celebrated on April 21. p. 381.

[37] Ibid., pp. 236–237.

[38] Ibid., p. 213.

Atreus and told him to stipulate with Thyestes that Atreus should be king if the sun should go backward; and when Thyestes agreed, the sun set in the east."[39] In describing the same event, Ovid refers to "Phoebus [the sun] wresting his car about turned round his steeds to face the dawn."[40] In its third unusual motion, then, the sun goes backward.

Euripedes' *Orestes* also confirms the event, giving us important information. He describes again:

> . . . the deadly feud
> Which turned the sun's wing'd chariot round
> Changing his course toward the dawn;
> While Zeus guided to a different path
> The seven tracks of the running Pleiads.[41]

At this point, the man on the street is wondering how the stars of the Pleiades, or any others, could be seen if the sun were shining. Naturally, they could not have been. Euripedes is telling us something, but we cannot grasp it. One is reminded of a high school math problem in which information necessary to write an equation is not given, but already a matter of common knowledge. To the ancients, it was common knowledge among those who had lived through the recent visits of Mars that planetary catastrophe occurred at the vernal equinox.[42]

Euripedes is telling us that the sun is in the constellation of the bull, Taurus, where the Pleiades are located, and a catastrophe is occurring as the sun reverses its motion, moving toward Aries. What we have here is not an anomaly in rotation but an apparent motion of the sun against the background of stars. Of course it is the Earth that moves. And we moderns, without the common knowledge of the ancients, must await further indication that this event occurred at the Spring Equinox.

Now any astronomer worth his salt is feeling his blood pressure rise. The motion described appears impossible; and the vernal equinox is known to have been in Taurus back in the 3rd and 4th millennium B.C. At least we assume this to be true since it has been retro-calculated to that time and position. Moreover, the equinox does not move suddenly—unless, that is, one accepts certain records of the ancients that indicate that it does.

In reading *Worlds in Collision*, it is difficult not to be struck by the number of times the equinox comes into focus. Like a bad penny, it keeps coming up. Catastrophic events seem to occur close to or at the equinox, and if we are to believe the ancients, the equinox itself can move suddenly. Velikovsky was alert to the problem,[43] which becomes central within these pages.

The fourth unusual motion of the sun occurred in 687 B.C. on the evening of the destruction of Sennacherib's army by a consuming blast. As it happened this event occurred during the first night of Passover, observed at the time of the vernal equinox.[44] This coincides with the sudden motion of the sun described in II Kings. "And Isaiah the prophet cried unto the Lord; and he brought the shadow ten degrees backward, by which it had gone down in the dial of Ahaz."[45] This remarkable quotation refers to two catastrophes, but is misleading in that there is the

39 Apollodorus, *The Library,* Epitome II, pp. 10–13.

40 Ovid, *The Art of Love,* Book I.

41 Euripedes, *Orestes,* trans. Phillip Vellacott (Baltimore, Penguin, 1972) p. 335.

42 This would account for the extraordinary accuracy of Biblical prophecy during this period.

43 Velikovsky, *Worlds in Collision,* pp. 344–347.

44 Ibid., pp. 231–232.

45 II Kings 20:11, also see Isaiah 38:7–8.

implication that the two motions somehow canceled each other out. In fact, they did not, as will presently become clear.

There is reason to suspect that this fourth motion of the sun is echoed in the Mahabharata, India's great epic of war.[46] In the Drona Parva of the Mahabharata,[47] Arjuna, the champion of the Pandavas, has vowed to slay Jayadratha, ally of the enemy Kauravas, before the sun sets. Time runs out for him, however, and the sun is about to go down. The divine Krishna, Arjuna's charioteer, shrouds the sun so that Jayadratha, aware of Arjuna's vow and believing the sun to have set, will reveal himself. Arjuna then fights and kills Jayadratha, after which the darkness is withdrawn by Krishna and the sun shines and presently sets again.

The distance between the Middle East and India gives a reasonable fit between events which took place late in the day. Added to this, the Mahabharata War was fought in the interval between the Dvapara and the Kali Yuga, the two most recent world ages of India.[48] These would correspond well with the Age of Heroes and the Iron Age, the two most recent ages of Greek tradition. Both sets of world ages would have been separated by the "twilight" of the 60-year period of the Earth-Mars Wars.

The great Earth-Venus Wars of the 15th century B.C. give us one more glimpse of sun and Moon in an unusual motion. Following Velikovsky, it was at the second visitation of Venus that Joshua asks sun and Moon to stand still so he will have time to rout his enemies.[49] That the heavenly bodies complied has provided grist for the mills of science and religion time out of mind. For Velikovsky, this passage was an indication that a catastrophic event was taking place, and he soon found a corresponding period of darkness across the ocean in Mexico in the Annals of Cuautitlan.[50] Here we need only distinguish the direction of the sun's motion. Since the day was extended, the apparent notion of the sun was backward. Now planetary catastrophe can be accompanied by an obscured sky, but the unusual motions of the sun mentioned above were sufficiently visible to record. Thus far, we have found but one forward motion of the sun, that which occurred at the death of Ahaz. The sun seems inclined to an apparent backward motion at planetary catastrophe, which is, in fact, the natural direction of its apparent path as viewed from Earth over the long precessional spin. We will return to this subject presently.[51]

Velikovsky always maintained that the terrestrial pole once pointed to the Great Bear, Ursa Major,[52] one of the best-known configurations of stars in the sky. Orthodox science rejects this view, in part because the constellation does not lie on the precessional path of the north celestial pole, a circle in the sky described around the pole of the ecliptic with a radius of 23.4 degrees. This figure reflects the tilt of Earth while the circle reflects Earth's top-like, 26,000-year precessional spin.

For the terrestrial pole to have pointed to Ursa Major implies a major catastrophic

[46] Artur Eisenberg, "Dating the Great Mahabharata War," *Kronos* II, 3, pp. 56–64.

[47] P. C. Roy, trans. *The Mahabharata of Krishna-Dwaipayana Vyasa,* ed. H. Haldar (Calcutta, Oriental Publishing Co., 1955) Vol. VI, pp. 343–347.

[48] Giorgio de Santillana and Hertha von Dechend, *Hamlet's Mill* (Boston, Gambit, 1969) p. 78.

[49] Joshua 10:12–13.

[50] Velikovsky, *Worlds in Collision,* p. 45.

[51] These apparent backward motions are sudden precessional spins, discussed in "Time Travel on the Precessional Clock" and elsewhere.

[52] Velikovsky, *Worlds in Collision,* pp. 310–311.

movement. It would also have put the Earth in something closer to a 40-degree tilt than the 23.4-degree tilt it presently enjoys. Having the Great Bear at the pole would have many implications, but one of the most critical is that those on Earth would see the sun travel much farther north and south. This change of the terrestrial tilt is reflected in the myth of Phaethon, how he drove his father's sun chariot and created havoc by going off the path.

A lucid account of this myth exists in Plato's *Timaeus*. From the mouth of an Egyptian priest we hear the following story. ". . . once upon a time Phaethon, son of Helius, yoked his father's chariot, and, because he was unable to drive it along the course taken by his father, burnt up all that was upon the earth and himself perished by a thunderbolt—that story, as it is told, has the fashion of a legend; but the truth of it lies in the occurrence of a shifting of the bodies in heaven which move around the earth, and a destruction of the things on earth by fierce fire, which recurs at long intervals."[53]

In the myth of Phaethon, we confront one of the more puzzling aspects of myth, difficult to grasp with our linear habit of mind. In man's continuing efforts to bring Heaven to Earth, certain myths become epichronic, adding layers of meaning down through the ages. The Phaethon myth comes into focus every time the Earth tilts and the sun's path changes. In the version of Nonnos,[54] as well as the "Battle of the Stars" in the Fifth Book of the Sibylline Oracles, references to several different catastrophes are brought together, while an account of Hyginus has Saturn being thrown out of the sun chariot.[55] We will unravel the story as we move back through cyclical time.

Because sun and Moon are the only ones left in the sky of what was formerly a company of highly visible gods, we find ancient texts obscure. When we read "sun," we naturally assume that it is our sun in the present order of the heavens. Velikovsky's genius lay in his realizing that there have been different orders of the cosmos. Since he brought myth and legend into focus as a way of illuminating past ages, the work of disentangling suns, stars, planets, and comets has proceeded apace, but outside the present paradigm.

Sun and Moon were often held to have been members of a triad including Venus, which was much brighter in past ages: Q're, Ashima, and Anatha;[56] or Jehovah, Ashima Baetl, and Anatha Baetl.[57] The names do not inspire confidence, however. Venus is present as Anatha, but Ashima sounds very like Shamash, who was Saturn.[58] In fact, the crescent symbolism which brings the Moon to mind has also been associated with both Venus and Saturn.

Now that we begin to understand that the ancients were as concerned to record sky events as we are to write and rewrite history, we can move into the cyclical past with more confidence. The ancient media of myth and legend, art and ritual, operates to save the phenomena and carry it down through the ages. The mutability of symbol allows it to attach information and bring it forward.

Because the story of the mother goddess is so tightly knit into historical catastrophism,

[53] Plato, *Timaeus,* trans. Cornford (New York, Bobbs-Merrill, 1959) pp. 9–10.

[54] Nonnos, *Dionysiaca,,* Book 38, lines 290–434.

[55] Hyginus, *Poetica Astronomica,* II.42.

[56] Robert Graves,*White Goddess,* p. 264.

[57] Ibid., p. 366, a reference to the cult at Elephantine in Egypt in the 5th century, B.C. A baetl is a cone-shaped rock which relates to the world mountain.

[58] M. Jastrow, "Sun and Saturn," *Review of Assyriologie et d'Archeologie Orientale* 1909, Vol. 7, pp. 163–178.

we will give a brief summary in the following chapter, both to illuminate and give form to the present survey. The fortunes of the mother goddess are in disarray; the deterioration in the situation of woman, nature, and the goddess is faithfully reflected by that of the whole of mankind. Today's epithet of choice, some form of "fornicate your mother," "chinga tu madre," etc. echoes through the culture. The celestial death of the goddess, brought to Earth, has operated to bring us to this remove.

Sun within Moon.

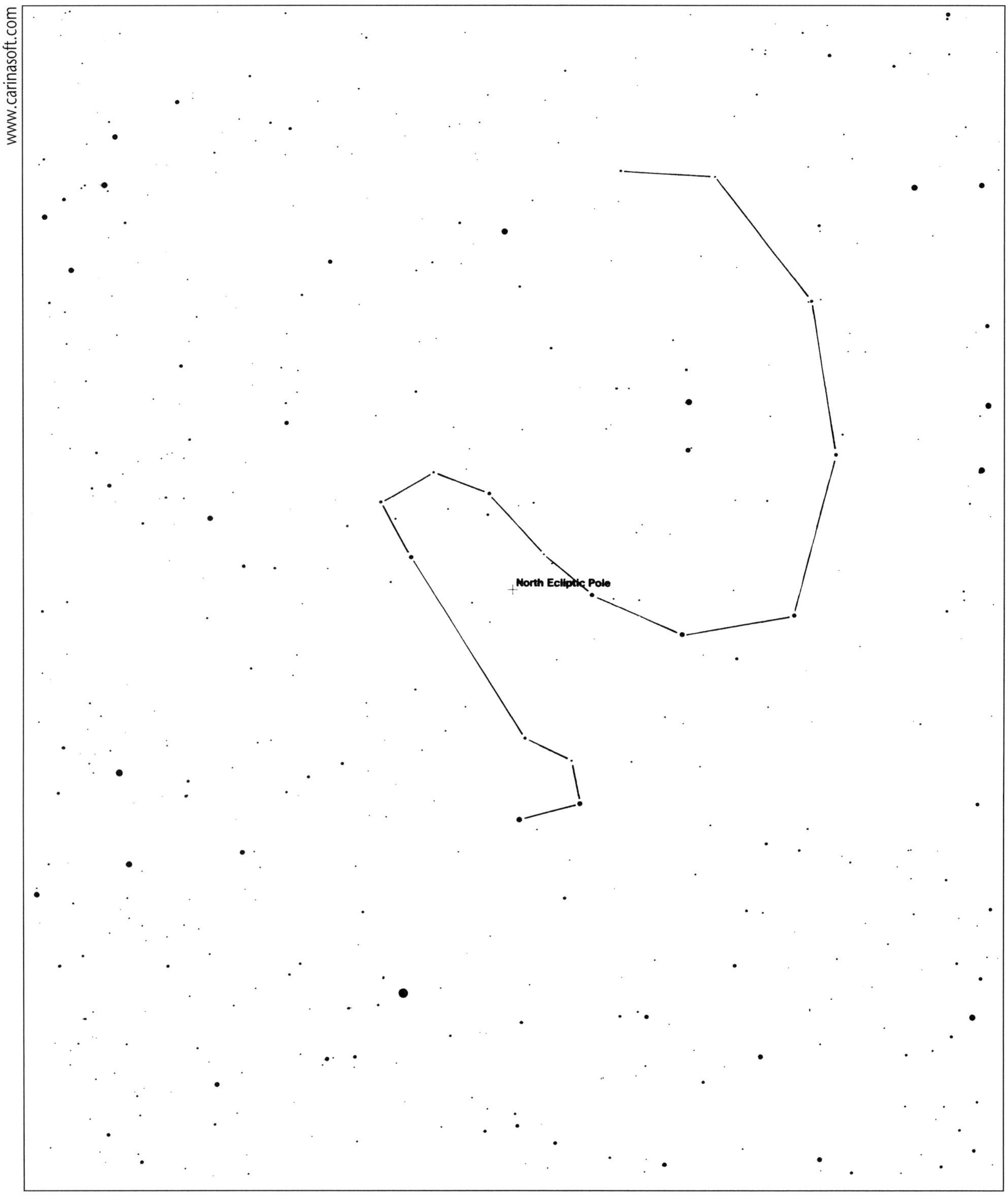

The Constellation Draco, the Babylonian "yoke-star of the sky,"
and source of the omega symbol commonly associated with Venus.

3. Life and Death of the Goddess

But we will certainly . . . burn incense unto the queen of heaven, and to pour out drink offerings for her, as we have done, we, and our fathers, our kings, and our princes, in the cities of Judah, and in the streets of Jerusalem; for then had we plenty of victuals, and were well, and saw no evil.

– Jeremiah 44:15–17

Worship of the goddess is one of the oldest forms of religion that can be traced. Figurines, wide-hipped, emphasizing her generative powers, go back as far as the Gravettian period,[1] while Orphic fragments tell us that the triple goddess was honored even before Uranus, the heavens.[2] And certainly the concept of triplicity continued to exert power over the mind of man down to the establishment of the Catholic Trinity. In the beginning of the current set of cycles, however, despite the many names she has acquired, the goddess was one.

The mother goddess, whom we would come to know in one aspect as Venus, ruled the skies from a beginning in the Golden Age. It is a difficult concept for Westerners, extending godhood to the feminine, but such was the case in our own tradition in past millennia. Jewish women, with the approval of their husbands, baked cakes in honor of Venus, the Queen of Heaven, in the 7th century B.C.[3] just as the Sumerian women did for Inanna even earlier.[4]

The appearance of Venus in the sky in her various incarnations has changed so much over the millenia that it behooves us to note the shapes in which she has shown herself. The sky goddess was watched carefully throughout the ages and her manifestations recorded, so that it is possible to trace her history. "Velikovsky's comet" is best known from *Worlds in Collision,* but more recent workers in the field have continued research on the history of the planet, extending her appearance farther into the past.

Toward the end of the Golden Age of Saturn, where our story begins, the mother goddess was understood as the band surrounding Saturn,[5] then much closer to the Earth. The bright wheel in the heavens is recalled in the name of the Celtic goddess Arianrhod, "silver wheel," while many Mediterranean goddesses are associated with the wheel. Some of us will recall the circular yoni, still symbol of the female orifice in the culture of India. David Talbott has identified Saturn and his surrounding band as the enclosed sun of Egyptian hieroglyphics, Ra in his Aten.[6]

Then as now, Saturn had a discernible pattern of rings, but highly visible and widely celebrated in ancient media in the number

[1] E. Adamson Hoebel, *Anthropology: The Study of Man,* 4th ed. (New York: McGraw Hill, 1966) p. 182.

[2] Robert Graves, *The Greek Myths,* 2.b.

[3] Jeremiah 44:15-17.

[4] S. Langdon, "A Hymn to Ishtar," pp. 32–37.

[5] Talbott, "The Great Mother," *The Saturn Myth,* pp. 80–89.

[6] Ibid., pp. 61–71.

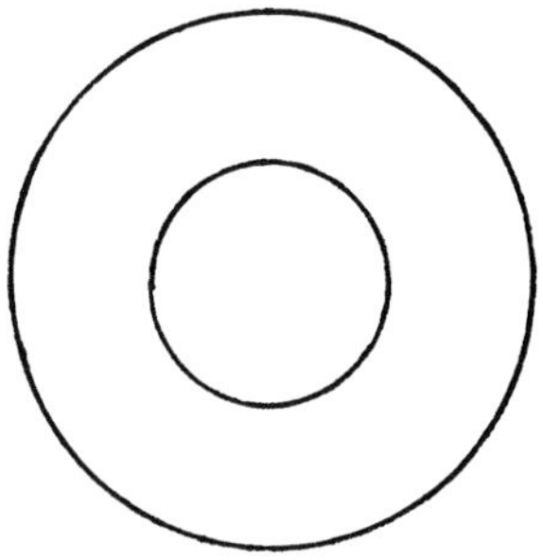

The enclosed sun of Egyptian hieroglyphics: Ra in his Aten.

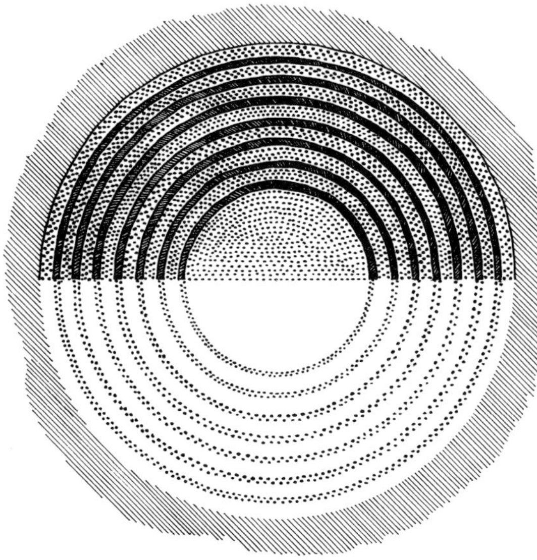

The ancient Saturn's illuminated crescent rings.

seven.[7] Among a host of images that recall the rings is the great serpent Ophion coiled seven times around the Universal Egg,[8] the sevenfold tiara of Shamash,[9] the seven walls surrounding the underworld, and certain cities such as the seven-walled Uruk and Nineveh. Pallas, a common epithet of Athena, the Greek Venus, is the number seven.[10] When the planet and its rings were lit on one side by the sun, they were seen as illuminated crescents: the wings of the seven angels that stand before God,[11] the seven birds (Vucub Caquix) of the *Popol Vuh*,[12] but commonly in the Middle East, seven sets of bovine horns.

The *Egyptian Book of the Dead*,[13] echoed by Egyptian art, depicted one bull and seven cows, clearly showing the feminine aspect of the sevenfold encircling band.

Crescent horns and the cow image would continue to attach to Venus down through the ages. When the primeval Saturn withdrew to a tiny point of light in the sky, the Egyptians began the ritual of leading a cow seven times around a temple. The ceremony was called "the Seeking for Osiris [Saturn]."[14] In the *Odyssey*, when Nestor wanted to sacrifice a heifer to the goddess Athena, he summoned a goldsmith to gild her horns first.[15] Sacred cattle were sometimes branded with another enduring mark of Venus, the omega sign, explained below.[16]

Toward the end of the Golden Age, the primeval Venus appeared to unwind from the Saturn Configuration,[17] looking like a lock of

7 For a period toward the end of the Golden Age, Saturn apparently had nine rings. Myth reflects this change as Orpheus adding two strings to his harp, the omega-shaped harp of Venus as hanging goddess.

8 Graves, *Greek Myths*, 1.b.

9 P. Gossman, *Planetarium Babylonicum* (Rome: 1950) p. 188.

10 John O'Neill, *Night of the Gods*, p. 48.

11 Ibid., p. 955.

12 See analysis in the chapters "Saturn Myth" and "Saturn Configuration."

13 E. A. Wallis Budge, *Egyptian Book of the Dead* (British Museum, 1895; New York, Dover, 1967) p. 239.

14 Plutarch, *Isis and Osiris*, 372.52. This is also echoed in the tawaf of Islam, walking around the Kaaba seven times. *Kronos* XII, 3, p. 15. Also see O'Neill, *Night of the Gods*, p. 125.

15 Homer, *Odyssey*, Book III.

16 Ilse Fuhr, "On Comets, Comet-like Luminous Apparitions and Meteors," *Kronos* VIII, 1, p. 49.

17 The appearance of this motion will be amplified in the chapter on the Saturn Configuration.

The bull displays round horns; the cow's horns have an omega shape (from the tomb of Nefertari in the Valley of the Kings).

One bull and seven cows (from a papyrus in the Berlin Museum).

hair or a snake and hanging for a protracted period, long enough for a "hanging goddess" to be mythologized. Greek myth records this as a revolt in Heaven led by Hera. She was punished by being hung from Heaven by golden bracelets (Saturn's rings) with an anvil tied to each foot.[18] Other hanging goddesses include Aphrodite, Artemis, Helen, Ariadne, Phaedra, Erigone, and Polycaste.[19]

Hanging snake (Book of the Dead, Plate XIX, Chapter LXXXI, Papyrus of Nebseni, British Museum, 9900).

[18] Graves, *Greek Myths,* 13.c, 23.b.

[19] Kerenyi, *Dionysus* (Princeton 1996) pp. 106, 158. Greek ritual played out the hanging or swinging of the goddess in the Charila and the Aiora.

Eventually the bright coil unwound into the elemental dragon or cosmic serpent, and when it detached from the configuration and went on its way, it left behind a pattern for one of the early groups of stars to be marked in the sky, the constellation Draco. One of the least prominent constellations, its shape traced out only by third- and fourth-magnitude stars, few people could find it in the sky today. We may thank the modern dream and myth factory, Hollywood, for better acquainting the public with Draco as the computer-animated dragon-hero of the film, *Dragonheart,*[20] properly placed in the sky at its conclusion.

The Egyptians knew the constellation Draco as Tau-ert, Isis-hippopotamus, who leads a procession of so-called northern gods on the ceilings of many tombs,[21] and Isis was the Egyptian Venus.

The Egyptian constellation Tau-ert, our Draco.

To the Babylonians, Draco was "markasu rabu sa e-si-kil-la," "great bond of the pure house,"[22] recalling the goddess as the band around Saturn. Curiously, Draco is not enumerated among the known constellations of the Babylonians, but its stylized symbol enjoys a prominent position on boundary stones, often after the three primary representations of the deity. It is sometimes represented as a snake.

Omega symbols from Babylonian boundary stones of 12th century B.C. (Boundary stones of Mardikapaliddina, Berlin Museum; the "Michaux Stone," Biblioteque Nationale; boundary stone of Malishipak; facing 106 MAR vol. V.)

[20] *Dragonheart* (Universal City Studios, Inc., 1996) produced by Rafaella de Laurentiis, directed by Rob Cohen, written by Charles E. Pogue.

[21] Draco is the model for the blue hippotamus, without which no collection of Egyptian art is complete.

[22] Eric Burrows, book review of Vol. V of *Mythology of All Races, Journal of the Royal Asiatic Society,* 1933 (London, JRAS) p. 699.

It is the shape of Draco which is important here. We Westerners see only a curved celestial dragon, but the ancients saw Draco as an omega sign and the yoke of a draft animal. Burrows and some other scholars concur with the ancients as to the representation on the boundary stones. "Since the object naturally suggests a yoke, and was listed as such by Hinke, it may reasonably be taken for a stylized representation of the constellation (Draco) which was actually called the Yoke, and which must in fact have been seen as a somewhat omega-shaped yoke." [23] But one has only to open a celestial atlas to be struck with the resemblance. The Sumerians called Draco "musir-kesda," meaning "yoke-bound." [24]

Venus appears as a dragon in Mesoamerica, sporting the same omega-shaped sign. The symbol can be identified in the tail of many images of Xiuhcoatl or Quetzalcoatl, the feathered serpent.

"Celestial dragons frequently have Venus glyphs on their heads . . ." remarks Thompson.[25] The "diving god" wears a Lamat glyph for a head in the Dresden codex. As we will see, the Venus glyph, Lamat, appears to have evolved from the omega-shaped sign.

Like so many other gods and heroes, Venus presently made the trip to Hell recorded in the Descent of Ishtar and that of Inanna,[26] and subsequently established a pattern of appearances in the skies of Earth. This was her brilliant cometary period, described in the "Hymn to Ishtar" as ". . . the mighty queen of heaven," "the pure flame that fills the heavens." [27] Also, repeated many times, she is ". . . garment of the sky." [28] With such an appearance, it is no wonder Venus was never enumerated with the planets by the ancients. The goddess as Saturnian band has not been forgotten, however, for in the Sumerian ritual described in this hymn, seven cakes are baked for Inanna, echoing the seven rings around Saturn.[29]

Venus is well recorded as the sickle-shaped Inanna symbol of cuneiform tablets, and "the sickle-shaped Typhon [with whom Velikovsky identified Venus] is referred to in countless mythic tales." [30]

The feathered serpent.

[23] Ibid., p. 699.

[24] Ibid., p. 699.

[25] J. E. S. Thompson, *Maya Hieroglyphic Writing* (Norman: Univ. of Okla. Press, 1960) p. 219.

[26] Hell, as we shall presently discover, has a location. It is at the south pole of the ecliptic, just as Heaven was formerly perceived at the north pole of the ecliptic. Thus both will be capitalized herein.

[27] Langdon, "A Hymn to Ishtar," pp. 18–19.

[28] Ibid., pp. 26–30.

[29] Ibid., p. 32.

[30] Fuhr, "On Comets," p. 59, note 2.

Variations of the Inanna symbol.

CODEX NUTTALL

CODEX DRESDEN

CODEX MADRID

Knotted serpents.

At the time of the planetary interactions with Earth, Venus-Typhon changed its appearance to a "spiral-shaped fiery knot," as described by Pliny.[31] The knot is represented artistically on both sides of the Atlantic; a knotted serpent in the Codex Nuttall of Mesoamerica,[32] others on the heads of goddesses in the Codex Madrid and the Dresden Codex.

The square knot of two intertwined omega symbols can be found in the Mediterranean world. This latter knot can be understood to symbolize the two visits of Venus. Interestingly, the double visit of Venus is also represented in the New World by two omega signs, but drawn somewhat differently. Cult knots are found in large numbers in Crete.[33]

The square knot was known to the ancients as a Hercules knot. The hero-god often tied his lion skin around his neck in such a fashion. We will explore the tight relationship of Hercules and his patron Athena (Venus) at some length in subsequent chapters. In modern times the square knot is sometimes known as a love knot.

We will not dwell on the catastrophic destruction visited on our planet during the

[31] Pliny, *Natural History,* II, 91.

[32] Codex Nuttall (New York, Dover, 1975) plate 75.

[33] Hans Baumann, *Lion Gate and Labyrinth,* trans. Stella Humphries (New York, Random House, 1967) p. 124.

Double Venus symbols.

The Hercules knot: intertwined omega symbols.

Earth-Venus wars. The burning hail, boiling rivers and seas, vulcanism and eruptions of lava, giant waves of translation and protracted periods of darkness have been dealt with in *Worlds in Collision.* Velikovsky identified Earth changes in *Earth In Upheaval,* some now synchronized with the 15th as well as the 8th and 7th centuries B.C.[34] We will have much to say about other critical phenomena associated with the Venus cataclysms, however. Velikovsky brought up important material in the section on "Changes in the Times and the Seasons," wherein the Papyrus Anastasi IV complains, "The winter has come as summer, the months are reversed and the hours are disordered."[35]

It was during the Earth-Venus Wars that the proto-planet earned her reputation as a warrior goddess, wreaking destruction on Earth but also exhausting some of her cometary substance. Her two approaches to Earth caused Venus to change both her appearance and her orbit; the unruly comet became a tame planet, our beautiful Morning and Evening Star. During the 700 years of the Age of Heroes, what was left of her cometary train was described as "horns" (during her crescent phase) or "hair" (when seen at superior conjunction).

Despite all the changes in the appearance of Venus during successive world ages, the omega or yoke-shaped symbol continued to be closely associated with her. It would become "hair" around the head of a woman, the "coiffure hathorique"[36] of the literature (a "flip" to moderns). Fuhr cites the extraordinary prevalence of this hair fashion in the pictorial art of all the ancient cultures of the Near East and Mediterranean region.[37] But the style reaches to more northern cultures as well.

Beginning with the first appearance of the omega symbol in the Isin-Larsa period, Fuhr traces it through the Middle Bronze Age to the "indispensable eyeglass" in the graves of Scythian males.[38] The "spectacle" spiral and spiral-headed pin or toggle pin[39] are found in the Indo-European trade axis of the Age of Heroes, from Magagnose in France to Mohenjo-Daro on the Indus River in India.

[34] *Science Digest,* August 1970, reported on page 48 that Mt. Rainier erupted 3,500 years ago (the time of the Earth-Venus wars) and filled two river valleys with deposits to depths of 700 or 800 feet.

[35] Velikovsky, *Worlds in Collision,* pp. 119–124.

[36] Venus is Hathor, the Egyptian cow goddess.

[37] Fuhr, "On Comets," p. 47.

[38] Ibid., p. 47.

[39] The spiral symbol, echoing Venus' spiral departure from the Saturn Configuration, is often attached to the omega sign. It will be discussed in "Killing the Bull."

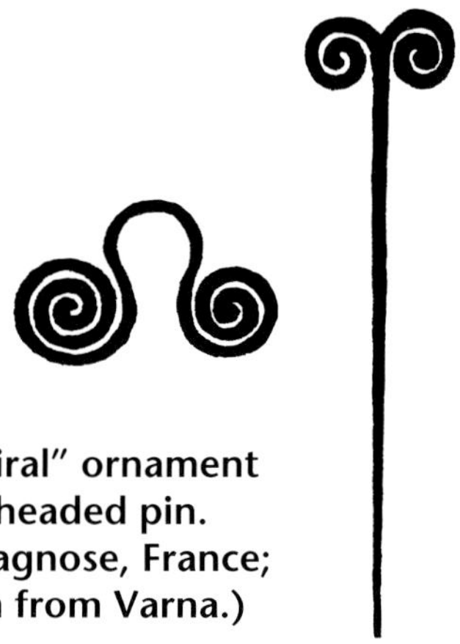

"Spectacle spiral" ornament and spiral-headed pin. (SS from Magagnose, France; S-headed pin from Varna.)

The widespread ingot torques,[40] sometimes take the omega form and can even be found around the neck of the goddess.

The coiffure Hathorique, above from Denderah, below from the Waldalgesheim tomb.

Goddess and omega torque, the goddess from a bronze caldron from Rynkby, Denmark. (Metal ingot torque p. 77 Dayton found throughout Europe at Ur and Kahun. Goddess wearing torque on bronze caldron of Rynkby, Denmark, 1st century B.C.)

[40] Dayton has argued convincingly in his monumental *Minerals, Metals, Glazing and Man* that these torques are the money of the period. The torque-necklace became an attribute of the goddess Freya. It was called "Brosingamene" in the Anglo-Saxon epic, *Beowulf*.

The omega spiral can be found in Egypt today, worn by some women as an amulet against miscarriage.[41] This type of protective amulet is called an apotropaion, reflecting the ancient and still widely held belief that what opens wounds can close them, and what hurts can heal. The appearance of comets was widely believed to cause miscarriage. Interestingly, omega is the last letter of the Greek alphabet, properly placed to signify an ending.

The use of the omega symbol in Mesoamerica is also widespread. Many of us are familiar with the Aztec Calendar stone, the monumental sculpture that was dug up in Mexico City where the earlier Tenochtitlan had been reduced to rubble by Cortez. It has been studied for generations and found to be full of coded information about history, religion, and astronomy, much concerned with the cycles and even transits of Venus.[42] The circular stone is wreathed with two cosmic serpents, and in the middle is the Ollin sign of the present Fifth Sun of Mexico.

This Ollin sign with its central motif has been variously depicted elsewhere. Most instructive is the representation on the Malinalco drum where the Venus omega sign surrounds the central orb exactly like the hair of the European "coiffure hathorique."

The handsome bas-reliefs of Cosumalhuapa in Guatemala figure the omega symbol in so large a form, in the background as it were, that it could easily be overlooked. Used in the same manner as the Venus "coiffure hathorique," the omega fills the background around the heads of several of the depicted gods and is repeated in smaller forms.

Venus symbol from the Malinalco drum.

Bas-reliefs from Cosumalhuapa (detail from #5; detail from #11).

[41] Winifred S. and A. M. Blackman, "An Ancient Egyptian Symbol as a Modern Egyptian Amulet," *Annuaire de L'Institut du Philologie et D'Histoire Orientales,* Vol. III, 1935 (Brussels) p. 93.

[42] Raul Noriega, *Piedra del Sol* (Mexico: Editorial Superacion, 1955) part III, pp. 431–445. Periods of Venus are also found on other monuments discussed in this book.

Early in the 20th century, the perceptive Zelia Nuttall saw these sculptures as visitations of the gods.[43] The replacement of the left hands of many of the subordinate figures with animal heads suggest totemic groups tributary to the left-handed Huitzilopochtli, war-god of the ancient Mexican pantheon, whom Velikovsky identified as Mars. These bas-reliefs must have much to do with the star wars of the 8th and 7th centuries B.C.

A thoughtful examination of the Venus glyph, Lamat, reveals a quincunx, four omega symbols around a central dot or star.

Venus Lamat glyphs (from Thompson, *Mayo Hieroglyphic Writing,* Figure 7).

The quartering of anything is not unusual for the Mexicans, who also multiplied many of their gods by four. Except that reading the central star as a numeral could make this symbol a five, a number that will presently come into focus. As Dechend suggested in the introduction to *Hamlet's Mill,* we must begin to count if we wish to proceed.

Velikovsky saw the battles in the *Iliad* as taking place against the planetary wars between Venus and Mars in the first half of the 8th century B.C.[44] It was Hesiod's Age of

Wood carving with Venus symbols (carved wooden teponazthli drum, Axzotla, Pueblo).

Heroes, a time of widespread trade as well as "wine and warfare." Athena-Venus, the warrior goddess, championed the invading Achaeans while Ares-Mars supported the defending Trojans. Rose and Vaughan have developed computer models based on Mars having been an inner planet before the Venus-Mars Wars, and Venus on an orbit close to the one it follows now.[45]

Athena won both of her confrontations with Ares in the 5th and 21st book of the *Iliad.* This should not surprise us, given the greater size and mass of the planet Venus. In the second battle, it is significant that Athena struck Ares with a boundary stone, suggesting that the orbit of Mars was changed at that time.

The interaction between Venus and Mars was perceived as a battle in the *Iliad*, but other cultures saw it differently. When Gilgamesh encountered Ishtar, it was seen as unrequited love. "Gilgamesh washed out his long locks and cleaned his weapons; he flung back his hair from his shoulders; . . . When Gilgamesh had put on the crown, glorious Ishtar lifted

[43] Zelia Nuttall, *Fundamental Principles,* p. 153 (589).

[44] *Worlds in Collision,* pp. 243–250.

[45] Lynn Rose, "Could Mars Have Been an Inner Planet?" *Pensee,* Vol. II, 2 (Portland, Ore.: Student Academic Freedom Forum, 1972) p. 42.

her eyes, . . ."[46] The "hair" of Gilgamesh, his "crown," suddenly comes into focus. Like Ishtar, we are struck by a change in his appearance.

The encounter with Venus affected Mars; he acquired a part of her cometary train. But Gilgamesh refuses Ishtar just as Arjuna refuses the heavenly Urvashi in the Mahabharata. It is a common theme, the hero refusing the goddess, because it happened first in the sky.

Seneca, in "de cometis," wrote, "Here is the explanation which is given by some ancient authors. When a planet enters in conjunction with another, they confound their lights into one light, and they have the appearance of an elongated star . . . the interval which separates them is illuminated by both of them and transforms into a trail of fire . . ."[47] Of course Seneca is recounting what happens at a real meeting in space.

In their encounters in the first half of the 8th century B.C., Venus lost more of her cometary substance while Mars gained flaming "hair," becoming the Babylonian fire god. In many ways the goddess set the style, with hair as fashion and hair as strength. Witness the biblical story of Samson, or the false beards worn by Egyptian pharaohs. But now Venus would grow weaker and dimmer as the centuries passed. Eventually the Romans would celebrate the cult of Venus Calua, the bald Venus.[48]

Between the 15th and 8th centuries B.C., Venus appears to have described a synodic period very close to its present one, eight Earth years to five synodic periods of Venus.[49] The eight-year period was celebrated in many cultures, often divided in half as four-year periods since the cycle divides easily at either inferior or superior conjunction. The most important of the Greek festivals which followed the Venus period was the Heraea, held every four years, in which 16 women wove a robe for Hera,[50] commemorating the ambrosial robe that clouded the atmosphere of Earth after the first visit of Venus.[51] Other festivals following the Venus period were also in place, some connected with the bear, probably celebrated throughout the Age of Heroes.

The battles between Venus and Mars did not appear to directly affect the Earth; for us, it was a simply a phenomenon visible in our skies. Reconstructing the chain of events requires a little imagination, an identification with ancient people who have just witnessed an impressive celestial display involving the warlike, terrible Athena-Venus and the hitherto unremarkable (during the previous age) Ares-Mars. All eyes would be on the warrior goddess, expecting the worst, and anyone with a nodding acquaintance with Homer could guess what the ebullient Greeks would do. They would schedule funeral games for the next likely timeframe connected with Venus, games that would become the Olympics. Velikovsky put this together in *Worlds in Collision.*

Now this is not a leap of fancy, honoring an astronomical event with games. Here we must

46 Gilgamesh Epic, N. K. Sandars trans. (Baltimore: Penguin, 1960) p. 85.

47 Seneca, "de cometis."

48 George Dumezil, *Archaic Roman Religion,* p. 422.

49 Both Babylonian and Maya astronomers allowed Venus a 90-day period of invisibility at superior conjunction, however, which does not agree with the 50-day invisibility at present. Anthony Aveni, *Skywatchers of Ancient Mexico* (Austin: 1981) pp. 187, 327.

50 Frances Cornford, "The Origin of the Olympic Games," *Prolegomena and Themis,* Jane E. Harrison (New Hyde Park, N.Y. Univ. Books) pp. 229–231.

51 Velikovsky, *Worlds in Collision,* p. 170.

remember that all foot races and chariot races that are run on a circular track came originally from the sky,[52] so that to the ancients there was a connection between the two. The Circus Maximus in Rome contained altars of the planet gods, while the chariot race at Olympus ran for 12 rounds, suggesting the yearly path of the sun through the zodiac. The starting point was at an altar of the Twins.[53] Races were run counterclockwise, as the planets "run" around the sun as seen from above.[54]

Scholars have argued persuasively that the Olympic Games celebrate a funeral (not for Earth, however),[55] or constitute New Year festivities.[56] Both arguments tend to support a catastrophic view since catastrophe is both an ending and a beginning. A problem the Greeks were faced with was one of scheduling, for the date of choice at the fall equinox was already occupied by the ladies.[57]

The Heraea was a much older, immovable festival, dating at least from the first visit of Venus, but also honoring the second in the period of its celebration, half of the eight-year Venus period. The foot races between young women probably gave its name to the month Parthenios, in which the festival was held.[58] That the Heraea was celebrated on the first of the month at the new Moon[59] tells us that we are in the Age of Heroes, though close to the end of it, and also sheds light, or we might say darkness, on the various New Moon feasts. "A Hymn to Ishtar" says that ". . . she decrees fates for her Land. Monthly, at the new moon," and later ". . . at the dark of the moon . . ."[60] In making use of the darkness at the new Moon, the Heraea might also have been honoring the pattern of darkness after both visits of Venus.

The Olympic Games had to defer to the Heraea, then, and were held either two weeks before or two weeks after the older feast.[61] The logical, rational male principle would incline toward the full Moon and its useful light. And after the Earth-Mars Wars, when the new order of the heavens manifested, there would be a constant struggle to keep sun, Moon, and equinox in sync to honor the previous order. The Olympic Games would continue to be a movable feast,[62] as would our Easter, imposed as it is on pagan festivities that antedate the Earth-Mars Wars.[63]

Not long after the interaction between Mars and Venus, Greek astronomers would be able to assure the people that Venus was not a threat. It was Mars that would stir up the heavens, change the number of days in the year, tilt the pole with a resulting change of climate, shake up the cycle of the Moon so the months had to

[52] Livio Stecchini, Appendix to *Secrets of the Great Pyramid* (N.Y.: Harper & Row, 1971) p. 295.

[53] Cornford, "Olympic Games," *Prolegomena and Themis,* p. 228. The Inca also saw Gemini as a beginning, calling it "creation time." Santillana and Dechend, *Hamlet's Mill,* p. 248.

[54] For a more comprehensive treatment of the catastrophic origins of the Olympic Games, see the chapter of that name.

[55] James Frazer, *The Golden Bough* (London, Macmillan) ed. 3, p. 92 ff.

[56] Cornford, "Olympic Games," p. 216.

[57] Why the Heraea was celebrated at the fall equinox rather than that of spring will be addressed in a later chapter.

[58] Cornford, "Olympic Games," p. 228.

[59] Ibid., p. 228.

[60] Langdon, "Hymn to Ishtar," pp. 20, 36.

[61] Cornford, "Olympic Games," p. 230.

[62] Our modern calendar has been calculated into the past for our convenience, yet Pliny, 23–79 A.D., was still dating by Olympiads.

[63] L. Boemler, "Asherah and Easter," *Biblical Archaeology Review* (May/June 1992), p. 66.

be kept by direct observation. But after the 60-year period of the Earth-Mars Wars ended in –687, the solar system would compose itself, falling into place for the Iron Age. The Moon settled into its present cycle, and an interesting resonance was observed. The eight-year Venus cycle coincided closely with 100 lunations.[64]

The catastrophic events of the chaotic period between world ages posed terrible problems for the religions of the times as they struggled to digest celestial events. Throughout the previous age, the goddess had been both powerful and heavy-handed. In many cultures, fecund Mother Earth was fertilized by the blood and bones of kings who suffered ritual castration and death after a one-year reign and the necessary fertility rites.[65] But as the once-incandescent Venus cooled and steadily grew dimmer in the sky, the power of the goddess waned and the king-consorts of the Earth Mother grew stronger.

Mythology tells us that in the first Olympic Games, one king replaced another in the chariot race, Pelops replacing the dead Oinomaos.[66] This, among many other myths and legends, was the source of the yearly cycle which required a kingly sacrifice, but the length of the cycle changed as a result of the Venus-Mars Wars of the first half of the 8th century B.C. The four-year Venus time frame that devolved on the Olympiad apparently extended the kingly reign to four years as well. Mars has a "twin" aspect which would also influence arrangements in the earthly kingdoms. The first kings of Rome, the twins Romulus and Remus, would divide an eight-year reign between them.

As might be expected, the extension of the kingly reign from one to four years gave the consorts of the goddess more time to consolidate their power. They grew less willing to give up either their power or their lives.

When it was noticed that the eight-year Venus cycle coincided closely with 100 lunations, an equivalence was established between Venus and the Moon that facilitated transfer of myth and symbol from the former to the latter.[67] Now when we see crescent and horn symbolism, it brings only the Moon to mind.

It was only a matter of time before the lunar nodes and the 19-year cycle of eclipses were noted, and at that time the sacred king reached for the longer time period of 19 years.[68] Eventually the kings claimed divine right, but it was Mars who provided the celestial pattern, the aggressive masculine prototype which came to dominate the mythology and psychology of the present Iron Age.

So much for the transfer of focus from Venus to Moon as well as a certain confusion between Athena and Aphrodite, both aspects of the goddess. Now we understand why there were 50 priestesses in the temple (½ Venus cycle = 50 lunations) and why they were fathered by a hero or god. Priam, for instance, had 50 sons, only 19 of whom were legitimate (following the number of the metonic Moon cycle).[69] Now the patriarchy comes down with a vengeance, Hera, Demeter, and Persephone forced into subservience to Zeus, Poseidon, and Hades. The endless philanderings of the male gods, doubtless in the missionary position of dominance, were a ploy to consolidate the new regime. Zeus was even

[64] The Olympic Games were held alternately every 49, then 50 lunations.

[65] Graves, *Greek Myths,* pp. 17–19.

[66] Cornford, "Olympic Games," p. 214.

[67] Graves, *Greek Myths,* 25.2. Myths relating to Athena were transferred to Aphrodite.

[68] Ibid., p. 19.

[69] Ibid., 158.7.

declared to be the father of several sets of triple goddesses.

So offensive did the new regime become to contemporary Greeks that the 5th-century B.C. playwright, Aeschylus, took it upon himself to give the head of the Olympic pantheon his comeuppance. Aeschylus puts the following words into the mouth of Prometheus, suffering torments for having brought fire to man, "He is a monster . . . I care less than nothing for Zeus. Let him do as he likes."[70] Campbell contrasts this reaction with that of Job in the Bible who also suffered torment by God, but without complaining.[71] Against this latter background, the Jews suppressed the feminine principle, and the later Church Fathers virtually eliminated it, an action which would eventually cause the Church much grief.

The fixation of the European psyche on the mother goddess would eventually raise the mother of Jesus to unprecedented heights as the medieval cathedrals to the Virgin attest. Unfortunately, Mary has suffered from the dualistic vision we have imposed on the world, and lacks the sexuality that would make her a whole woman. But at least she cared for her child. There is usually a striking transition from matriarchy to patriarchy when the care of the young king devolves on someone other than his mother: Romulus and Remus suckled by a she-wolf, infant king raised by shepherds, etc.

Of the many celebrations of the 8:5 period of Earth years to the synodic periods of Venus, a persistent one was held in Egypt at the temple of Isis. The heliacal rising of Venus was the occasion for a festival and procession inaugurating the four-year Egyptian year. Now this rising is a movable feast, tied as it is to a slightly irregular astronomical alignment. The Egyptians had already added the five extra days attributable to the Earth-Mars Wars to their calendar and were preparing to deal with the other ¼ day in the Canopus Decree of –238 by having the calendar follow the brilliant star Sirius. Star time is the correct way to fix the calendar, but this particular star was an even more perfect choice as its appearance on Earth was commensurate with the Earth year of 365¼ days.[72] Its heliacal rising also happened to occur close to the time (within six weeks) of the annual flood of the Nile, a phenomenon that had probably been used by practical farmers for centuries.

Unfortunately, the Canopus Decree stepped on the toes of the priests of the temple by specifying that the feast of Isis be nailed down to a particular date. Such was the devotion to Isis-Venus and her movable feast that the calendar reform never did take root. These were times when confrontations between church and state were usually won by the former. It was left to the Romans to institute a leap year to keep the calendar in order. Though the Canopus Decree was written in three languages, it contributed to a confusion between Isis and Sirius that is still with us.[73]

At the time of the Spanish Entrada into the New World, two Venus festivals were in place there. The Atamalqualiztli took place every eight years close to the autumnal equinox,[74] reflecting the present 8:5 ratio of Earth years to synodic periods of Venus. The Toxiuhmopilli,

[70] Aeschylus, *Prometheus Bound,* 937.

[71] Joseph Campbell, *Myths to Live By* (N.Y.: Bantam, 1973) p. 82.

[72] The slight motion of Sirius against the vault of stars comes close to negating that of the precession.

[73] For a discussion of the problem of the Canopus Decree and Sothic dating, see "Astronomy and Chronology," supplement to *Peoples of the Sea,* by Velikovsky.

[74] Thomas A. Joyce, *Mexican Archaeology* (London, Boston: The Medici Society, 1914) pp. 73–75. Also J.E.S. Thompson, "A Correlation of the Mayan and European Calendars," *Field Museum of Natural History Anthropological Series,* Vol. XVII.

the binding of the sheaves of years, took place every 52 years, commemorating the period of Venus in previous millennia, both before and between the catastrophes she visited on Earth in the 15th century B.C.[75] A bas-relief on the Platform of Venus in Chichen Itza illustrates both the 52-year bundle and the 8-year synodic period.

Venus symbols from the Platform of Venus in Chichen Itza.

The Spanish soon put an end to the outward display of much Indian culture, but a curious myth of the feminine has survived which has been aired so many times in popular fiction that it is worth putting to rights.

When the Aztecs were approaching the environs of Tenochtitlan, the original site of Mexico City, the will of their sun and war god Huitzilopochtli was construed to mean that his sister and co-regent Quilaztli should be left behind in Malinalco. She was a very heavy presence possessing four names: Woman-serpent, Eagle-woman, Woman-warrior, and Woman of the Underworld, names which strongly suggest Venus.

As the Aztecs settled in Tenochtitlan and consolidated their power, it was presently noticed that there was a certain lack of the feminine principle in their world order. They set out to remedy this situation by inviting the daughter of the ruler of Culhuacan to become their queen. She yielded to their request but was subsequently killed and flayed by her erstwhile subjects, and a youth dressed in her skin.

Still unsatisfied, the rulers of Tenochtitlan arranged to have a male of the line of Culhuacan become their ruler. This individual came with his wife to the Aztec capital, where she withdrew into political seclusion. He began to rule in the year One Flint (1364 A.D.) with the title of Woman-serpent (Cihuacoatl), one of the names of Quilaztli. From that day forward the title seems always to have been borne by a man, who performed the bulk of human sacrifice and acted as an alter-ego of the actual ruler.[76] In such fashion was what was seen to be a necessary balance between masculine and feminine preserved.

The Aztec creation of Snake-woman is somewhat reminiscent of the installation of the Goddess of Reason in Notre Dame Cathedral during the French Revolution. Maude Oakes has called the latter a "repression of the true spirit of the feminine in favor of a false masculine intellectual ideal."[77] Both Aztecs and French were struggling with a problem not yet solved, achieving the proper balance of masculine and feminine in society. After bringing Heaven to Earth over recent millennia, we find ourselves sadly out of order in this respect.

75 "The Fifty-two Year Period," *Worlds in Collision*, pp. 152–153.

76 Nuttall, *Fundamental Principles*, pp. 61–63.

77 Maude Oakes and Joseph L. Henderson, *The Wisdom of the Serpent* (New York: George Braziller, 1963) p. 23.

In the following chapter we will begin our journey into the cyclical past, consolidating Velikovsky's work on the Earth-Mars Wars with an examination of the same celestial phenomena recorded by the ancient Olmecs of the Gulf Coast of Mexico.

Dedication Stone of the Templo Mayor in Tenochtitlan, Mexico City (Las Calendarias Prehispanicos, Figure 19).

The two different crossed bands of Olmec iconography.

4. Olmec Crossed Bands: An Analog of Time

Great were the descriptions and the account of how all the sky and earth were formed, how it was formed and divided; and the measuring-cord was brought, and it was stretched in the sky and over the earth, on the four angles, on the four corners, as was told by the Creator and the Maker, the Mother and the Father of Life, . . .

– *Popol Vuh*

The Olmec crossed bands element, St. Andrew's Cross, or Maya sky glyph is a symbol common to Mesoamerican culture. It is figured liberally throughout Mexican and Mayan codices, painted on pottery and chiseled in stone from Central America to the Yucatan and across the breadth of Mexico. Even where writing was not well developed, as in the culture of the Olmecs of the Gulf Coast, we still find frequent representation of the X of the crossed bands.

Since the civilization of the Olmecs has been pushed back to the 1st millennium B.C. and earlier,[1] we may properly examine it for evidence of the catastrophic events described in Part II of Immanuel Velikovsky's *Worlds in Collision,* the problematical Earth-Mars encounters of the 8th and 7th centuries B.C. The Olmecs, often referred to as the mother culture of Mesoamerica, initiated the number and writing system of the Maya and ancient Mexicans, and left a rich iconography that yields to catastrophist interpretation.

We know the Olmecs of the Gulf Coast of Mexico from their sculpture, a stolid-looking people with faces set in a permanent scowl. Yet there is something striking about the binding of the skull, the elaborate headgear, the incising on the countenance of statues suggesting tattooing on the face. The Olmecs seem very much focused in their heads. This is particularly evidenced by the V-shaped cleft indicated on the heads of art and statuary at about the location of the fontanel where one can see the pulse beat in tiny babies. One is reminded of the Hopi keeping their kopavi open to listen to the messages of the gods. The cleft in the skull is not limited to man but is found in representations of birds and animals as well, suggesting that the Olmec had a connected attitude toward nature.

A people who place a statue on top of a volcano[2] and build a model of a volcano at one of their ceremonial centers[3] would probably not allow a catastrophic event to pass unmarked. We will use the X of the crossed bands to turn the tumblers in the lock of the Olmec iconography. As we will see, the various symbols of their civilization tend to illuminate each other when they are considered together.

[1] *Science,* July 12, 1957, p. 72. The first radio carbon dates from La Venta were 1154 to 174 B.C., plus or minus 300.

[2] San Martin Pajapan.

[3] La Venta.

Like their fellow man around the world, without the distraction of artificial light, the Olmecs watched the lights in the sky. What is more unusual, they represented themselves at this activity in at least two pieces of sculpture that have been recovered.[4] Why a statue of a man looking upwards? Their attention must have been caught by the same alien skies and catastrophic phenomena recorded on other continents. Certainly the lights in the sky were the source of their first symbols.

Olmec skywatcher from La Venta Parque Museo, Villahermosa, Mexico.

The Olmecs had no written language as we know it. They were just beginning to use symbols for communication, the same number system later used by Maya and Mexica: a dot for one, a bar for five. Michael Coe notes four symbols used by the Olmecs, all of an astronomical nature,[5] that were later transmitted to Maya hieroglyphics: the U sign believed to be the Moon, the four-petalled kin flower of the sun, a primitive four-pointed star sign thought to be Venus, and the X of the crossed bands element. There is at least one more, the omega symbol of Venus,[6] sometimes written so large on monumental sculpture that it is easily overlooked.

The first three symbols must provoke wonder at the way in which they are represented. Perhaps the U sign is most like the crescent of the Moon, although any spherical body that happened to get into a certain position between Earth and sun would exhibit a crescent. Venus has a crescent phase, easily seen with a good pair of binoculars. Certainly the sun by no stretch of the imagination resembles the four-petalled kin flower, and a four-pointed star might represent any of the brighter stars or planets in the heavens.[7] The fourth symbol, the X of the crossed bands, is generally believed to be a celestial symbol because it is so often represented in the skyband, a length of distinctive signs that can accompany codices and monumental sculpture. It is read "sky" or "snake,"[8] "center" or "zenith," tan when its function is phonetic.[9]

Straight lines are unusual in nature, and we look to the sky in vain for crossed bands. An early Western observer perceived an X in the sky, although he would have pronounced it differently. In *Timaeus*, Plato tells how the demiurge constructed the universe out of two circles which he joined "in the form of the

4 Monument 56, La Venta Park Museum in Villahermosa, Tabasco: standing figure, hands behind head, looking almost straight up. Another statue from La Venta at the museum: seated figure looking up at a 45-degree angle.

5 Michael Coe, "Early Steps in the Evolution of Maya Writing," *Origins of Religious Art and Iconography,* edited H. B. Nicholson (Los Angeles: UCLA Latin American Center Publications, 1976), p. 111.

6 Ilse Fuhr, *Ein Altorientalisches Symbol,* Chapter V (Weisbaden: Otto Harrassowitz, 1967), also *Kronos* VIII, 1, pp. 38–52.

7 However, see David Talbott, "The Saturn Thesis," *Aeon*, Vol. VI, No. 3, pp. 31–37 and his video "Remembering the End of the World."

8 The words are homophonic, caan and can. The number 4 is also kan. Linda Schele and David Freidel give chan. *A Forest of Kings* (New York: Morrow, 1990) p. 416.

9 David Joralemon, "The Olmec Dragon: A Study in Pre-Columbian Iconography," *Origins of Religious Art and Iconography,* edited H. B. Nicholson (Los Angeles: UCLA Latin American Center Publications, 1976) p. 66.

letter X."[10] These circles are understood to be the celestial equator and the ecliptic.[11]

An Olmec, if he was like other Indians of the New World, probably saw himself at the center of his world, feet solidly on the ground, his existence braided snugly into the cycles of the Earth. He had strong feelings about the pole, or a point eight degrees west of it,[12] and the courts and earthworks of his ceremonial centers generally display orientation to the cardinal points. The X that he perceived would have been formed by the sun on the horizon at its extreme solstice positions.

A diagram by Villa-Rojas in Appendix A of *Time and Reality in the Thought of the Maya* [13] helps us hook up the X with reality.[14] He refers to a culture other than the Olmec, but the concept works for all. It is how the Meso-americans represented the path of the sun in the sky. The X connects the solstice positions of the sun at its rising and setting, the most extreme northern and southern points that it reaches, thereby expressing an angle. In many cultures, the solstices are considered to be gates.

The X extends across time already, for the sun can only be in one place at a particular moment. Aveni, the archaeoastronomer, puts it thus: "The four arms of the St. Andrews Cross signify the four houses of the sun in the sky, two in the east and two in the west."[15]

The diagram gives us a low, flat X, the image of the crossed bands as it is most commonly encountered. Deceptively simple in appearance, it contains a powerful piece of information that underlies our world: seasons,

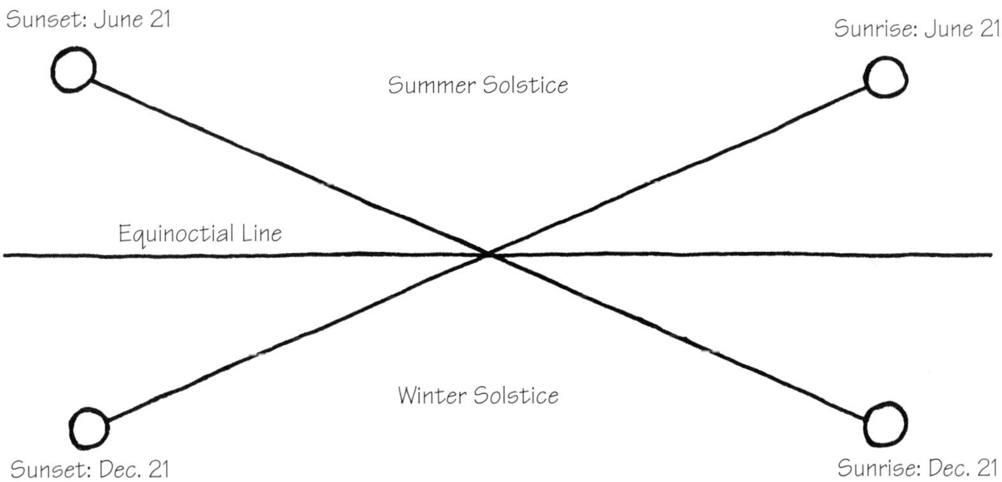

The four houses of the sun, gates of the solstices.

[10] Plato, *Timaeus*, 36.

[11] F. M. Cornford, *Plato's Cosmology* (Indianapolis: Bobbs-Merrill, 1975) pp. 72–93.

[12] The main complex at La Venta is oriented eight degrees west of north. The northerly orientation of the oldest civilizations has been traced around the world by Zelia Nuttall in "Fundamental Principles of Old and New World Civilizations," Archaeological and Ethnological Papers of the Peabody Museum, Harvard University, Vol. 2. John O'Neill has illuminated the problem in *Night of the Gods* (London: Harrison & Sons, 1893), while the underlying phenomenon may be examined in David Talbott's *Saturn Myth* (New York: Doubleday, 1980).

[13] Miguel Leon-Portilla, *Time and Reality in the Thought of the Maya*, 2nd edition (Norman: Univ. Oklahoma Press, 1988) Appendix A.

[14] Here the diagram is redrawn.

[15] Anthony Aveni, *Skywatchers of Ancient Mexico* (Austin: Univ. Texas Press, 1980) p. 156.

climate, and weather, the very angle of the ecliptic which defines our Tropics of Cancer and Capricorn, our Arctic and Antarctic Circles.[16] Now orthodox thought finds this moderately interesting, but if one is a historical catastrophist and has noted X's in Mesoamerican culture with different angles, the information becomes literally Earth-tilting.

Crossed bands are figured abundantly on Olmec statuary, monuments, and pottery. We find it on headgear, mythical animals, and birds, on the chest of the half-human, half-jaguar babies that appear to occupy the center of the Olmec mythos. Oddly, the X can be found incised within the eye of a human or other creature, suggesting that it was something that could be seen, perceived with the eye. And, as is also apparent in the other Mesoamerican cultures, the X comes in at least two forms: the low X describing the present ecliptic, and a taller version in which the lines approach being perpendicular to each other, almost as if they were inscribed in a square.

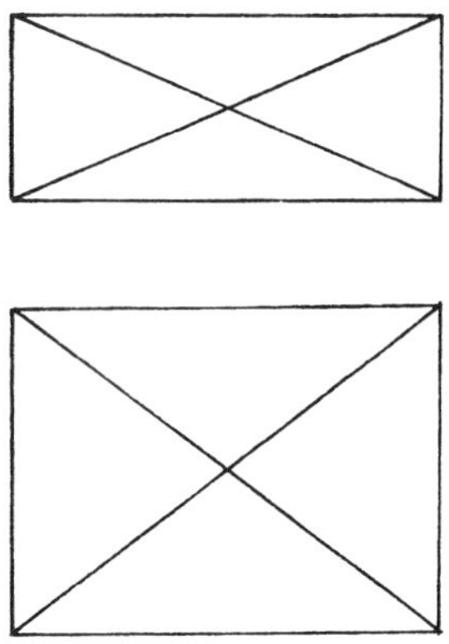

The two different X's of the Olmecs.

Again, it is a matter of taking the ancients at their word, in this case, their symbol. The Olmecs are clear on having two different X's in their art and sculpture. If the angles are not always perfectly drawn, neither have they been carefully copied. There is no getting around it. If the low X gives us the angle of the ecliptic in the present age, then we must begin to deal with the other, taller X of the Olmecs which is giving us an entirely different angle.

Here we must depart from the conventional view of history. Velikovsky followed Greek myth in claiming the north terrestrial pole once pointed to Ursa Major, the Great Bear, "utraque sicca," always dry, never dipping below the ocean.[17] He felt this indicated a catastrophic tilt of the terrestrial axis at some time during the Earth-Mars Wars. In fact, ancient Mexican myth echoes the same tilt in the legend where Tezcatlipoca jumped down from the pole and had his foot bitten off by the earth monster.[18] The suggestion is that the constellation was visible year round, then suddenly moved to a point where a part of it dipped below the horizon, below the sea in a maritime culture. In the land of the Olmecs, this event was seen to occur when the jaguar, Oceotl,[19] jumped down to Earth and began attacking mankind.

For us here on the planet experiencing such a catastrophic tilt of Earth, the old celestial pole would appear to go down, the new one go up; Ursa Major moved away from the pole while, presumably, Ursa Minor moved toward it.[20]

[16] The angle of the ecliptic, 23.4 degrees, registers only at the equator. The apparent path of the sun widens as one moves north or south. At Mexico City it forms an angle of about 26 degrees.

[17] Homer, Ovid, and many other classical writers refer to a time when the Great Bear never dipped beneath the sea. Aratos in *Phaenomena* wrote of Ursa Major, "By it on the deep Achaians gather where to sail their ships."

[18] Nuttall, *Fundamental Principles,* pp. 8–9. The myth is fairly common in the literature and Tezcatlipoca is often represented artistically without his foot.

[19] Oceotl, jaguar, is the name of Ursa Major in Nahuatl. Brinton 1893:56, Forstemann 1904:568.

[20] The exact motion of the celestial pole will be dealt with in a later chapter.

When we translate the difference in the angle of the two Olmec X's to the sky, the resulting positions of Earth open interesting possibilities for speculation.

The crossed bands suddenly appear as an elegant coding of information, two different X's indicating two different world ages. This allows us to use the different X's to penetrate the Olmec mythos further.

Our Big Dipper is one of the most prominent constellations in the sky, the only one with which many people are familiar. To the Romans, it was the Great Bear, Ursa Major. To the Hebrews and Phoenicians, it was Dubh, also bear. Although the "jumping down" of Tezcatlipoca was caused by the catastrophic movement of the terrestrial pole, the constellation itself was called Oceotl, the jaguar, and so it was apparently known to the Olmecs who have represented the jaguar many times.

It is the were-jaguar babies that are the focus of our interest, however, and it would appear that they were engendered when Oceotl jumped down to Earth and began attacking men and fornicating with women. Jaguars are occasionally seen attacking men in Olmec iconography and there is some badly damaged sculpture indicating carnal interaction between jaguar and woman. Matthew Stirling has hypothesized that the were-jaguar babies are the product of this unnatural union.[21]

This interaction between jaguar and woman, between Sky and Earth has all the earmarks of a powerful cosmic event. The

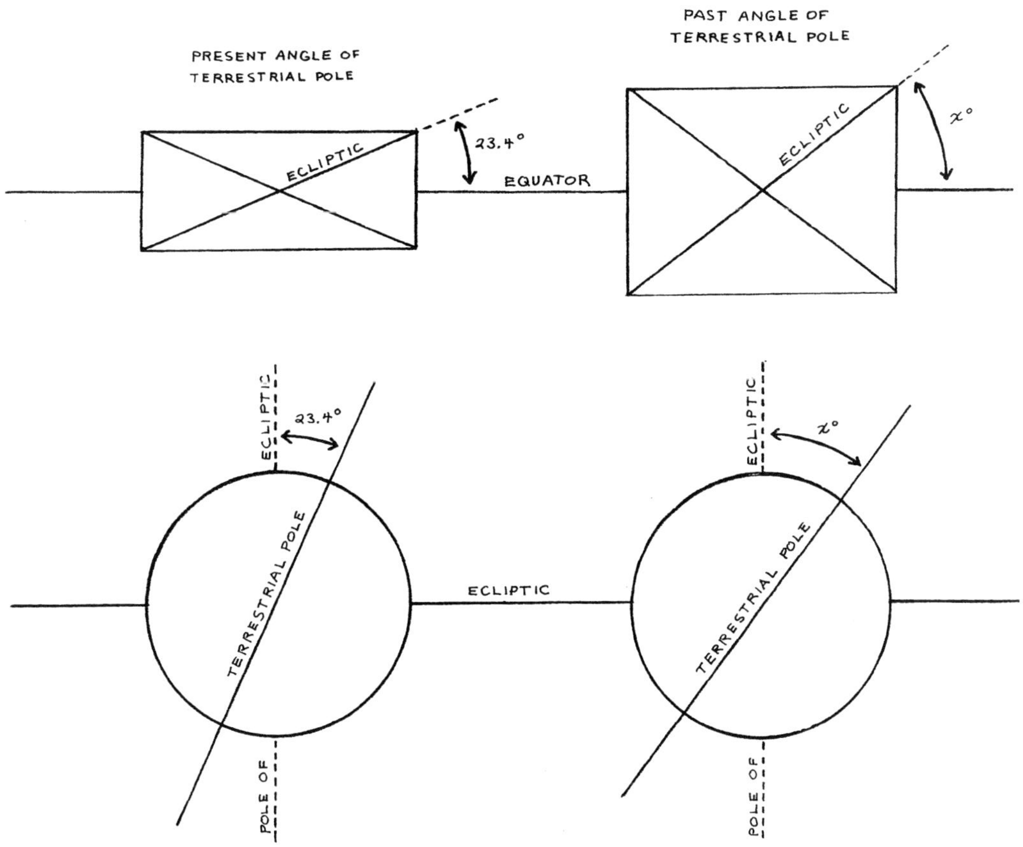

Earth tilts indicated by Olmec X's.

[21] Matthew Stirling, "Stone Monuments of the Rio Chiquito, Veracruz, Mexico," Smithsonian Institution, Bureau of American Ethnology, bulletin 157 (Washington, D.C.) pp. 1–24.

jaguar in the sky, Oceotl, moved down from the pole and attacked Earth. We recall Eliade, "... for archaic man, reality is a function of the imitation of a celestial archetype." The Olmecs knew exactly what to do; they would bring Heaven to Earth, recording the event as perfectly as they could. But there were a number of jaguar babies recorded in a way that suggests a series of events.

Here I interrupt the narrative to make what may appear to some to be a departure from logic and reason. Although much lip service is paid to the desirability of a more holistic world view, the dreams, impulses, and intuitions that lie at the heart of creative work and give it much of its zest and excitement rarely make an appearance in serious discussion on the frontier of knowledge. The following is a record of events followed by a dream that helped me loose the knot of the Olmec mythos. The background is important, as will become apparent.

✦ ✦ ✦

Early in December of 1992, while reading through a stack of books and papers on the Olmecs, I was suddenly struck with a strong impression of the approaching holidays. It was time to prepare for the celebration of the last polar personage of our culture, Santa Claus. Yielding to impulse, I picked up the phone and made arrangements to make Christmas cookies with two of my sisters. Later, on checking my recipe file, I realized that my favorite recipes were missing and made a note to call one of my sisters the next day to check on ingredients.

That night, I was delighted to meet another sister, deceased several years, in a dream. She gave me a cookie recipe, very simple, just five ingredients, and I struggled to commit it to memory in order to bring it back from the dream and share it with my other sisters. Then I was awake with the first ingredient clearly in my mind: five pounds of flour. Impossible. It was too large a quantity, and I soon realized that it wasn't a recipe at all, but one of those highly edited fragments that dreams present you with when you are busy and they are determined to get through.[22]

The "five" was clear enough, though not yet resonating with waking reality, but it was days before I accepted the rest of the connections. In the word play common to dreams, "pounds" was to be understood as "blows" or "hits," "flour" to be "flower." The message my sister had given me was that there were five blows to the flower, the four-petalled kin flower which was one of the recognized astronomical symbols of the Olmecs. The kin flower does not remotely resemble the sun, which it supposedly represents. For some time I had recognized it as a quartered symbol, many of which had been found at the pole by other researchers.

It was an interesting dream, but only a beginning. Now the work would commence, checking it against reality.

✦ ✦ ✦

As the civilization of the Olmecs continues to be unearthed, we have begun to perceive a series of monumental sculptures, generally considered to be altars, at the major centers. To date, the best-preserved series is at La Venta in the state of Tabasco. Although the site fell victim to the search for oil, many of the sculptures have been removed to the Museum Park at Villahermosa where they may be seen today. Perhaps the most remarkable of the monuments is the altar of the quintuplets.

The altar is massive, of a blocky, rectangular shape, with a figure emerging from a niche in front, holding a baby across his arms. On the sides of the altar are bas-reliefs: four more figures, two on each side, each holding a were-jaguar baby. The monument is badly disfigured, particularly the baby in the center, but it is

[22] Jeremy Taylor, *Dream Work* (New York: Paulist Press, 1983) pp. 46–49.

Altar of the quintuplets.

generally supposed to be another were-jaguar infant, engendered, following Stirling, when jaguars attacked women.

Here we will discard the idea of quintuplets and assume that these five were-jaguar babies were born serially, over a period of time, and represent the catastrophic events of the Earth-Mars Wars as described by Velikovsky. Beginning with this hypothesis, we can examine other Olmec sculpture and art to see if they fit the pattern. The altar of the quintuplets is already speaking to us in catastrophic terms, for the niche from which the central priest and were-jaguar infant are emerging is surrounded by a giant omega symbol, usually associated with Venus. We begin to suspect this middle infant, the third of five, of having done the most serious damage.

The monument must have been erected after the catastrophic events had transpired, at a time when their serial nature was well understood. But there would have been a chaotic interim period for the Olmecs. Working back through time, we will enter this twilight zone which Western man has not yet made sense of. The Olmecs, with an apparently unshakable confidence in the natural order, have done a workmanlike job of describing it.

First, we will trace the “five-element”` further, for the culture of the Olmecs emphasizes it strongly. David Joralemon has done some of the work for us in his treatment of Olmec art and iconography, recognizing the four dots and central bar of Olmec work and grouping them together for our convenience.[23]

[23] David Joralemon, “A Study of Olmec Iconography,” *Studies in Pre-Columbian Art and Archaeology,* Vol. 7 (Washington, D.C.: Dumbarton Oaks, 1971) No 85. Also see “Olmec Dragon,” pp. 37, 41.

Elsewhere four similar objects are placed, two and two, on each side of a central figure.[24] The handsome Las Limas sculpture in the archaeological museum in Xalapa, Vera Cruz, is a good example, a seated figure holding a were-jaguar infant across his lap. The detail is next to impossible to photograph, but four birdlike heads are incised on the knees and shoulders of the seated figure. Four other symbols bracket his mouth, and four more his forehead.[25] The were-jaguar infant's chest is emblazoned with the X's of two different crossed bands.

Were-jaguar infant held by Las Limas figure.

Some years after Joralemon published his opus, another example of the five-element was unearthed at Cacaxtla, a painted figure with two Mexican Star signs on either side.[26] The central figure holds the fifth Mexican Star in his left hand[27] while the five-element is repeated in the five points of the Mexican Star and the unusual five-lobed skirt worn by the figure in which the Venus omega glyph is centered.[28]

Painted figure from the Temple of Venus at Cacaxtla.

In further examining Joralemon's work, the upright bar with two dots on each side can be seen as a shorthand. It can be read as nine, following the Maya convention of a dot for one, a bar for five, but in this instance it is probably meant to be a symbol. The five-dot element of the quincunx, where one dot is centered within four others, is also well represented in Olmec culture.[29]

The bar itself receives considerable attention from the Olmecs. The monumental basalt sculpture that Franz Blom and Oliver La Farge found atop the volcano San Martin Pajapan represents a seated figure wearing a jaguar mask and holding a cylindrical bar in

[24] Ibid., No. 5, 12, 22, 89, 228.

[25] Ibid., No. 10.

[26] John B. Carlson, "America's Ancient Skywatchers," *National Geographic*, Mar. 1990, p. 98.

[27] This suggests the "left-handed" Huitzilopochtli.

[28] Additional examples of figures displaying the five-element can be seen in "Mural Masterpieces of Ancient Cacaxtla," George E. Stuart, *National Geographic*, Sept. 1992, pp. 120–136.

[29] "The quincunx is the most frequently occurring sign in the Meso-american symbolic language." Laurette Sejourne, *Burning Water* (Berkeley: Shambhala, 1976) p. 91.

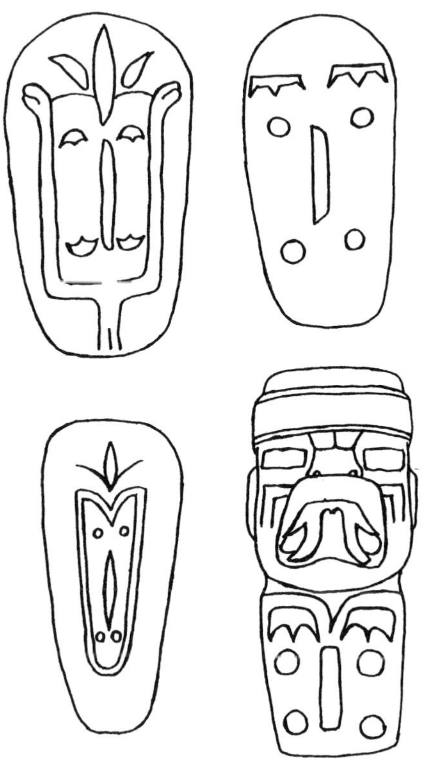

The five-element in Olmec writing, celts from La Venta, an hacha from the Gulf Coast (bottom right).

The sculpture found on top of the volcano San Martin Pajapan.

its hands. The attitude of the figure is very like that of the priests (for so they are assumed to be) holding the were-jaguar babies.[30]

Drucker found an almost identical statue broken into pieces at La Venta,[31] where an earthen volcanic cone over 100 feet high was constructed at one end of the ceremonial center. Monument 8 at La Venta also depicts a figure sitting cross-legged, holding what is left of a bar in its hands. Monument 11 from San Lorenzo holds one as well.

The bar, in numerical notation, represents five; the way it is held recalls the way the were-jaguar infants are held in the altar series, suggesting that it, too, was a result of catastrophism. But the bar, in the hands of a creative people, has not finished its development.

At Chalcatzingo in the highlands south of Mexico City is a series of rock carvings in relief.[32] In petroglyph 1, a figure known as El Rey holds a bar and is seated on a block. Both bar and block are actually S-shaped figures molded to a block shape. We know now that the Earth has tilted, for the S-shape represents the blue worm, Xonecuilli,[33] as the ancients knew Ursa Minor.[34] In fact, the crossed bands of the present world are inscribed on top of the cave-like area in which El Rey sits.

[30] The right hand of this particular statue holds the bar from below, the left from above.

[31] Robert F. Heiser, "New Observations on La Venta," Dumbarton Oaks conference on the Olmec (Washington, D.C., 1968) p. 35.

[32] David Grove, *Ancient Chalcatzingo* (Austin: Univ. Texas, 1987).

[33] Jorge Enciso, *Designs from Pre-Columbian Mexico* (New York, Dover, 1971) pp. 101–103. Also see Enciso, *Design Motifs of Ancient Mexico* (New York, Dover, 1953) pp. 55–60.

[34] Sahagun, *History,* Book I, VII, cap 3. Nuttall claims that the recurved scepters of the culture, hecaxonecuilli, are a representation of Ursa Major at the pole. The recurved scepters are a widespread phenomenon. See O'Neill, *Night of the Gods,* pp. 431–432.

Most important, three Venus omega glyphs are attached to the construction that serves the figure for a hat, indicating that three catastrophes have taken place.

It is worth mentioning here that because of the focus on the pole throughout Mesoamerica, one occasionally sees a kingly figure seated on a jaguar throne, and the little jaguar stools are still common in the markets around Lake Atitlan in Guatemala and elsewhere. This would reflect the jaguar, Oceotl, at the pole throughout the Jaguar Age (Hesiod's Age of Heroes).[35] Thus putting El Rey on a Xonecuilli throne after a catastrophic tilt of Earth can be easily understood. The king now sits on the Blue Worm, or Ursa Minor. Later, jaguar and eagle (Oceotl and Cassiopeia) would be perceived as juxtaposed near the celestial pole, providing the foundation for the two great warrior groups of the Aztec.

Petroglyph 1 at Chalcatzingo.

[35] Two Olmecs with headdresses pay homage to a jaguar in "New Light on the Olmec," George Stuart, in *National Geographic*, Nov. 1993.

The blue worm, Xonocuilli, Ursa Minor (from Jorge Enciso).

Petroglyph 4 at Chalcatzingo.

The El Rey bas-relief is the last in a series which appears to describe a change of climate, fertile ground for consideration after a considerable tilt of the terrestrial pole. Jorge Angulo V describes the sequence of reliefs as showing an increase in rainfall which he assumes to be a seasonal event,[36] but which I believe to have been a result of the catastrophic tilt: increase in the size of clouds, the amount of rain, the growth of squash plants. Meanwhile a series of zoomorphic animals, crouched on the S-symbol of the blue worm, Xonocuilli, look up at the clouds, communicating in breath or speech scrolls.

Petroglyph 4 shows two jaguars attacking two men. A cartouche over the eye of the upper feline contains the crossed bands of the present world. We are indebted to Angulo for describing a Venus omega element in the ear of this creature,[37] difficult to see in the accompanying plate. Again, the low angle of the X indicates a catastrophe that occurred after the tilt of the pole. These two jaguars with their low-angled X's probably represent the last two catastrophes of the Earth-Mars Wars.

The altar series at La Venta probably best represents what we would think of as mainstream Olmec thought. Of a total of seven "altars" recovered, only five have niches and emerging figures, only two have were-jaguar babies.[38] Altar #4 in the literature now claims our attention.

The figure of a priest sits in a niche directly under the huge, stylized mask of a jaguar, within the mouth of which are the crossed bands of the present world. Four familiar symbols appear to be striking on the edge of the border of the niche, four omega symbols with long, streaming tails. Although they are considered by some to be vegetation symbols, we believe they mark four visits of a comet-like apparition.

This altar must have been erected after the fourth catastrophic event, worrisome because the jaguar had moved from the pole on the third event. Would he continue to attack mankind, to move further in the sky? There

36 Grove, *Ancient Chalcatzingo,* pp. 133–141.

37 Ibid., pp. 145–146.

38 Soustelle, *The Olmecs,* pp. 40–41.

Altar #4 at La Venta (front view).

Altar #4 at La Venta (side view).

is no were-jaguar infant in this sculpture. Instead, the priest holds the end of a rope that extends around the corner of the altar and is attached to a human figure in bas-relief on the side. Certainly a people who have just seen their world come loose from its moorings would find a rope a comforting symbol. I suggest the figure is representative of the jaguar Tezcatlipoca, no longer at the pole but tied or yoked to it, circling around it, missing one foot because it now dips below the horizon. We will return presently to the image of a figure tied to the pole, since it was used in ritual by the ancient Mexicans.

Fortunately, it is an ill wind that blows no one good. A bulldozer, working to create an airstrip to service the drilling operation at La Venta, accidentally dug up one of the most evocative sculptures produced by the Olmecs. It is in bas-relief, a figure of a priest with his medicine bag in his hand, sitting attentively, surrounded by a snake, a protective earth monster in this case.[39] This helps us imagine the situation of the priest when he is underground, before he emerges from the niche beneath the altar.

There is a tablet above his head, on the top of which are inscribed two square X's descriptive of the previous age. The first two hits on the Earth have already taken place, but the terrestrial pole has not yet tilted. The bag held by the priest is also marked with at least one, possibly two square X's,[40] as he is apparently keeping score. Thus the priest waits within the

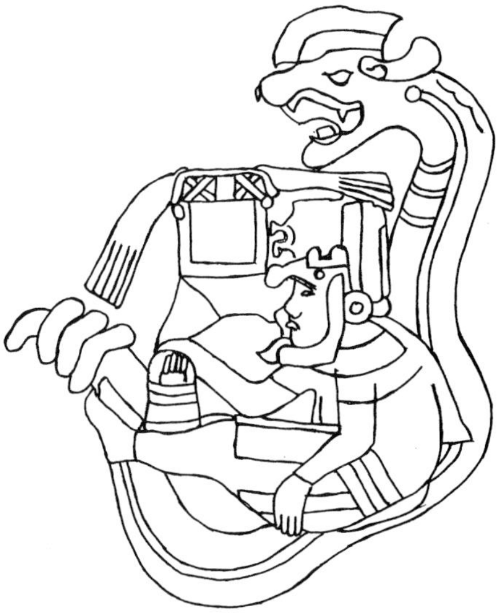

Bas-relief of a priest surrounded by a snake.

[39] This snake is often considered to be threatening.

[40] The X's on the bag are impossible to see both on the sculpture and in most photographs of it. See the carefully lit photograph in *Art of the Maya* by Henry Stierlin, plate 23 (New York: Rizzoli, 1981). It may be that the images can still be recovered through rubbings or latex cast.

Earth, in expectation of a comet-like visitation, not knowing what the shape of the heavens will be when he emerges, but perhaps hoping it to be another square X. Alas, such was not the case. Following Olmec iconography, the pole tilted at the third in a series of five catastrophes.

Another of the monumental sculptures of La Venta will help us see the problem the Olmecs were faced with. A large Olmec figure sits cross-legged (see photo on page 40), leaning forward in the attitude common to Olmec statuary.[41] On the torso are figured two different crossed bands. As on the chest of the jaguar baby held by the Las Limas figure, the X of the present world age is represented above the X of the previous one. The symbols of both ages must be honored because of the serial nature of the catastrophes and the uncertainty which accompanied them.

There is a celt, a ceremonial ax from Las Tuxtlas published by Pina Chan in 1964, which speaks to us in a more human way. The effigy wears a crown with five elements, the center one bearing the crossed bands of the previous age. The other elements in the crown are similar to those on the face of the Las Limas icon. The eyes of the effigy on the celt are both incised: the right one with the crossed bands of the present world, the left one with a wavy line. I offer a reading of barrenness for the wavy line. It can be understood as the edge of a flint, probably a proto-sign for the day sign Etznab, which came to be expressed as crossed bands of wavy lines.[42] Etznab, flint, is the sign of barrenness and destruction.[43]

A ceremonial celt from Las Tuxlas.

The effigy sees a changed world, decidedly different from the one expressed on his crown. The angle of the path of the sun has changed. The world appears barren and destroyed, just as one would expect after a catastrophic event. We should not be surprised to see two tears depending from his right eye.[44]

The characteristic downturned mouth and occasional tears of Olmec art probably owe much to the change in the heavens indicated by the two different angles of the crossed band element. What we are talking about here is the catastrophic end of the Fourth Sun of the ancient Mexicans, Ocelotonatiuh, the Jaguar Sun. There is some conflict in ancient media on the order of the five suns enumerated by the Mexicans, but the Olmec speak clearly and powerfully about the most recent one.[45]

The "massive offerings" at La Venta can be more easily understood in a catastrophic

[41] Monumental sculpture No. 77 at La Venta Park Museum, Villahermosa, Tabasco, seen at the beginning of this chapter on page 40.

[42] Etznab is a Maya day sign. The corresponding day sign in the Mexican calendar system is a picture of a flint.

[43] A similar face can be seen in the Borgia Codex (Plate 53 of the Diaz and Rodgers restoration, page 5 of the five directional trees), where the left eye has a flint drawn over it.

[44] This celt is sometimes drawn with tears in both eyes.

[45] The ages of Ehecatl, Quiahuitl, Atl, and Oceotl are not always given in the same order in the mythology of the ancient Mexicans, nor does the jaguar always occupy the same position in monumental sculpture.

context. Michael Coe describes one pit 50 × 61 feet on the sides, 24 feet deep, filled with over 1,000 tons of rough serpentine slabs, a material highly prized by the Olmecs. Over this was placed a green mosaic set in yellow and orange soils, a jaguar mask of 485 shaped blocks of serpentine. The mosaic, a work of art not made for the eyes of man, was then covered with a layer of adobe blocks and clay.[46]

Three jaguar masks were found at La Venta, two of them identical, all buried along with other offerings of jade and serpentine.[47] What we see here is a herculean effort to bury the Jaguar Age, to propitiate the Earth, to make it cease from the movement in which the Age of Ollin was born.[48] One of the jaguar masks can be seen at the La Venta Museum in Villahermosa.

Although the Olmecs used the omega glyph customarily associated with Venus to depict the appearance of the cometary body that periodically visited them, I agree with Velikovsky that it was Mars they were dealing with. He identified Huitzilopochtli, the Aztec war god, as Mars even before the old Mesoamerican cultures were understood to be extant during the 8th and 7th centuries B.C.[49] There are two representations in Olmec art of a configuration that suggests Mars, a doorway or opening with three bars around it. One of these has tall crossed band elements on either side. A linear reading would be "an opening of 15 (years) between two events."[50] This gives the interval of Mars' closest approaches to Earth, echoing the periodicity of its catastrophic appearances in our skies.[51]

A massive jaguar mosaic at the La Venta Museum in Villahermosa.

[46] Michael Coe, *America's First Civilization* (Washington, D.C.: Smithsonian, American Heritage, 1968) p. 61.

[47] One slightly different mask was centered in what is now suspected to be a ball court. Friedel, Schele, Parker, *Maya Cosmos* (New York, Wm. Morrow, 1993) p. 134.

[48] Ollin means movement, motion or earthquake, from the same root as the word for rubber, olli.

[49] Velikovsky, *Worlds in Collision*, pp. 250–251.

[50] This image is from a pottery vessel from Las Bocas. Another similar image gives "an opening of 25 years," a close fit between the Venus and Mars catastrophes.

[51] At present, the close interval of Mars occasionally ratchets up one synodic period, making it 17 years.

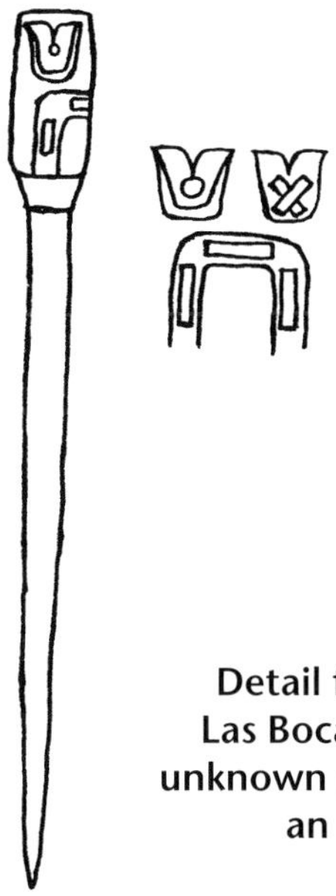

Detail from a pottery vessel from Las Bocas, and a jade perforator of unknown provenence, both suggesting an opening of 15 (years).

Confusion between Venus and Mars during the Earth-Mars Wars will be examined in the chapter on the Mexican star sign. The records of destruction in the Olmec civilization are not perfectly clear; some appear as natural disasters, others as purposeful. Workers in the field are well aware of the problem, and it is to be hoped that they will soon be ready to examine the evidence from the viewpoint of historical interplanetary catastrophe.

For moderns to grasp the ancients' orientation in space has always been problematical. Now that we begin to understand that time is also an issue, the nature of the confusion clears somewhat. In the following chapter we will observe the ancient Mexicans turning from north to east to honor the Fifth Sun, Nahui Ollin, and their new sun and war god who will eventually find expression in Huitzilopochtli.

The top section of Stela 10 at Yaxchilan features an omega-shaped skyband that describes some of Earth's recent catastrophic history.

5. The Shape of World Ages

The theory about "how the world began" seems to involve the breaking asunder of a harmony, a kind of cosmogonic "original sin" whereby the circle of the ecliptic (with the zodiac) was tilted up at an angle with respect to the equator, and the cycles of change came into being.

We do not think it is an accident that this originator of time [Xpovos/Kronos] begins with an X, representing the obliquity of the ecliptic in Plato's Timaeus.

– Giorgio de Santillana, Hertha von Dechend, *Hamlet's Mill*

The authors of *Hamlet's Mill* have overlooked the Fall, but they are honing in on what the X we have been discussing is all about. Rather than tilt the enduring ecliptic, however, the tilt is better seen as one of Earth. The X, of course, remains the same. As we learned in the previous chapter, in each age it is the angle formed by the X which is important.

The X of the crossed bands, then, though it cannot be directly perceived, comes from the sky in each age and shapes life on Earth, handing down the path of the sun, our seasons, climate, and weather. It has not been so for us Westerners, however, for we have not understood it. What is the X to us?

It is the deepest symbol for the unknown in the culture, the quantity for which we solve equations when we first learn algebra, the mark we make for our name when we know nothing of writing, the secret place on the map where the treasure is buried. It is the least-used letter of the alphabet. In point of fact, K, C, and Q do its work, and in Latin all questions begin with Q as does the word itself in English. X is our symbol for the unknown. Unless we are conscious of doing geometry, "X marks the spot" where two lines cross rather than telling us anything of angles. The X stands between us and a valuable body of knowledge, the parameters of the recent ages that man has lived through, the shape of world ages.

It was while reading Leon-Portilla's *Time and Reality in the Thought of the Maya,* meditating on the X drawn by Villa Rojas, that the crossed bands of the Mesoamericans began resonating with the tilt of the terrestrial pole against the ecliptic. Some of you may have noticed that other parts of one's body are often wiser than one's head. My fingers began ruffling back through the pages to a drawing of the top of Stela X at Yaxchilan (shown on page 56), the Maya city on the Usumacinta. There is an unusual skyband on the bas-relief, actually in the shape of an omega sign, and it has three crossed bands on it, each one drawn at a different angle. The old god of time, Kinh, was pictured just above it, wearing a cartouche around his neck with an X inscribed on it.

At some point in time the skyband began to make sense. The classic Maya had created an overview of time as they understood it, a record of world ages going far into the past. And to what do we owe this remarkable record? During the classic period of the Maya, the elderly Shield Jaguar fell in love with a young woman named Lady Evening Star. Their portraits appear in cartouche on either side of the top of the stela,[1] which is a tribute to their

[1] Linda Schele and David Freidel, *Forest of Kings* (New York: Morrow, 1990) fig. 2:12, p. 86.

May-December marriage. We owe much to whoever chose Venus as a namesake for this young woman.

Like other men in other cultures, the Maya were not above such puffery as claiming ancestors in the dawn of time or even in the sky. When the time came to memorialize this important wedding, it would have been only natural to call in an astronomer-historian to devise a skyband recording the historical appearances of Venus to honor Lady Evening Star.

Reading between the two rather comical double Venus glyphs at either end of the sky-band, left to right and past to present, are: the four-petalled kin flower, the day sign Akbal for darkness, an extremely high-angled X that we have not seen before, the day sign Lamat for Venus, Akbal again for darkness, two unusual signs of a circular nature, the X of the Jaguar Sun, and the X of the present Age of Ollin.

Recapitulating, we have: polar sun, darkness, unknown age, Venus event, darkness, pattern A, pattern B, the X of the Jaguar Sun, the X of the present sun. It is not unlike the sequence of events that Velikovsky has reconstructed in *Worlds in Collision.* He also felt that a different world order preceded the Venus events, and that is exactly what is described by the first high-angled X.

The omega-shaped skyband of Stela X agrees with Maya records of successive world destructions in the *Popol Vuh*, and with Aztec ritual of binding the sheaves of years in honor of the destructive Venus who brought darkness. Finally, it agrees with the two crossed bands of Olmec culture. But these master astronomers and calendar makers, the Maya, often have more than one string to their bow.

Many students of epigraphy as well as members of the general public have always been charmed by a handful of remarkably ancient dates on Mayan monuments, some going back millions of years. Perhaps something is behind them, but they are not generally regarded as a realistic description of the past. These dates might be described as a "leaning into time" for the sake of it. But the dates that cluster 4,000 or 5,000 years ago are something else again, probably real but very shadowy events not well understood at present.

As the Mayan glyphs continue to yield their secrets, they reveal images of the past somewhat contrived by the political arrangements of the Classic Maya. In an effort to legitimize their regimes, exaggerated claims were made to relate individuals from the Classic era to ancient legendary figures. History was projected into the past, but this only means that the Maya are more like us than we have hitherto been prepared to believe.

On the one hand, the Maya give us Stela X at Yaxchilan with a clear record of catastrophic change and the shape of previous ages; on the other, they retrocalculate smoothly in 360-day years back to 3114 B.C. according to the Thompson correlation. One has to keep in mind a powerful kingship during the Classic period telling the priests and artisans what to put on the monument. The astronomers and ahau had different agendas. We must continue to read between the lines. It is entirely possible that other fragments of the catastrophic record are buried in the mass of self-aggrandizing Classic Mayan history.

Anyone who visits the earliest sites of civilization on the Mosquito Coast, the Peten of Guatemala, or the Gulf Coast of Mexico must ask themselves the same question. With all the beautiful areas available in these countries, why did early Maya and Olmec cultures choose to settle here? Insufferable heat and noxious insects are only the beginning of the unpleasant conditions. Archaeologists who have worked in these areas to study ancient cultures, often at risk to their health, deserve our deepest appreciation and gratitude.

As it happens, historical catastrophism has an answer to the question: the climate was different. With the sun moving farther north and south, following the angle of the square crossed bands and spending much less time over the tropics, the low-lying areas near the coast would have been the choice ground on which to live.

The Olmecs brought their culture reasonably intact through the chaotic period of the Earth-Mars Wars only to watch it broil under the sun of Ollin. The heat must have seemed devastating as the new sun hung closer overhead. There would have been gradual migration toward the mountains of Central America, toward the high central plateau of Mexico where a later civilization would have its focus. The migration now appears as an extension of Olmec influence as more and more artifacts in their characteristic style are dug up in the high plateau of Mexico and the mountains of Mexico and Guatemala.

That the later Mexican culture had its roots in that of the Olmecs cannot be doubted. The famous Stela C at Tres Zapotes with its controversial date[2] illustrates this perfectly. The top and bottom half of the monument were dug up at different times and reside in different places, somewhat obscuring the connection. The bottom half, representing the Olmec culture and the end of the Jaguar Age, is in the Archaeological Museum in Mexico City; the top half, representing the new age of Ollin, is still in Tres Zapotes where it was dug up. It would make better sense the other way around.

Joined together, the two halves of Stela C depict the transition between the ages clearly. The Olmec mask on the bottom is as we have come to expect: the downturned mouth, tears running down the cheeks, the cleft in the middle of the forehead. A wineglass image

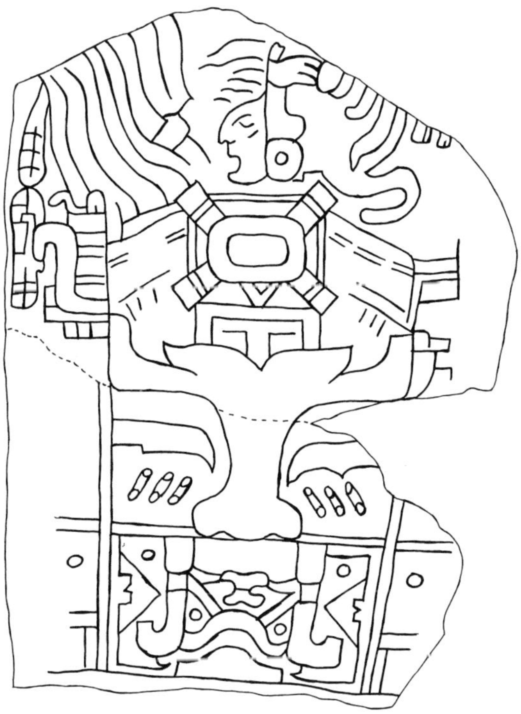

Stela C at Tres Zapotes.

runs up from the nose.[3] The image of a man on top grows directly out of the forehead cleft and wineglass shape below. The figure of the man is centered above the image of our old friend, the crossed bands element, the arms of the X broadening out to look more like Ollin. The angle reflects the present age, but something unusual is about to happen.

At some time between the flowering of the Olmec culture and that of the Aztecs, the Mesoamericans broke some of their ties to the pole with its former jaguar and present blue worm and began to look east to the rising of the new sun of Ollin, Huitzilopochtli. The move involved turning the crossed band element 90 degrees, at which time it transformed into Ollin, the sign pictured on the

[2] Scholars have addressed the question of whether the Olmecs were working from the same zero date as the Maya. Stela C has been dated at 31 B.C. Placing it one baktun earlier would give a date of 425 B.C., which would seem more reasonable.

[3] The wineglass image can also be traced as a symbol of catastrophe though not as easily as the Venus omega sign. See *Olmec Iconography*, No. 10, 22, and 192. Also Gerald Berjonneau et al., *Rediscovered Masterpieces* (France, Editions Arts, 1985) image 46 with its four flanking wineglass symbols.

The Aztec calendar stone, the sign of Ollin reflecting the angle of the ecliptic at Mexico City.

Aztec calendar stone. The smaller angle would then be at the top of the calendar stone and elsewhere, the emphasis on the east where the sun rises.

That the calendar stone reflected the angle of the ecliptic at Mexico City was well understood by Nuttall and others.[4] Because her work has become unfashionable and is rarely referenced, younger scholars have to reinvent the wheel.

It is difficult for us to understand the importance the ancient Mesoamericans attached to the crossed bands element expressed in the Ollin sign, unless it is juxtaposed to the four earlier suns that were world ages, in each of which time had a different shape. At stake is nothing less than the sacralization of the world. The new sun requires ceremony and ritual to reorient space and make it sacred. Temples, houses, and fields must be situated so they can be assimilated to the center of the world, so they will be "real." Franz Tichy has written a paper showing that the work was done.[5]

The pyramids and fields around Mexico City and Puebla have been found to be oriented to the sun of Ollin and the angle of the Aztec calendar stone. In their attempt to obliterate the Mexican civilizations, the Spanish razed their ceremonial centers and erected churches

4 Nuttall, "Fundamental Principles," p. 13.

5 Franz Tichy, "Order and Relationship of Space and Time in Mesoamerica: Myth or Reality?" Mesoamerican Sites and World Views, a Conference at Dumbarton Oaks, ed. Elizabeth P. Benson (Washington, D.C.: Dumbarton Oaks, 1981).

on their foundations, faithfully following the solar orientation of the native religion. Their behavior in this instance must have provided any of the surviving Indian cognoscenti with a few bright moments, seeing that the sun of Ollin continued to be honored.[6] Tichy has elucidated the system as using the solstitial points to keep the world and the calendar in order.

Thus the X of the crossed band element is as well adjusted to time on a grand scale as is the analog clock of Western technology on a small one. By looking at the angle of the "hands" of the X, we see not what time of day it is, but what age we are in.

Since Velikovsky has identified Huitzilopochtli as Mars on the basis of his warlike aspect, an examination of the myth may prove helpful. The sun and war god appeared late on the scene in Tenochtitlan, arriving with the Aztecs in the wave of migrations that took place in the Americas in the 13th century A.D. He would necessarily have been integrated into the existing pantheon with some care, something which must always be kept in mind.

As a warrior, Huitzilopochtli was impeccable. Engendered by a ball of feathers that fell into the garments of his mother, the grisly Coatlicue, she of the serpent skirt; born with shield and weapons intact, ready to make war; the left-handed hummingbird cut a wide swath with a swordlike a ray of light. He annihilated his brothers and sisters, chasing some of them out into a lake before killing them.[7]

If Huitzilopochtli is Mars, included in the family of the solar system as long as records and myth exist on the subject, this spectacular birth and immediate warlike behavior must have occurred far in the past, followed by a long period of peaceful behavior during which he caused no comment. It was only in the 8th century B.C. that he resumed a warlike career.

Hummingbird, huitzitzilin, means "the resuscitated."[8] The sense in which Huitzilopochtli could be understood to be resuscitated is the same sense in which Mars became "outrageous" at the time of the mythology of the *Iliad*.[9] Both resumed warlike behavior after a long period of quiet.

There is a link between Olmec iconography and Aztec ritual which comes into focus here. At the time of the Conquest, the Aztecs celebrated a festival known as Tlacaxipehualiztli, during which a strange ritual was dramatized. A combat took place between a prisoner, tied by a length of rope to the center of a large circular stone, and five warriors.[10] Often referred to as gladiatorial combat, it was rather a ritual execution. The prisoner could move freely over the circular stone but was armed only with toy weapons. A wooden sword is mentioned in some accounts, sometimes a macana or club set with feathers instead of razor-sharp obsidian.

Any of the five warriors, armed with conventional weapons, could have dispatched this man with little trouble, but this was a ritual, an instance of ancient media containing valuable

[6] Solar orientation of churches is not unknown in Western tradition. Certain churches in England are oriented to the sun rising on the day of the saint whose name attaches to the structure. Lockyer in *Dawn of Astronomy* references Gilbert Scott, "Essay on Church Architecture," *Builder*, Jan. 2, 1892.

[7] Lewis Spence, *Myths of Mexico and Peru* (Boston: David Nickerson, 1913) p. 70. Spence says that Huitzilopochtli occupied a place in the Aztec pantheon similar to that of Mars in the Roman. He mentions that Huitzilopochtli's shield bore a quincunx of five tufts of down, visible in some artistic representations.

[8] For the warrior-hero to be resuscitated, renewed, or reborn is a conception reaching far into the past.

[9] Velikovsky, *Worlds in Collision*, pp. 243–250.

[10] Bernardino de Sahagun, *A History of Ancient Mexico*, Book II, Calendar Festivals, second month. Also see T. Joyce, *Mexican Archaeology* (Boston, Medici Society) p. 66.

information. The warriors approach the prisoner one at a time, fight for a period, and then withdraw. It is the fifth warrior, fighting left-handed, who finishes off the victim.[11] Subsequent to this unequal combat, the corpse is flayed, and the skin worn by another. Accounts vary as to the disposition of the flayed skin, but it was definitely considered to be a disguise or transformation.

Modern scholarship has reduced this fascinating account to a celebration of the change of seasons,[12] but it also resonates closely with the five serial catastrophes described by Olmec iconography. The prisoner represents the old Jaguar sun, Oceotl, tied to the pole which he no longer dominates, helpless before the serial onslaught of the left-handed Huitzilopochtli. The flaying of the victim and the donning of his skin by another represents the metamorphosis of the sun from Jaguar to Ollin.

Traces of this five-aspected warrior occur elsewhere. The divinity Macuilxochitl (five flowers) was celebrated at Xochilhuitl (feast of flowers) which lasted five days. Offerings were made of five corn tamales and bread was baked in the S-shape of Xonecuilli, Ursa Minor.[13] The association of Mars with a flower is actually cross-cultural.

As well as the Olmec described the catastrophic events of the 8th and 7th centuries B.C., it is disappointing to find that the later Mesoamericans have left us an obscure, even confusing record. The quincunx representing the five visits of Mars is visible on the shield of Huitzilopochtli, but is also attached to Venus, becoming the day sign Lamat! I am convinced that the Mesoamerican astronomers of the 8th and 7th centuries B.C. were perfectly clear on what happened in the skies. Problems of an orthodox nature may have intervened to confuse the record or, what is more likely, a holistic world view left an imperfect linear record. The Mexican star sign is the key to how well the Mesoamericans grasped the cycles of Venus and Mars. We will analyze it in a later chapter.

For now, we leave the Jaguar Age to the Mesoamericans and cross the ocean to where the most recent age was perceived as Hesiod's Age of Heroes. There the constellation at the pole has a more familiar name, the Great Bear, Ursa Major.

---

[11] Laurette Sejourne identifies the fifth warrior as Huitzilopochtli. *Burning Water* (Berkeley: Shambhala, 1976) p. 158. But as Mars, each of the warriors represent Huitzilopochtli.

[12] Catastrophic myth often expresses seasonal change, but this is only one level of meaning.

[13] Sahagun, *Historia,* Book I, Part II, cap. 14 (Santa Fe, Univ. Utah) pp. 31–32. In cap. 3 of I, VII Ursa Minor is described as S-shaped.

www.carinasoft.com

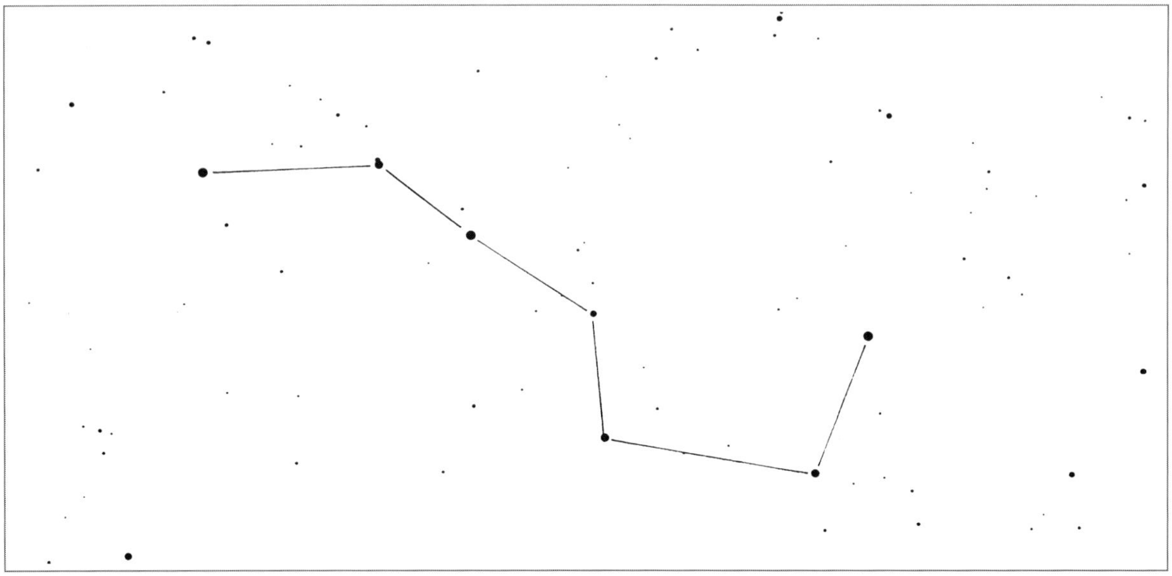

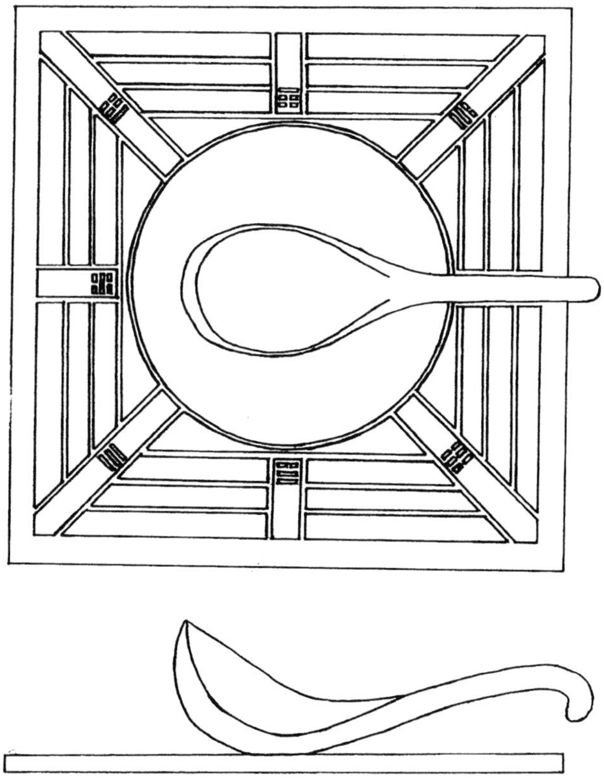

The Chinese lodestone spoon rotated on the Earth-plate as Ursa Major rotated at the pole during the Age of Heroes (Jaguar Age). The handle of the spoon (dipper) always pointed south.

6. Ursa Major at the Pole

. . . Great Bear, nicknamed the Wain, which always wheels round in the same place . . . the only one which never bathes in Ocean's Stream, . . .

– *Odyssey,* Book V

Bear stays in the same place and rotates on the same point.

– Corpus Hermeticum

No longer did the stars in the Bear, moving in a circle fastened around his hips, dance up high near the northerly Pole, but moved to the southwest and wet their feet in the unfamiliar Ocean in the Lake of Hesperia.

– Dionysiaca of Nonnos

And the Wain, which was never bathed in the sea, shall be plunged beneath the all-engulfing waves.

– Seneca, *Thyestes*

In reading the Classics, we occasionally stumble on a reference to the Great Bear, "always dry," "never dipping beneath the sea." Or again we might find a description of the constellation moving suddenly so that it does get wet. Those of us who live in the USA have difficulty spotting the constellation on autumn evenings because it lies so close to the horizon. In the latitudes of ancient Greece and Rome the bear would dip below it. In a maritime culture, below the horizon means below the sea.

The statements at the head of the chapter, often couched in poetry, make no sense to a modern. Chaucer, Spenser, and Bryant, as well as many of the ancients, have all written of Ursa Major in these terms. But then poetry is akin to myth, spun from the same elusive filaments. Robert Graves speaks of proleptic and analeptic thought, the ability of the poet to access past and future.

Velikovsky made it clear in *Worlds in Collision* that the reason the Great Bear never dipped beneath the sea was because it was at the pole during a previous age. It "plunged beneath the all-encompassing waves" when the terrestrial pole tilted. The telling information he assembled in *Worlds in Collision* supporting this vision has been little noticed, even though it is a case in which the lines are clearly drawn. Orthodox astronomy is fond of retrocalculating the position of the pole, but nowhere does the circle of the precession of the pole enter Ursa Major. There is no getting around it. We must side with either the poets or the astronomers. In this chapter and several that follow, we will access ancient witnesses and ancient media that support the poets.

The Great Bear, Ursa Major, is a highly visible configuration of stars. Its present position in the northern hemisphere in the circle of perpetual apparition, and the brightness of its stars in an easily recognizable "dipper" shape, have made it a well-known constellation throughout the present age and doubtless much earlier. Both the constellation and its individual stars have names conferred by the oldest civilizations. The Chinese called Ursa Major the "Government," while the lucida, brightest star in the constellation, was called "Pivot of the Heavens."

In contrast, despite its present position at the pole, Ursa Minor, the Little Bear or Little Dipper, is otherwise unremarkable; the pole star itself is difficult to spot without reference to the "pointers" at the mouth of the Big Dipper. There could never be visual confusion between the two constellations. Nevertheless, since two different world ages and two different sky-bears are involved, the habit of the ancients of referring to the polar constellation as simply "the bear" makes it difficult to differentiate between the two. The Mithraic scholar Roger Beck quotes the work of Gundel: "In most contexts it is virtually impossible to tell which constellation is intended."[1] Hence the paucity of solid reference to Ursa Major at the pole.

In these pages we will follow Velikovsky's lead and pursue the Great Bear, Helice (Ελικη), and learn why Greek navigators (ελικωτασ) called themselves after the constellation.[2]

✦ ✦ ✦

Here I beg the reader's indulgence, interrupting the discussion with a personal experience balanced by a dream, both of which gave strong direction to the work in progress. I began work on the bears in the sky, half expecting a dream as often happens. Instead, an unusual waking event transpired, followed by a dream a week later.

In preparation for a camping trip, a well-provisioned metal cookbox stood outside the door of our mountain cabin one night toward the middle of September. Hearing a noise, I got out of bed and went outside, promptly falling over a litter of food and pots, backward into the now empty cookbox so that my hips were wedged into it. Bruised and in a vile humor, I extricated myself, repacked the box, and set a heavy rock on it, assuming I was dealing with a raccoon.

A few minutes later, back in bed and hearing more noise, I got up to run the varmint off. But something slowed me down. Peering through the window of the kitchen door with one hand on the doorknob and a broom in the other, I gradually perceived a large black bear not six feet away, pushing the cookbox around in a search for food. Immediately a different frame settled around the event. This was a real bear!

The dream bear arrived a week later, a huge golden-brown grizzly followed by six other smaller bears of the same color. Backing away from this imposing apparition, I fell down, whereupon the huge bear came up and sat squarely across my hips until I woke up.[3] In retrospect, there was a feeling similar to that of pregnancy. I soon realized that the work before me involved gestating, birthing, and nurturing a vision of the bear. But there were now two bears to be dealt with: the majestic, mythical bear in the sky, and the flesh-and-blood bear, smelly, untidy, and mischievous, but with oddly engaging ways. It is, moreover, from the living symbol that we can learn the qualities that caused it to be raised up to the sky.[4] As they did to Dante, ". . . the Muses nine point out to me the Bears."

✦ ✦ ✦

[1] Beck, "Interpreting the Ponza Mithraeum," *Journal of Mithraic Studies,* Vol. II, No 2, p. 143, note 81.

[2] The relatively stable center of the North Pole provides orientation.

[3] Polar figures are frequently represented as seated, resting, immobile. Nuttall, *Fundamental Principles,* p. 36.

[4] "The psychological approach to the enigma of the symbol, the design to extract from it the secret of its depth, cannot but fail, if the searching intelligence refuses to acquiesce to the chance of being taught something by the living aspect of the object under its attention." Campbell, Joseph, "The Dilettante Among Symbols," Heinrich Zimmer, *The King and the Corpse* (Princeton Univ. Press, 1956), p. 4.

The bear is too large, too long-lived a mythic figure to do justice to within these pages: old enough to preoccupy Neanderthal man as a cult figure, young enough to be the favorite stuffed toy for a child to take to bed for the night. Although the bear is sometimes mentioned carelessly in company with other large predators, in truth he enjoys a special position vis-à-vis humanity, a claim on our good will that outrages the purely rational concerns of orthodoxy. The Park Service in Yellowstone and other parks have their hands full keeping visitors from trying to relate to resident bears, some of whom are grizzlies. The bear is Gentle Ben and Winnie the Pooh, more recently Muffie Van der Bear. He enjoys the same nickname as one of our most popular presidents, Teddy.

The bear has an enormous range; it is found on all continents in the northern hemisphere in the temperate zones and farther north. With the development of cultural anthropology, it was gradually realized that the bear was a cult animal over most of this area, supported by a mass of ritual and ceremonialism.[5] The bear was a member of the family, sometimes nursed at the human breast, as well as friend, pet, and sacrifice. In death he could be a messenger to the gods. Many so-called primitive societies claim descent from the bear.

Campbell finds an instance of mythological thinking in early man who buried his dead in a way that suggests they were readied for the next world. He mentions "... a number of chapels in high mountain caves, where cave bear skulls, ceremonially disposed in symbolic settings, have been preserved ..."[6] and goes on to report modern instances among the hunting and fishing peoples of the far north who follow similar practice. In fact, the bear was known long before he appeared at the pole, but it is only since then that he has enjoyed ritual importance.

Thus far we have been finding phenomena in the skies brought to Earth, but the bear falls into a different category, an earthly creature raised to the sky. In dealing with the flesh-and-blood bear, our feet are on the firm ground of common knowledge with no need for bookish reference. The bear loves honey,[7] lives in a cave in the mountains, and climbs trees. He has resemblance to mankind: sitting, standing, using his paws. He can be encouraged to perform something like a dance.

When night first fell on Earth after the last catastrophic visit of Venus, the stars gradually became visible. The handsome new configuration of stars at the pole required a name. Who first saw a bear at the pole? Myth enters the linear world at its own pace, splashing into time like a stone patterning the surface of a pond. The physicist David Bohm posits a folded implicate order unfolding into the world we know.

When the seven bright stars appeared at the pole, they had to have some connection with the previous configuration of stars, for that was the way the ancients perceived the world. The last figure to sit at the pole had provided the sweet manna or honey-lash that had fallen to the ground between the two visits of Venus. A bear would have been after the honey presumed to be there. He must have slept in a cave in the world mountain that once stood at the pole, climbed the world tree that connected Heaven and Earth, and now he would dance

[5] A. J. Hallowell, "Bear Ceremonialism in the Northern Hemisphere," *American Anthropologist* XXVIII, 1926, pp. 1–175.

[6] Joseph Campbell, *Myths to Live By* (New York, Viking, 1972) p. 31.

[7] Honey-eater is the taboo name of the bear in all Slavic languages. Rhys Carpenter, *Folk Tale, Fiction and Saga in the Homeric Epics* (Berkeley, repr. 1962), p. 124.

around at the pole throughout the newly dawned age.[8]

The zodiac, belt of animals, had been loosened from the waist of Mother Earth by Venus and slipped into a new position. But now it was securely fastened with a new sky in place. Accustomed to finding a father god or mother goddess at the pole, mankind accepted the bear as the new ruler of the heavens.

The new sky of the Age of Heroes signaled the development of new records in primeval media. Since Venus had placed the bear at the pole, the four-year, one-half Venus cycle was the interval at which it was celebrated.[9] The bear's cave, symbol of the cosmos, is the entrance to another world, and the bear himself a messenger to the gods. The bear cult of the Ainu, some of whom still live on Japan's northern islands, is one of the best known.

Bears are taken as cubs and reared in a family, only confined when they grow too big and rough to play with the children. At the age of four, ceremonies are arranged and the bear is given an important charge: "... we are holding a great festival in your honor. Be not afraid. We will not hurt you. We will only kill you and send you to the god of the forest, who loves you ... You will ask god to send us, for the winter, plenty of otters and sables, and for the summer, seals and fish in abundance. Do not forget our messages."[10] The bear is mercifully dispatched in the belief that, after delivering his messages, his soul will return and become a bear in his old haunts. His hide is removed with head and paws attached and arranged so that he is the guest of honor at a subsequent banquet of his own flesh.

In his "Cult of the Sleeping Bear," Rhys Carpenter examines several immortality cults which perform ceremonies analogous to those above. In the Salmoxis cult, recorded among the Thracians by Herodotus, the bear has become the deity. Deathlike sleep was insufficient to the purpose, however, and so strong was the belief in immortality that humans were regularly sacrificed. "Every four years they chose one of their number by lot and, after instructing him in their various wishes and needs, send him away as a messenger to Salmoxis. And this is how they send him. While some of them group themselves, holding three javelins, others seize the feet and hands of him who is to be dispatched to Salmoxis, and swinging him up in the air let him fall on the spear points."[11] This action is tantamount to skewering the messenger on the world axis or cosmic column.[12]

The Father of History, Herodotus, continues on his way, questioning Greeks on the Hellespont and the Black Sea about Salmoxis and learning that he had been a slave of Pythagoras of Samos. "After gaining his freedom, he amassed considerable wealth and returned with it to his native land ... and there he built a banqueting hall in which he entertained the leading citizens and in the course of the feasting set forth his doctrine that neither himself nor his guests nor yet their children's children should die, but should come to that very place and there should live forever in

[8] Priestesses called arktoi (bears) performed arkteia (danced like bears) at Brauron and Munychia in Attica. John O'Neill, *Night of the Gods,* p. 945.

[9] The Lydians recorded the connection between the bear and Venus in the sanctuary of Artemis-Anaitis. Graves, *Greek Myths,* 116.j.

[10] Frazer, Golden Bough, quoted in Rhys Carpenter, "Cult of the Sleeping Bear," *Folk Tale, Fiction and Saga in the Homeric Epics* (Berkeley, Univ. of Cal.) p. 112. Also see Campbell, *Myths to Live By* (New York, Viking, 1972) p. 32.

[11] Herodotus, *Histories,* quoted in Carpenter, *Folk Tale,* p. 113.

[12] That the cosmic column should have three points relates to the trident of Poseidon, which will be discussed in a later chapter.

enjoyment of every happiness. But all the while he was saying this, he was engaged in making for himself an underground chamber; and when it was completed, he disappeared from among the Thracians by descending into the underground chamber, and there he abode for three years. The rest lamented and mourned for him as dead. However, in the fourth year he reappeared among them; and thus they were confirmed in what Salmoxis had told them."[13] By the time of Herodotus, of course, Ursa Major is no longer at the pole, and the Father of History, busily translating ancient media into modern, is already losing touch with the previous world age.

In China, the tenure of the Great Bear at the pole was recorded in a more playful style. Orient and Occident have different ways of thinking, well expressed by the magnetic compass from which both take direction. While our indicator of direction points north, theirs has historically pointed south. Though a moment's reflection tells us that either would serve the purpose of orientation, the Chinese model is elegant in that it has a celestial archetype, the turning constellation of Ursa Major at the pole with its south-pointing tail (see illustration on page 64).[14]

We recall that the brightest star in the constellation, the lucida at the mouth of the dipper, was called by the Chinese, "Pivot of the Heavens." Other ancient descriptions have the pole between the bowl and handle of the dipper.[15] In the 700 years of the Age of Heroes, assuming the same rate of precessional motion, the pole would only have moved about nine degrees of longitude. With the constellation in position at the pole, it is immediately apparent that the bowl of the dipper would appear to rotate while the long handle would sweep the sky around it, always pointing south.

The Chinese saw the dipper as a spoon which, brought to Earth, had an unusual function. The spoon was short, with a high, raised handle, and carefully balanced so that it could not only sit but spin on a point on the bottom of the bowl. Thus it could be spun on a smooth surface and the handle of the spoon used as a pointer in divination. There were two different ways in which this was done.

The diviner's board "... was composed of two boards or plates, the lower one being square (to symbolize the Earth, thus called the ti phan); and the upper one being round (to symbolize Heaven, hence called the tien phan). The latter revolved on a central pivot and had engraved on it the 24 compass points, ... It always bore, engraved at the center, a representation of the Great Bear."[16] The double board could be used by itself, spinning the Heaven plate on the Earth plate with the central figure of the Great Bear giving direction by its tail, or a spoon representing the constellation could be spun on a single Earth plate.

"Exactly how it was used has long been forgotten, and all that we can be sure of was that the heaven plate was turned on its axis in imitation of the imaginary passage of the Great Bear's tail around the horizon according to the seasons."[17] As the eight trigrams of the I Ching were placed at compass points and intermediate points, the board could probably

[13] Herodotus, *Histories,* quoted in Carpenter, *Folk Tale,* p. 114.

[14] The old-style Western indicator of north on a map, the fleur-de-lis, also has a celestial archetype, the same triple-pointed symbol that caused the Salmoxis victim to be impaled on three spears. See the discussion of Poseidon's trident in "Killing the Bull."

[15] See Egyptian star map in later chapter.

[16] Joseph Needham, *Science and Civilization in China* (Cambridge University Press, 1962) Vol. 4, part 1, Physics, p. 262.

[17] Needham, *Science in China*, Vol. 4, part I, p. 268.

have been used as a simple way of consulting the oracle rather than by throwing yarrow stalks or coins. The Tao Te Ching contains one of the first mentions of the diviner's board.[18]

Needham feels that the Heaven plate with the representation of the constellation came before the model of the spoon. Since the Chinese appeared to be working toward a compass which would require a freely moving pointer, this would be a sensible order of events. He has witnessed the performance of a modern lodestone spoon (natural magnetite) on a bronze plate and is satisfied as to its south-pointing ability. He accepts the rotating Great Bear on the Heaven plate though aware that it actually revolves around the pole.

That a model came down from the heavens should be rather expected than not at this point. Needham claims "... that the tail of the Great Bear may be unhesitatingly denominated the most ancient of all pointer readings, and in its transference to the 'heaven plate' of the diviner's board, we are witnessing the first step on the road to all dials and self-registering meters."[19]

The Chinese are consistent in placing their spoon, the Northern Bushel or Ursa Major, at the pole and generally at the center of things. In warfare they carried four banners figured with their four great directional constellations: Warrior, Dragon, Tiger, and Bird. In the center a fifth banner was carried depicting "The Pointer of the Northern Bushel," Ursa Major.[20]

Like Needham, another careful scholar, Roger Beck seems not at all troubled by an image of Ursa Major at the center of the stage. Neither of these men, of course, perceive the issue under discussion here, and both have other fish to fry. Ursa Major is pictured at the center of the zodiac at the Ponza Mithraeum. Beck sees this as validating the Mithrasliturgie, a section of the Greek Magical Papyri. One of the spells gives the point of view. "Bear, Bear, you who rule the heaven, the stars, and the whole world; you who make the axis turn and control the whole cosmic system by force and compulsion; I appeal to you, ..."[21]

It is worth mentioning that the bear is addressed as Taurian in this particular spell, confusing to an orthodoxy that knows little of astrological signs, much less that the bear should have one. But a constellation at the pole throughout the Age of Heroes is important enough to have her birthday noticed, in this case the Spring Equinox at the commencement of the year. The bear is a Taurian because, as we will presently see in the chapter on Mithraic mysteries, the Spring Equinox was in Taurus throughout the 700 years of the Age of Heroes. Again we part company with orthodox science, for they have retrocalculated an Age of Taurus to 4000 to 1800 B.C.

The shoulder of a bull, an Egyptian image of Ursa Major, is mentioned in the Mithrasliturgie: "... a god immensely great holding in his right hand the golden shoulder of a young bull." The text is unequivocal. "This is the Bear which turns heaven around."[22] A number of investiture scenes in Mithraic iconography repeat the image. It has been argued that the image of Messiah holding seven stars in Revelations 1:16 owes something to Mithraic imagery. Certainly the stars of the bear were a significant configuration long before the existence of the seven churches of Asia to which orthodoxy refers the image. Suetonius

[18] Needham, *Science in China*, Vol. 4, part I, p. 268.

[19] Ibid., p. 266.

[20] O'Neill, *Night of the Gods*, pp. 903, 907.

[21] Greek Magical Papyri, PGM VII. 686–702.

[22] Ibid., PGM IV. 695–705.

mentions that Augustus was supposed to have had birthmarks in the same pattern as Ursa Major.[23] Truly the oil of anointing drips down from the bear on Christian and pagan alike.

The Egyptian image of Ursa Major as the shoulder or thigh of a bull bears further examination. In hieroglyph, it is sometimes pictured on top of a platform, a symbol which the nome of Letopolis used as its standard.[24] The platform itself is distinctive, for it frequently supports other figures which we will come to recognize as polar: a seated figure or the eye of Ra.[25] The message to be conveyed was that of a god resting on the axis column, in this case Ursa Major at the pole.

The Maya are on record as bringing out the measuring cords when the heavens change, and the Egyptians were similarly concerned with orienting Heaven and Earth. "On account of the stretching of the measuring cord, the Egyptians were called by the Greeks αρπεδοναπται [harpedonaptai],[26] whose art Democritus boasted of having acquired."[27] In an inscription at Abydos, the goddess Sesheta participated in the orientation ritual, addressing the king. "The hammer in my hand was of gold, as I struck the peg with it, and thou wast with me in thy capacity of Harpedonapt. Thy hand held the spade during the fixing of its [the temple's] four corners with accuracy by the four supports of heaven."[28]

The Bull's Thigh, Ursa Major, is sometimes included in the description of the ritual. The temple of Hathor at Denderah was aligned as follows: "The living God, the magnificent son of Asti [a name of Thoth], nourished by the sublime goddess in the temple, the sovereign of the country, stretches the rope in joy. With his glance toward the ak [the middle?] of the Bull's Thigh constellation, he establishes the temple-house of the mistress of Denderah, as took place there before." Again, "Looking to the sky at the course of the rising stars [and] recognizing the ak of the Bull's Thigh constellation, I establish the corners of the temple of her Majesty."[29]

The word ak implies the central driving point of the pole, since ak or ag comes from the Sanskrit aj, "to drive." Its mention in the inscription is a ritual reference to an earlier dedication "as took place there before." Another inscription at Edfu recalls the time when Ursa Major was at the pole. "I have grasped the wooden peg; I hold the handle of the club; I grasp the cord with Sesheta; I cast my face towards the course of the rising constellations; I let my glance enter the constellation of the Great Bear; . . . I establish the four corners of thy temple."[30] The presence of Sesheta in these ceremonies of alignment is particularly apt, since as Venus[31] she realigned Heaven and Earth on two occasions, on the second actually placing Ursa Major at the pole.

The sides of Egyptian pyramids are aspected toward the cardinal points of the compass: north, south, east, and west. It is thus that the

[23] Suetonius, *Augustus,* 80. Quoted by Beck, "Ponza Mithraeum," p. 124.

[24] Wainwright, "Letopolis," Journal of Egyptian Archaeology XVIII, 1932, pp. 170–171.

[25] The concept of a resting god is also examined in *The Saturn Myth*, pp. 42–59, 158–161.

[26] The harpe of Saturn appears to have transmuted into both a directional indicator and measuring instrument. It will be treated presently.

[27] Norman Lockyer, *The Dawn of Astronomy*, p. 175, quote from Nissen.

[28] Ibid., p. 175.

[29] Ibid., p. 176. Lockyer does not grasp the meaning of ak.

[30] Ibid., p. 179.

[31] David Talbott & Ev Cochrane, "The Origin of Velikovsky's Comet," *Kronos* X, 1, p. 38.

Egyptians assure us (those who came after) of their ability to locate their markers on Earth. Once we understand this proof of terrestrial orientation, we can look to various shafts in the pyramids and to exterior temples for orientation in the cosmos. The temples of Hathor (mentioned above) and Isis in Denderah are both oriented 18 degrees off the cardinal points, suggesting that the Egyptians may have left us a message there.

Europeans who came to the north temperate regions of the New World after Columbus were surprised to learn that Ursa Major was called "bear"[32] by many Indian tribes. "Its stars [Ursa Major] seem to have been called the Bear over nearly the whole of our continent when the first Europeans, of whom we have knowledge, arrived."[33] Native Americans claim bear, badger, eagle, and many other creatures in their totems, but it is the Bear Clan of the Hopi which is dominant in this, their Fourth World. Jung has been quoted by a former colleague as saying that the American Indian is a deep archetypal figure for Americans.[34] Native Americans relate to bears, and so, perforce, do those who came after.

Before the eradication of the matriarchy, the bear was feminine. Spells in the Greek Magical Papyri address her thus: ". . . greatest goddess, ruling heaven, . . ."[35] ". . . universal queen . . ."[36] "O queen of mortals . . ."[37] Zeus's impregnating Callisto and raising her and her son up to the sky as bears is a late patriarchal myth, but there is a germ of the Bearson legend in it. Presently the Bear will transform into a male and engender his own new age. We will examine this phenomenon in the following chapter.

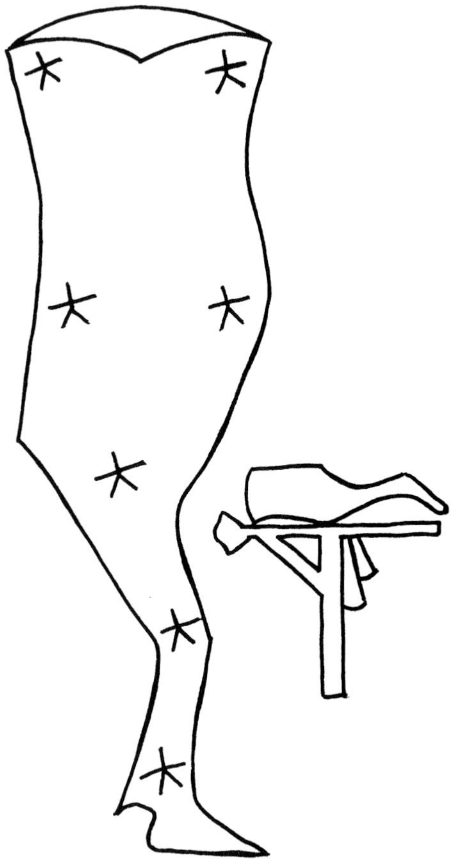

Ursa Major, the standard of the Egyptian nome of Letopolis.

[32] Europeans have their own traditions of the bear, reflected in the names of cities such as Berne, Berlin, and Dublin. (Bear is dubh in Hebrew and Phoenician.) *Epigraphic Society Occasional Papers,* Vol. 17, p. 203.

[33] Stansbury Hagar, "The Celestial Bear," *Journal of American Folklore* (N.Y. Houghton, Mifflin) Vol. VIII, April–June, 1900, pp. 92–103. The great southern branch of the Comanche, the Pehnahterkuh, were "Honey-Eaters." T. R. Fehrenback, *The Comanches* (New York, Knopf, 1974) p. 320.

[34] See video series, "Remembering Jung."

[35] Greek Magical Papyri, trans. Hans Dieter Betz (Chicago, 1986) PGM IV, 1300–1310.

[36] Ibid., PGM VII, 695.

[37] Ibid., PGM LXXII, 15–20.

The legendary Bearson.

7. The Bearson: Seminal Legend of Western Civilization

The most we can do is dream the myth onward and give it modern dress.
– C. G. Jung

When a historical catastrophe causes transformation, realigning Earth and Sky, the myths that have sustained the culture must change. Mother Earth must give birth anew, fertilized by a new consort, a new Father Sky.

A remarkable parallel exists between the way Old and New World perceived the end of the Age of Heroes or Jaguar Age in the 8th and 7th centuries B.C. In both worlds, the interaction between Earth and Sky was seen as a generative event, the image at the pole, Ursa Major or Oceotl, fertilizing the human. As the Jaguar is father to the were-jaguar of Olmec iconography, so the Bear is father to the Bearson of folk tale.

A German scholar, Friedrich Panzer, has developed an unusual thesis, that the ultimate plot of *Beowulf*, an epic of 12th century Britain, is a widely known folk tale which he calls the Bearson.[1] To do this he collected hundreds of variants of the story in almost every part of Europe and several areas in Asia and collated them.[2] In "The Folk Tale of the Bear's Son," Rhys Carpenter takes Panzer's formulation of the story, but goes on to show parallels with the *Odyssey*![3] Using material extraneous to Homer, he brings a bear connection into clearer focus. Truly the bear names and stories cluster thickly around Odysseus and Carpenter's argument is telling.

Reading through Panzer's formulation of the Bearson, I was struck with a certain similarity to the events which terminated the Age of Heroes. The cast of the narrative, however, is also a beginning, the beginning of the Age of Iron. Using the translation of Carpenter, we will analyze the story and make comparisons.

"A wedded but childless woman meets a bear in the forest, is taken to his cave, and remains there to bring forth a son to him. Very exceptionally, the child is born of a she-bear and begotten by a man." The bear is Ursa Major and the cave a symbol of the cosmos, a dark, mysterious region at the pole where Earth and Sky are connected. The woman is symbolic of the Earth, the bear of the constellation at the pole.

"So strange a child shows traces of his animal origin, is ugly or hairy, or has bear's ears." This creature is the new mythic hero, the offspring of the recent catastrophic interaction between Earth and Sky. His connection with nature is visible, cannot be overlooked.

"The bear prevents his captives from escaping by closing the entrance to the cave with some large object, usually a boulder." The old connection between Heaven and Earth, the pathway to the gods at Ursa Major, has been closed.

---

[1] Friedrich Panzer, *Studien zur germanischen Sagengeschichte*, Vol. 1.

[2] Stith Thompson claims this is one of the most popular tales in the world, though he considers it to have traveled by diffusion. *The Folktale* (New York, Holt, Rinehart, 1946) p. 33.

[3] Rhys Carpenter, *Folk Tale, Fiction and Saga in the Homeric Epics*, chapter VII.

"Mother and son escape from the bear's den thanks to the Bearson's attainment of sufficient strength to roll the stone aside. (In some versions, he kills the bear, his father.)" The new polar opening to the cosmos is revealed to be at Ursa Minor.

"Mother and son return to the human world where the husband of the woman adopts the child as his own. He is usually well along in years and cannot hope for other offspring." Bearson's connection to the old cosmic order is recognized.

"The child grows up to be prodigiously strong, though sluggish and indolent and generally thought of no account." Since the constellation at the pole was generally given the credit for turning the cosmos around, Bearson would be seen as strong; since Ursa Minor moves little itself, he would be sluggish. That Bearson was thought to be of no account would reflect the general lack of prominence, the dim stars of Ursa Minor.

"Before starting on his adventures, Bearson acquires a marvelous weapon which he learns to wield with consummate skill." This weapon represents the polar axis to begin with, and the sword or dagger is the weapon of Mars. The development of metallurgy with its focus on warfare is well underway at this period in Europe. It is the obvious prerequisite for technology.

"Bearson, starting on his adventures, acquires companions endowed with astonishing physical capacities. These perform feats of prodigious strength; tearing out trees by their roots, removing mountains, diverting rivers, mastering waters." These physical changes suggest the catastrophes of the Earth-Mars Wars.

"Main adventure now follows. Bearson and his companions come upon the House in the Woods. This is frequently sighted from a treetop.

"The house is empty of owner or occupant, but filled with food, often still cooking on the fire, and with everything in spotless working order. The visitors inspect everything, eat heartily and go to sleep." Carpenter calls this the Hansel and Gretel variant of the House in the Woods, and relates it to a scene in the *Odyssey*. Spotted from a tree suggesting its situation at the pole, the house has the flavor of an ideal Golden Age.

"The owner suddenly returns, discovering one of Bearson's companions in charge of the house. He is usually an old bearded dwarf, occasionally a giant,[4] and accosts his visitor politely, but soon turns truculent and manhandles him. One after another, on successive days, the company has to endure this treatment, until at last it is Bearson's turn." Here is an impression of successive catastrophic events, echoing the Aztec ritual of Tlacaxipehualitzli recounted above, where the five warriors attack an individual one at a time.

"When it is Bearson's turn, he seizes the demon—by the beard, if he has one—and overpowers him. Sometimes Bearson suspends him by his beard from the cleft of a tree; sometimes he cuts off his head or hews off an arm; in any event, the demon recovers himself and makes off to the underworld, whither Bearson follows him by the trail of blood." The demon, the planet Mars, is defeated and retreats to the underworld, problematical in both myth and legend. It can be the southern hemisphere, an objectified hell, or far outer space. As Bearson continues it is most like a fantasy Earth.

"Access to the underworld is usually down a deep well by means of a rope." Heaven, Earth, and the underworld have a common connection; the rope is a symbol of the axis column.

"The way to the underworld leads through fire and cold, by wind and water, beset by great darkness, but emerges in a true world beneath,

[4] Dwarf or giant is no mistake. See "Mars: The Magic Twin."

usually adorned with meadows, trees, and flowers." This is a transition from the coding of catastrophic events to a coda which is the hero journey familiar to us all: killing the monster (destruction of nature), freeing the princess (removing her to another prison), finding the treasure (ripping metals from the Earth).

As the story continues, the faithless companions find their way home, bringing the rescued princesses and maidens with them. In a favorite ending, Bearson arrives in time to punish his former companions, marry the youngest princess, etc. Readers of the *Odyssey* will recognize obvious parallels here. It is a triumph of the ego, but now his alienation is complete: from the natural world, the feminine, and his fellow man.

The incredible persistence of this legend over 2,700 years is illustrative of Velikovsky's hypothesis in *Mankind in Amnesia.* Catastrophic events have been repressed, but we are compelled to repeat or act them out in one form or another, probably until they are raised into consciousness. Carpenter puts it well. "Are such folk tales really so old? And if so, why are they so persistent? One could hardly class the plot of Bearson as edifying or even interesting. It is not a very good story and it is a thoroughly preposterous one. Why should it exert such a power over men's minds?" Why indeed?

Although the watershed of Western thought is generally located in the ideas of Plato and Aristotle, a case can be made for placing the actual break between Western Civilization and what came before to the cosmic interactions of the 8th and 7th centuries B.C. The presence of our own hero myth growing directly out of the catastrophic Bearson folk tale cannot be dismissed lightly.

The implications of the hero journey grow deeper. "The inward journeys of the mythological hero, the shaman, the mystic, and the schizophrenic are in principle the same."[5] In speaking of a schizophrenic journey, R. C. Laing asks, "Can we not see that this voyage is not what we need to be cured of, but that it is itself a natural way of healing our own appalling state of alienation called normality?"[6] The shaman and mystic are in control of their journeys; the hero is fighting his way through with style; the schizophrenic is drifting, at the mercy of his trip. We may ask ourselves if the present hero journey truly reflects the way man is, or is merely the psychological signature of the present age.

There are hints of a change in the cast of consciousness in the second half of the 1st millennium B.C. In *White Goddess,* Robert Graves discusses what he perceives as the decline of myth and poetic thought, distinguishing between poetry inspired by a feminine muse and the classical, more rational poetry in honor of Apollo. He quotes *Phaedrus* at length, calling the passage "One of the most uncompromising rejections of early Greek mythology . . ." And again, "The fact was, that by the time of Socrates, the sense of most myths belonging to the previous epoch was either forgotten or kept a close religious secret."

Campbell describes the change in *Myths to Live By.* "A radical shift of consciousness has bathed the universe and everything in it in a new, more brilliant light, blotting out the moon, and planets, and the other lights of the stars. And this new light, in the centuries then following, penetrated and transformed the whole world westward of Iran."[7] Does not this suggest the rise of the ego as we know it, rising above former, more profound connections?

[5] Joseph Campbell, *Myths to Live By* (New York, Viking, 1972) p. 237.

[6] R. D. Laing, *The Politics of Experience* (New York, Pantheon, 1967) chapter 7.

[7] Campbell, *Myths to Live By,* p. 77.

Bearson appears to be at the root of all the hero journeys, the going-off-to-change-the-world ideals which are the verbal iconography of the present age. They begin in myth and legend, develop in the theater, novel, and film, and will continue to be created until the culture has worked its way through. Alas, dragon slaying is still in vogue, but the princess is making efforts to free herself.

Consider Bearson's animal nature. Odysseus, as it happens, will presently prove to be a bearson despite all his bear connections having been carefully edited out by Homer. The notion of his hero having bear's blood in his veins was as offensive to Homer as it would be to any tidy housewife. Dirt, nature, particularly the animal nature present in man, can be very offensive to our princess. And yet it is this very nature that cries to be understood and integrated. It is not nature red in tooth and claw, but the intelligent environment suggested by human intelligence.

In Bearson, Ursa Major has suffered a sex change and is now male. Even though Venus twice bested Mars in sky-battle, it was Mars who came on to threaten Earth, providing the aggressive masculine prototype which would dominate both the mythology and psychology of the coming Age of Iron. The Hebrew prophets had a hand in this, interpreting the celestial wars as the heavy hand of the new patriarchal god. The Greeks added developing ideas of free will.

The pattern was brought down from the sky. The new ill-natured god would rule his people as mankind would rule the Earth, as man would rule woman, as the developing masculine-style ego would rule the psyche: in an alienated, unconnected way. The earthly model would be the ruthless kings of Babylonia and Assyria. The feminine would be submerged, the prototype of Sleeping Beauty. Mankind would embark on a new adventure, control of the Earth with the "marvelous weapon which he learns to wield with consummate skill," metallurgy followed by technology.

We might consider one reason why catastrophic events had to drop below consciousness. The true horror of interplanetary catastrophe is not that people will die, animal populations be destroyed, or the Earth itself be drowned, charred, or poisoned. These are acceptable parts of the equation of control in our present management of the Earth, as long as they go on slowly and at a reasonable personal distance. The true horror of interplanetary catastrophe is that we cannot control it.

The tight focus of our present ego-dominated, figure-to-ground consciousness simply shuts out whatever doesn't fit the agreed-upon paradigm. Psychic discomfort is relieved in game and story, in compulsive behavior and addictions. The dry records of cultures with a different cast of consciousness simply do not impinge on our world.

In sum, the Bearson folk tale expresses both the series of planetary encounters and Earth changes that initiated a new age and world view in the 8th and 7th centuries B.C., and the hero myth that we would use to penetrate it. The salvation of Western Civilization probably lies in those hairy bear's ears, the connection with Nature that we must acknowledge but which is beyond the scope of this paper.

✦ ✦ ✦

I had just finished "Bearson," as much as could be done in Guatemala, far from a library, clipped the handwritten pages together and put them in an envelope for return to the States. That night I had the following lucid dream.

I come awake in a swirling white fog, floating outside my childhood home. The thrill of lucid dreaming (as if awake, knowing one is dreaming) never pales. I think, "Wow! I'm going to fly!" And I do, swimming through the air as if it were water, turning on my back, having a wonderful time.

Then I consider having a sexual encounter, and immediately find myself in a dark, formal office with plush carpet, shelves of books, and polished dark furniture. There is a big, powerful-looking blond man in a dark blue suit behind a huge desk. He gets up and moves toward me without speaking or smiling, a very heavy presence. I wonder why I am not frightened, but move toward him. He turns his head to the side so I can see a black mask on the back of his head, big enough to cover his whole face were it in front.

He holds out a small model of a head in his left hand. It is pale, colorless, but a moment's scrutiny reveals that it is, though rough, a model of his head, square-looking, Teutonic. He is certainly not my type: thoroughly establishment, massive style, uncommunicative and humorless, but apparently he is the one. He picks me up, holds me against him, and, still fully clothed, we essay a face-to-face, upright sexual encounter which is interrupted almost immediately by my waking up.

This modern, archetypal male has put off his mask and is ready to be known, even allowed his head to be modeled in miniature and offered it to me. He is the successful hero at the end of his journey, alienated from the world and now trying to get in touch with the feminine. On another level, this man is my shadow, representing aspects of myself that I dislike and push below consciousness.

My willingness to know him, demonstrated by writing this paper, has evoked a response, the dream gift of the model of his head. This is highly significant, his greatest treasure, symbolic of the development of reason and logic, the great work that has been going forward during the Age of Iron. I also regard it as an affirmation of my reading of Bearson.

Sadly, the body of this archetypal man, representing feeling and intuition, remains unintegrated, still swathed in the persona of outer garments. Like the mask, these need to be removed before the polarities of masculine and feminine can merge. But happily, progress is being made.

✦ ✦ ✦

In the following chapter we will examine ancient media in a pictorial form, religious art which provides us with astronomical information as to the orientation of Earth and Sky in previous world ages.

The Mithraic tauroctony: science encoded in ancient art.

8. Mithraic Mysteries: The Road Not Taken

The study of Mithraism therefore provides us with insight into "the road not taken" by Western Civilization nearly 2,000 years ago—insight, that is, into an unrecognized part of who we are today.

– David Ulansey, *Origin of the Mithraic Mysteries*

Shall we write about the things not to be spoken of?
Shall we divulge the things not to be divulged?
Shall we pronounce the things not to be pronounced?

– Julian, "Hymn to the Mother of the Gods"

For the terrestrial pole to have pointed to Ursa Major, as described by the poets of Western Civilization as well as in ancient media, implies a major tilt of the axis of Earth within recorded history. This aspect of Velikovsky's work has never been seriously examined. The pole was in Ursa Major, and then, after the Earth-Mars Wars, it was in Ursa Minor on its present precessional path. Getting from there to here will prove even more problematical than we might suspect but, fortunately, help has been on the way for some time now. A number of astute scholars whose pleasure it is to penetrate the Mithraic Mysteries have, inadvertently, come up with an interim position for the Earth which fits our physical model well. How can this be? It is a matter of keeping an eye on the celestial vault.

The mystery religions have remained true to their name over the course of thousands of years despite vigorous gleaning of ancient records by capable researchers. Velikovsky has maintained that they would prove to be repositories of information on historical catastrophism, and so it proves to be. One can imagine the difficult situation of mankind during the Earth-Mars Wars. Cults and oracles would have been activated or come into being in the same way that the the Hebrew prophets began to prophesy. There was a terrible need to bring order out of chaos.

Celestial events had to be dealt with, assimilated, brought down to Earth. When a planetary catastrophe occurred, it was a comfort to believe that it was not completely arbitrary. It might, hopefully, be a pattern of happenings that had been foreseen and foretold. "The way of mankind was rough, but not completely dark."[1] The Sibyl's somber utterances were distributed by the Chresmologoi, tellers of oracles, professional verse makers who could provide an appropriate verse for any occasion.

But after the chaotic period 747 to 687 B.C., mankind went on in his generally upbeat way, making the best of things and trying for improvement while the catastrophic information began to drop below consciousness, but was preserved in large part by the mystery cults and oracles. When we think of these groups, cults of Trophonios and Salmoxis, Orphic mysteries, Sibylline Books, we think of limited, esoteric associations which preserved doctrine and knowledge thought to be important. But when we approach the Mithraic

[1] H. W. Parke, *Sibyls and Sibylline Prophecy in Classical Antiquity* (London & N.Y., Routledge, 1988) p. 7.

Mysteries, we are dealing with what can only be considered a major religion.

Arriving somewhat late on the scene in the first century B.C., Mithraism had spread over the known world by the time that Christianity was getting underway. In part because of its popularity with Roman soldiers, there were hundreds of Mithraic shrines throughout the Roman Empire.[2] Women were not welcome, a circumstance which must have contributed to the eventual displacement of Mithraism by Christianity. Of Mithraic religious practice little is known. There were levels of initiation to work through, a hierarchical system that would appeal to the military mind. At each level, one would be admitted to a higher rank, greater knowledge.

Students of history are rarely drawn to the study of Mithraism, so that the few who are enjoy their own special niche with small connection to the rest of the academy. Mithraic scholars research and publish, convene and discuss, and carry on their own internecine warfare with few looking over their shoulders. This situation may not continue due to their recent adventures into historical astronomy.

Mithraic studies have taken a turn that will prove useful to historical catastrophism on many levels. Although it was suggested by K. B. Stark over a century ago that the tauroctony, the slaying of the bull by Mithras, represented constellations in the sky, it was not until after the First International Congress of Mithraic Studies in 1971[3] that the idea of astral symbolism began to be reexamined.[4] The art and iconography of Mithraism is extensive, each cave-like shrine with its figures and zodiacs, but the heart of the religion can fairly be said to reside in the omnipresent tauroctony or bull-slaying.

A series of scholars have now dealt with the idea of a celestial origin for the tauroctony, several now in substantial agreement on a remarkable hypothesis: the figures in the tauroctony represent the celestial equator of a previous age when the Spring Equinox was in Taurus.

Now when orthodoxy talks about a previous age, they usually have in mind the slow, backward movement of the Spring Equinox through the 12 constellations of the zodiac. The complete circuit of the precessional clock takes approximately 26,000 years, 2,200 years per 30-degree sign, and each sign is thought to be an age. The Age of Taurus, for instance, is retrocalculated to about 4000 to 1800 B.C. Again we part company with orthodoxy. Within these pages, we call previous ages separated by historical catastrophe by the names the Greek mythographer Hesiod bestowed on them: golden, silver, bronze, Age of Heroes, and the present Iron Age.[5]

Origin of the Mithraic Mysteries is the best introduction to the subject available, well-referenced and readable, a valuable distillation of knowledge. Ulansey works his way through the celestial background of the tauroctony and Mithraism in general before revealing his thesis, that the Mithraic Mysteries were founded on the discovery of the Precession of the Equinoxes in 128 B.C. by Hipparchus, and the tauroctony is a retrocalculation of its position, a mystical display of knowledge. Whether a major religion of the past grew out of a phenomenon barely discernible within the lifetime of a man is highly problematical on

[2] David Ulansey, *Origin of the Mithraic Mysteries* (New York, Oxford Univ. Press, 1989) pp. 4–5.

[3] The Congress papers were subsequently published as *Mithraic Studies,* ed. John Hinnells (Manchester Univ. Press, 1975).

[4] Ulansey, *Mithraic Mysteries,* pp. 10–19.

[5] The five Ages of Hesiod equate with the five Suns of ancient Mexico, the Iron Age with the Sun of Ollin, the Age of Heroes with the Jaguar Sun.

the surface, but one begins to believe the concept might work with some moderns.[6]

Depiction of the tauroctony has produced some striking art. One instinctively senses a message in it. Without artificial light, the ancients would have been more familiar with the night sky and have recognized the major figures immediately. Bull and scorpion are easy constellations to pick out in the sky: Taurus at the Spring Equinox, Scorpius at the fall equinox. The dog, Canis Minor, was Procyon to the ancients, a dog star, the lesser one. The snake is Hydra, its dim stars slithering along 90 degrees of the celestial equator in antiquity. The crow is Corvus; the cup is Crater, as neat a cup as one could wish.

Cup and lion are often juxtaposed, but they are as a rule present only in Germany. Ulansey suggests the possibility that they could represent the solstices at Leo and Aquarius. But we will treat the cup where it is, clearly on the celestial equator, and return to Leo and Aquarius presently. Meanwhile, we must recognize exactly what is before us, a sculpture that is a star map. It is the wedding of science and art that Western Civilization holds to be either unlikely or impossible. The tauroctony, however, still speaks to us, but on more levels than we are accustomed to dealing with.

Those whose attention is not caught by art may be interested in the mythical adventures of a mighty hero. Although the Labors of Hercules have generally been understood to reflect a journey through the zodiac, the efforts of mythologists and astrologers to make the paths agree have met with limited

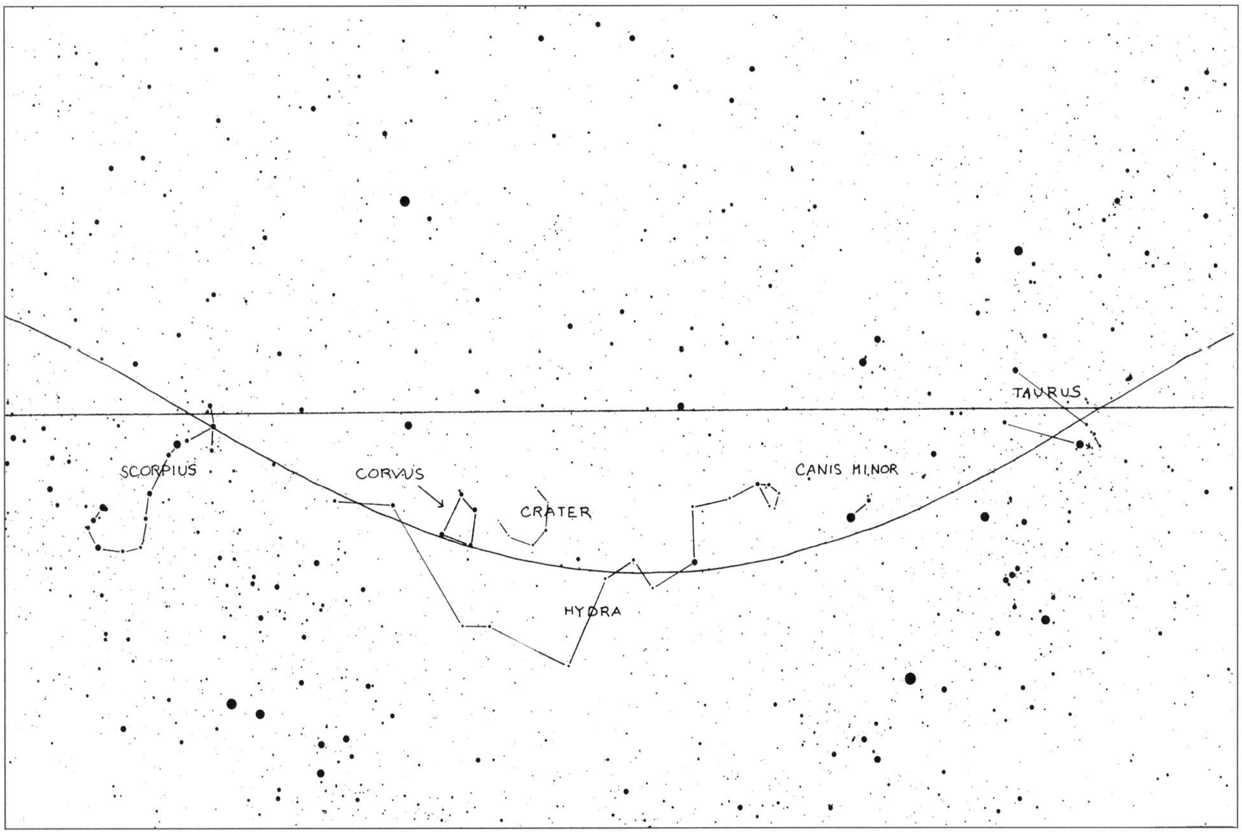

www.carinasoft.com

The Hercules Equator is described by both the Mithraic tauroctony and the Labors of Hercules.

[6] There is some quality in the supposed "catastrophic" changes of the precession as retrocalculated by science that engages the poetic leanings of orthodoxy. Witness the otherwise sensible and valuable *Hamlet's Mill.*

success. There is disagreement about the labors themselves as well as the order in which they occur, and the extensive mythology that surrounds Hercules can only cause confusion. However, the four unfamiliar equatorial constellations of the tauroctony fit easily, some elegantly, into the pattern of the Labors. Like that other hero to whom he is often compared, Gilgamesh, Hercules can be understood as the opener of a "new way."[7]

Corvus the crow would represent the Stymphalian birds, sacred to Ares (Mars), which Hercules has to drive away from their marsh. The birds breed in the desert[8] while Allen suggests the name "Bird of the Desert" for Corvus.[9] Hercules accomplishes his task with brazen castanets, a gift from Athena, who assists him throughout the labors.[10]

Hydra is a snake known to the ancients, identified by name in one of the labors. Hercules must destroy the Lernean Hydra, a monster born to Typhon and Echidne.[11] He masters the Hydra with the help of Athena.

Crater, the cup, is the golden goblet shaped like a water lily which Helius lent Hercules to sail in to Eritheia. Since the hero's "new way" is a reflection of the new path of the sun, the gift is appropriately from Helius, the sun god.[12]

The dog, Canis Minor (Procyon), is Cerberus, the Hound of Hell, whom Hercules had to bring back from the underworld,[13] again with help from Athena. Hercules and the dog was a favorite subject of Greek and Etruscan art.[14]

For easy reference, we will call the equator under discussion the Hercules Equator. It enjoys the distinction of being in place for only 15 years, between the third and fourth attacks of Mars. The other side of the Hercules Equator is also worth some attention, bringing obscure parts of the sky into focus: dark portions of Aries, the blazing Square of Pegasus, Aquila the eagle, and unremarkable sections of Serpens Cauda and Ophiucus.

The great Square of Pegasus, named after the winged horse, resonates well with another labor, cleaning the Augean Stables.[15] Hercules didn't muck out the stalls, but cleverly diverted two neighboring rivers to wash water through them, never soiling his hands. In both the round and the rectangular Denderah Zodiac, there are two fish figured near the Square of Pegasus. They would indicate the location of the two rivers, celestial ones, that Hercules diverted: ". . . the figures of the sacred rivers, Tigris and Euphrates, were seen in the bands of our Fishes."[16]

The small constellation Sagitta is mentioned by name in the Labors. It is the arrow which Hercules shot through the heart of the gryphon-vulture, thus setting Prometheus free.[17] Aquila was sometimes pictured holding the arrow, Sagitta, in its claws.[18]

---

7 Hercules' labors are echoed by those of Gilgamesh and the Irish Amathaon. Graves, *White Goddess,* p. 53.

8 Graves, *Greek Myths,* 128.

9 Allen, *Star Names,* p. 124.

10 Their intimate relationship is examined in "Mighty Hercules."

11 See the chapter on "Catastrophic Art and Monsters."

12 Graves, *Greek Myths,* 132.c.d.

13 The next visit of Mars will put the Hercules Equator in the "underworld," below the new equator.

14 Graves, *White Goddess,* p. 53.

15 Graves, *Greek Myths,* 127.d.

16 Weidner, *Handbuch,* 50, 52, referenced by Eric Burrows, "Some Cosmological Patterns in Babylonian Religion," *The Labyrinth,* ed. S. H. Hooke (New York, MacMillan) p. 61.

17 Graves, *Greek Myths,* 133.l.n.

18 Allen, *Star Names,* p. 350.

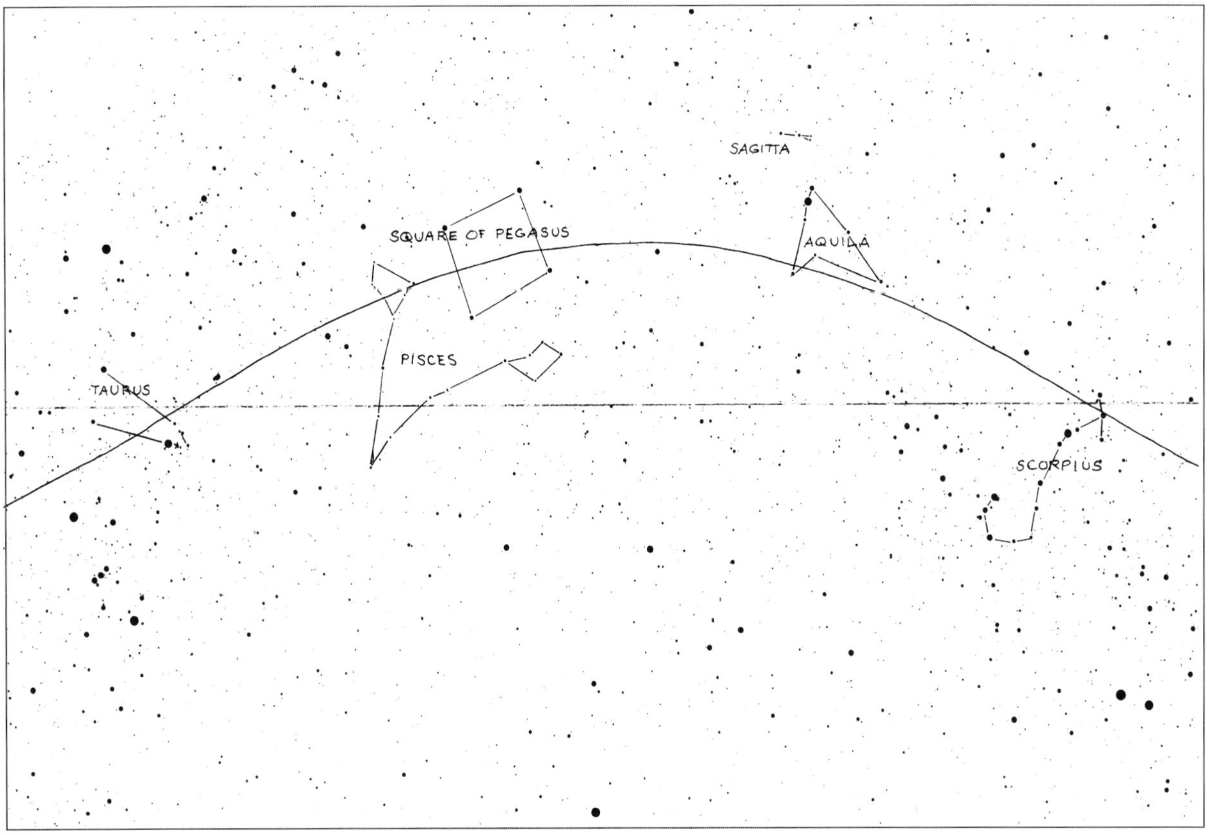

The Hercules Equator continues the Labors of Hercules on the other side of the sky.

Moving the celestial equator itself (or perhaps that of Mars) is suggested by the Labor in which Hercules has to obtain the golden girdle of Mars worn by the Amazon queen, Hippolyte. When belts and girdles of planets are moved, the results are catastrophic.[19] Whatever changes occurred here on Earth would likely have been reflected by corresponding damage to Mars. This labor could relate to the third interaction of the Earth-Mars Wars presently under discussion. Elsewhere, in a confrontation between Hercules and Ares, it is related that on the third lunge, Hercules pierced the armor of Ares.[20] Hercules is a twin aspect of Mars; he is opposed to him here, a defender of Earth who can be compared to Superman.

Mithraic iconography emphasizes the equinoxes. There are twin torchbearers, Cautes and Cautopates, who indicate the crossing of equator and ecliptic. They stand at either station: Cautes at the Spring Equinox with torch pointed upward, Cautopates at the fall equinox with torch pointed downward, each indicating the approaching season. They frequently stand with legs crossed, emphasizing the crossing of equator and ecliptic.[21] Mithraic doctrine insists on Mithras' seat being at both the pole and the equinoxes.[22]

[19] See belt-wrestling in the following chapter.

[20] Graves, *Greek Myths,* 139.b Also see Arjuna's arrow hitting the mark in the following chapter.

[21] The torchbearers are also connected with two particular signs: Cautes with Taurus and Cautopates with Scorpio. Ulansey, *Mithraic Mysteries,* pp. 62–64.

[22] Ulansey, *Mithraic Mysteries,* pp. 103–107.

Even a cursory reading of *Worlds in Collision* leaves one with the impression that planetary catastrophes seem to occur in spring and fall. Velikovsky believed they occurred at the equinoxes. The lesser mysteries of Eleusis, founded in honor of Hercules, were celebrated in the fall in the month of Boedromion corresponding to our September; the greater mysteries were celebrated in the spring in the month of Anthesterion.[23] Velikovsky mentions that the Flood of Deucalion (at the first visit of Venus) was also commemorated in the month of Anthesterion. The Babylonian New Year, Akitu, like that of the Jews and Phoenicians, was celebrated at either the Spring Equinox in the month of Nisan or the fall equinox in the month of Tisrit.[24] New Year's festivities indicate both an ending and a beginning.

It is puzzling but noteworthy that the Earth is likely to have a catastrophic interaction at the equinox. In discussion of the Olmec crossed bands symbol, we have focused on the angle of the ecliptic, but the lines of the symbol definitely cross at the equinox. And Plato has reminded us that the "motion of the same" and "the motion of the different" cross at the equinox.[25]

The record of the Hercules Equator, attested by the Mithraic tauroctony and the Labors of Hercules, is ancient science. Those who have done dream work and understand the way information is coded will understand it best; artists will grasp it easily. There is no attempt to be obscure. The idea is to deliver as much information as possible in a cohesive, layered model.[26] Whatever mystery exists is something we have thrown up in our own minds so we can work undistracted on a linear level.

Now that we have worked our way through the tauroctony and some of the Labors of Hercules, we will be able to grasp the next star map more easily. The Egyptians have left us a modest drawing, recovered a number of times, of their constellation deity, Tau-ert, holding a chain attached to Mesket, the Bull's Thigh or Ursa Major. A hawk-headed authority figure points to the connection with his rod or spear. There are sometimes slight differences in the rendering of the scene, but the seven stars of Ursa Major are always arranged in the same way: three above the bull's leg, four below it.

An Egyptian star map.

The image has sometimes been interpreted as the constellation being attached to the pole. This would parallel the image of Oceotl tied to the celestial pole in Aztec ritual. But Ursa Major is not just tied to the pole of Earth here. Livio Stecchini, the doyen of ancient measurement, has pointed out that the rod held by the hawk-headed man indicates the meridian which also passes through the pole of the ecliptic.[27] The Egyptians were familiar with the Precession of the Equinoxes and its pole, which in their system was located in the breast of Tau-ert. This odd-looking creature is a standing figure of a hippopotamus with

[23] Graves, *Greek Myths,* 134.b.

[24] Eliade, *Eternal Return,* p. 55. Also Budge, *Babylonian Life and History,* p. 211.

[25] F. M. Cornford, *Plato's Cosmology* (Indianapolis: Bobbs-Merrill, 1975) pp. 72–93.

[26] There are many other layers of the tauroctony to be explored.

[27] Peter Tompkins, *Secrets of the Great Pyramid* (New York, Harper and Row, 1971) p. 174.

pendulous dugs and distended stomach, sometimes figured with an alligator on her back. Tau-ert corresponds to our Draco, but may have more to say.

This little star map, then, tells us that Ursa Major is chained to the pole of the ecliptic. The unusual way the seven stars are represented—three above, four below the Bull's Thigh—only becomes clear when we notice the relationship of the seven stars of Ursa Major to the pole of the ecliptic on a celestial map. With the North Pole in Ursa Major and the same forces at work, we must assume a wider radius for the circle of precession. In drawing this circle so that three stars (handle of the dipper) are above it and four stars (bowl of the dipper) are below, the little star map begins to read out.

The "chain" held by Tau-ert, also attached to Ursa Major between the handle and the bowl, can also be drawn. There is little room for play here. A meridian can easily be drawn connecting: (1) the celestial pole when it was in Ursa Major (between the handle and bowl of the dipper), (2) the celestial pole at Alpha Draconis on the present circle of precession, and (3) the pole of the ecliptic!

Many of us have read that the North Pole was once at Alpha Draconis. It is one of orthodoxy's favorite stars for retrocalculation into

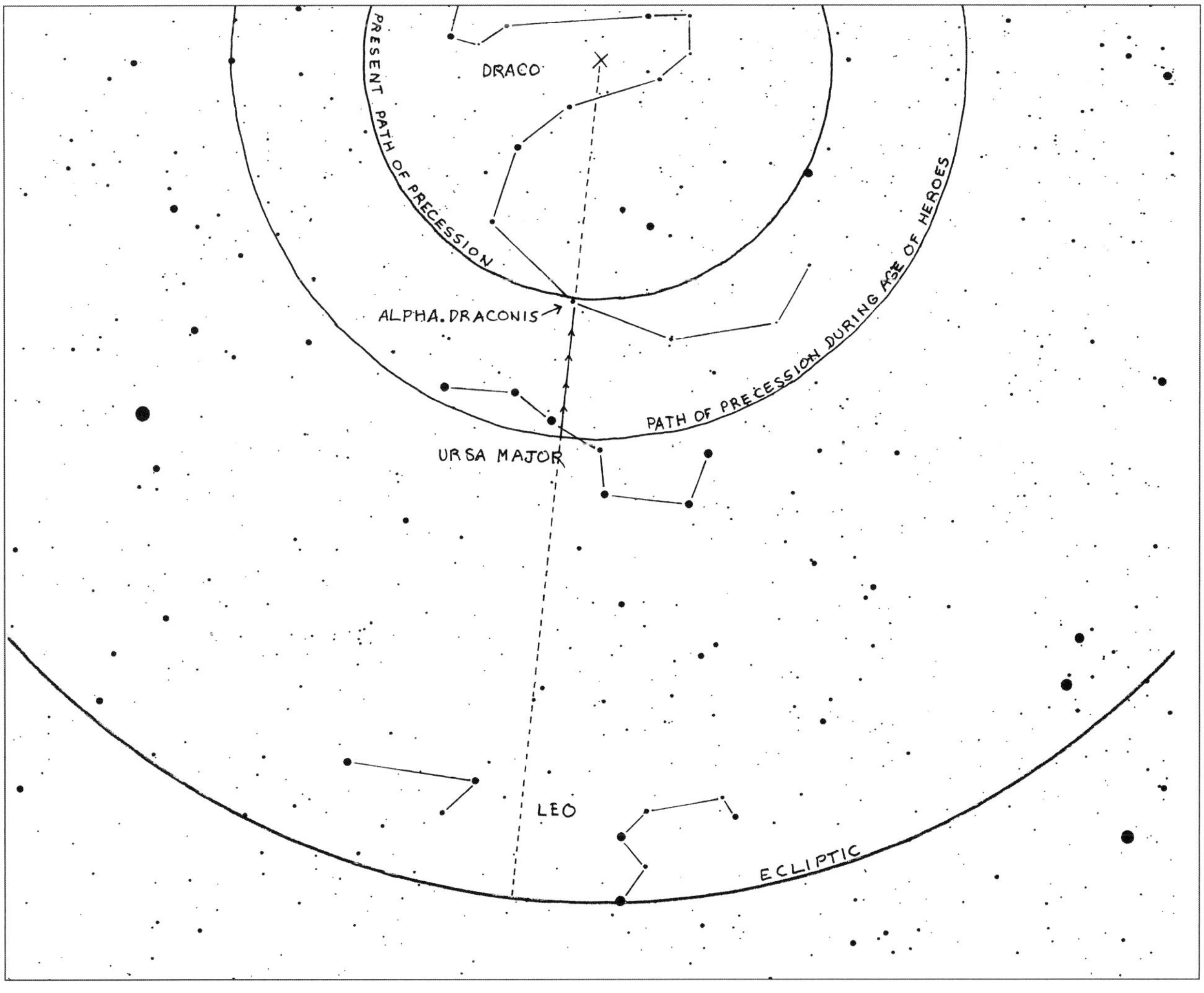

The solstitial colure of 717 B.C., Leo at the gates, the arrowed section marking the tilt of the terrestrial pole from Ursa Major to Alpha Draconis (detail from a 270-degree azimuthal projection).

the past. But historical catastrophists are interested in the star for other reasons. Velikovsky noted that ". . . in the days of Sargon II a new world age began,"[28] and Sargon expressed great interest in Alpha Draconis.[29] It has long been a source of speculation as to why a shaft in the Khufu pyramid should point to Alpha Draconis. If the star suddenly appeared at the pole after a catastrophic Earth motion, we could more readily understand the attention paid to it.

Thus, in drawing the line indicated on the small Egyptian star map, we are not just drawing a line on an astronomical map, but delineating the tilt of the pole! We suddenly discover why the seat of Mithras is both at the pole and at the equinoxes,[30] for the axis the Earth tilts on is at the equinoxes, exactly 90 degrees on either side of where the line of tilt could be extended to cross the celestial equator. The imaginary axis on which the Earth tilts pierces the sky at the two equinoxes, spring and fall, in this case, Taurus and Scorpius. But since axis is commonly understood to be the axis of rotation, we had better use the term of the ancients. The "hinges" on which the Earth tilts are at the equinoxes!

Happily, the Earth, after this tilt, has come to rest in a familiar setting. We rejoice with orthodoxy to see Alpha Draconis at the pole, particularly since it is, as it happens, the pole of the Hercules Equator, already well attested in art and myth. The angle of the ecliptic is approximately where it should be, and the Earth is on its catastrophic way through the Earth-Mars Wars.

Some other names for Ursa Major now begin to make more sense, echoing the Egyptian star map. The Babylonians called her "markas same,"[31] bond of Heaven; the Greeks called her "Omphaloessa" or umbilical cord. The message to be conveyed is that the constellation at the celestial pole is fastened to the pole of the ecliptic. Fortunately the Egyptians gave us the relative position of the stars of Ursa Major at the tilt.

Seneca's description of catastrophe in *Thyestes* gives the impression that all the constellations fall. Naturally, some would have had to rise. The zodiac did indeed fall as described, but only the half above the equator. The half of the zodiac below the equator would have risen. Nonnos' description of Hydra newly fallen to the Hercules Equator is particularly evocative. "Now the serpent scraped with his writhing belly the equator far away from the bear."[32] The constellations of the Hercules Equator do not relate to the celestial circle in the same way as zodiac signs, each equaling 30 degrees. Hydra followed the Hercules Equator for 90 degrees.

We may properly look to historical art to support our developing model of cosmic catastrophe. An outstanding example would be the Lion Gate at Mycenae, one among many lion gates. During the Age of Heroes with the Spring Equinox in Taurus, the lion Leo was at the gates of the solstice, on the ecliptic directly below Ursa Major. With historical dating constantly being revised downward, the Lion Gate will not suffer being pushed back to orthodoxy's putative Age of Taurus. After the tilt of Earth, the lion remained at the gate, but not for long.

---

[28] Velikovsky, *Worlds in Collision,* pp. 346–347.

[29] Allen, *Star Names,* p. 206 Orthodox history records two Sargons, but in fact there is only one, his persona divided by celestial catastrophe.

[30] Porphyry claims the equinoxes (plural ισημεριασ) as the seat of Mithras in "De Antro."

[31] Eric Burrows, S.J. "Some Cosmological Patterns in Babylonian Religion," *The Labyrinth,* ed. S. H. Hooke (New York, Macmillan) p. 62.

[32] Nonnos, *Dionysiaca,* W.H.D. Rouse trans. (Heinemann, 1940).

The Lion Gate at Mycenae, one among many lion gates.

Our review of Olmec iconography suggests five visitations of Mars, at the third of which the terrestrial axis was tilted. In the following chapter we pursue the episodes of the Earth-Mars Wars in epics and sagas of various cultures, some familiar, some little known to Western Civilization.

Ares and Aphrodite: Mars and the Moon.

9. Epics and Sagas: The Big Stories

None can escape the sky and what it bringeth.

– The Shahnama of Firdausi

There are, perhaps, a double handful of epic tales that have grown up in different parts of the world providing us with our earliest glimpses into the culture of literate mankind. They are once-upon-a-time stories, populated with heroes and heroines, gods and demigods, as well as fantastic creatures and objects that have their existence "in illo tempore" when the world was younger and more magical than it presently appears to be. Some of these epics were shaped by a unifying intelligence: the *Iliad* and the *Odyssey* of Homer, the great Iranian epic, the Shahnama of Firdausi. Others were collected and assembled: the Kalevala of Finland and the "Epic of Gilgamesh." Some, like the Mahabharata of India, have just grown. And grown and grown. Reading these epics is like entering a fantasy in living color, a film with special effects.

The epics have an open, easy style, but prove dense and layered. They persist, like water or love, because while we are reading them, some older, wiser part of us is looking over our shoulder, saying, "Yes, that is how it was. Remember. Remember." And sometimes we catch a part of it on the wing and make it our own, but always there is a felt motion, a resonance. The epics endure.

✦ ✦ ✦

The discovery of the public library was a major event in my young life. The pabulum which is presently served up to children had not yet been thought of, so I read straight through shelf after shelf of fairy tales, then branched out into legend and myth. How many times did I read that the hero "girded up his loins" to do battle? Even after I found out what loins were, it seemed a strange thing to do. It was a knot that would not come loose for many decades.

✦ ✦ ✦

Velikovsky found the structure of the planetary wars in the Bible, and much background and color as well as individual events in the epics of Homer. The Kalevala yielded catastrophic descriptions: "dreaded shades" enveloping the Earth, Earth sprinkled with red milk, hailstones of iron, etc. There were hints of this sort of material in the Mahabharata and the "Epic of Gilgamesh." Realizing that these tales referred to historical events, I determined to mine other epics for catastrophic grist. Matthew Arnold had lifted the story of Surhab and Rustam from the Shahnama for English-speaking readers, but otherwise it is little known. The legends of the early shahs were put together by Firdausi in the 11th century, and Firdausi is a poet akin to Homer.

It was Cyrus Gordon who used the term "belt-wrestling" for "sippu" in the "Epic of Gilgamesh,"[1] and belt-wrestling was also the style of battle in the Shahnama. In a fight, one would always go for the belt. Gustaham . . .

[1] *JNES*, 7, p. 264. Also see Appendix 25 in *Hamlet's Mill*.

"Thrust at his foeman's belt; the buckle held;"[2] "He thrust a deadly dart with all his strength, / Struck Gurdafrid upon the belt . . ."[3] Gurdafrid was a woman, but even a female warrior must gird up her loins with a big, strong belt to do battle. As a warrior she is an exception, a hold-over from the previous age when the goddess had more power. The Shahnama is decidedly patriarchal.

As a rule these heroes were fighting to the death, so that seizing hold of the belt of one's opponent does not seem a sensible way to dispatch him. But it was a form that had to be observed because it came from the sky. It is clearer in the Gilgamesh epic when Gilgamesh and Enkidu have a belt-wrestling bout, loosening "door posts" (cosmic axis) and the like.[4] The belts with which the heroes of the Shahnama firmly girded up their loins represented the belt of animals, the zodiac. Loosen the belt of your opponent and you send him off skewed at a new angle. Mars loosened the belt of Earth in their third interaction, tilting her about 16 degrees.

Another item of warlike accouterments was the lasso with which a warrior tried to catch his opponent "around the waist." It echoes the belt-wrestling while suggesting a hold between opponents, possibly a magnetic hold between planets.

The Shahnama is the story of a long, drawn-out conflict between Iran and Turan with the same kind of catastrophic imagery reflecting war on Earth and in Heaven as one perceives in the *Iliad*.

> Blood to the Fish and dust clouds to the Moon.[5]
> The White Div . . . showered stones and darts.[6]
> The whole world turned as black as ebony,
> The mountains shivered at the trumpets' blast,
> And earth was bent beneath the horses' hoofs, . . .[7]
> The atmosphere was darkened, earth was hidden—[8]

As in the Kalevala, there are catastrophic Earth conditions. When Fariburz made expedition to capture the castle of Bahman, he was defeated by excessive heat.

> Our loins are scorching underneath our mail,
> The bodies of our beasts of burden burn.[9]

And later:

> The rampart of the stronghold split asunder; . . .
> The earth became black as a negro's face,
> Sun, Moon, and Pleiades were lost to sight,[10]
> Immediately a cloud rose charged with hail.
> Hail charged with death;

2 Firdausi, *Shahnama,* Arthur and Edmond Warner trans. (London: Kegan Paul, Trench, Trubner & Co., 1906), Vol. 2, p. 113.

3 Ibid., Vol. 2, p. 133.

4 See "Heavenly Gates: The Solstices."

5 Firdausi, *Shahnama,* Vol. II, verse 304, p. 15. The fish are our Pisces.

6 Ibid., verse 329, p. 40.

7 Ibid., verse 382, pp. 84–85.

8 Ibid., verse 399, p. 96.

9 Firdausi, *Shahnama,* Vol. 2, verse 758, p. 406.

10 Ibid., verse 760, p. 408.

Full many a div was slain
And many venom-stricken fell to earth.[11]

Extracting an exact sequence of events from the epic would be a challenge, but all the critical happenings are there. A new order begins with Kai Khusrau. "For from today, new feasts and customs date, because today is born Shah Kai Khusrau."[12] His reign is 60 years, exactly fitting 747 to 687 B.C., four 15-year periods of Mars. When he says farewell to his warriors before his ascension into Heaven, "A cry rose from the army of Iran: The sun hath wandered from his way in heaven."[13] This echoes the tilt of the Earth in 717 B.C.

A remarkable claim of Kai Khusrau is the extent of his sovereignty. "The whole world is my kingdom, all is mine from Pisces downward to the Bull's head."[14] This comprehends both the constellation of Taurus and that of Aries! In the following chapter we will see how the Spring Equinox moved catastrophically from Taurus to Aries in 702 B.C. Kai Khusrau apparently saw himself as a sun king, Spring Equinox in Aries and Taurus, and that is the direction that kingship is taking.

Following the pattern of many hero-gods, Kai Khusrau retires into the mountains and ascends into Heaven. His five paladins accompany him to the very end although he warns them to go back, to leave the arid sands, for a storm of snow, will come the next day. They refuse to leave, their lances are covered with snow, and "they yielded up sweet life."[15] These paladins are Tus, Giv, Fariburz, Bizhan, and Gustaham. They are reminiscent of the five warriors that participated in the Aztec ritual of Tlacaxipehualitzli, but their battles have already been fought, with much belt-wrestling in the war with Turan. Giv's sable banner was charged with a wolf,[16] echoing the profuse wolf myths and legends surrounding Mars.

It is a bit of a surprise to find these heros playing polo for diversion and seemingly surrounded with elephants. "Elephantine" is a favorite descriptive word with Firdausi, and the animals are there in abundance. The scene is reminiscent of India, and one is reminded of Dayton's Indo-European axis of trade.[17]

The events that shaped and provided background for the epics came out of the sky, were visible to everyone, and were recorded in the ancient media of many cultures. The parallels between the Shahnama and the Mahabharata are startling enough to have been widely noticed.

The Hindus tell of four yugas or ages: Krita, Treta, Dvapara, and Kali, the present age being the degraded one of Kali. As described in the Mahabharata, the war between the Pandavas and the Kauravas takes place in the "twilight" between the Dvapara and Kali yuga,[18] i.e., in the period of the Earth-Mars Wars.[19] We have already discussed the anomalous movement of the sun in the Mahabharata. Although we shall see that it is not the same sort of motion as

[11] Ibid., verse 761, pp. 408–409.

[12] Ibid., verse 670, p. 326.

[13] Ibid., Vol. IV, p. 303.

[14] Ibid., Vol. II, verse 759, p. 407.

[15] Ibid., Vol. IV, p. 310.

[16] Ibid., Vol. III, p. 33.

[17] Dayton, *Minerals, Metals, Glazing and Man,* passim. Dayton has also dealt with the elephants in "The Lost Elephants of Arabia," *Antiquity,* Vol. 42, 1968, p. 42.

[18] Mahabharata, Roy trans., Vol. 1, p. 18.

[19] This would correspond to the period between Age of Heroes and Iron Age, or between the Jaguar Sun and the Sun of Ollin.

that described in the Shahnama where the sun actually changes its path, catastrophe is indicated in both instances.

The five paladins of the Shahnama are reflected in the five sons of King Pandu, the Pandavas: Yudhisthira, Bhima, Arjuna, Nakula, and Sahadeva. An incident early in this section gives us critical information. A neighboring king wishes to marry off his daughter and holds an archery contest to find a suitable husband for her. The five brothers decide to compete for her hand, each in turn attempting to bend a bow of enormous strength and hit a small target with an arrow. It is Arjuna, the third of the Pandavas, who pulls the bow and hits the target with the greatest of ease.

There is an unusual reflective resonance in the performance of Arjuna, the third brother, which mirrors Hercules hitting Mars on the third lunge in a previous chapter. When Earth is tilted at the third interaction with Mars, we naturally identify with our own planet as the injured one. In fact, the smaller Mars would likely have come out on the short end of any interaction.[20] Hesiod takes this view as quoted in a subsequent chapter, seeing Hercules, defender of Earth, as dominant throughout his bouts with Mars.

When the battles have been fought and the catastrophic interim between the Dvapara and Kali Yuga, with all its gloom and celestial apparitions, is at an end, the five Pandavas set out on the last leg of their journey. They, like Kai Khusrau and his paladins, also head for the mountains, but a particular one better connected to the pole, Mt. Meru.[21] Only the eldest, Yudishthira, enters Heaven in his mortal body. The others, with their joint wife Draupadi, must die first.

Santillana and Dechend discuss both Shahnama and Mahabharata in *Hamlet's Mill*, going over the ground with thoroughness and flair, also introducing Enoch and Quetzalcoatl into the pattern of events.[22] Quetzalcoatl's departure on a raft follows the example of the many gods who travel by boat through the sky, and certainly the Mesoamericans knew little of chariots. But Enoch's departure to Heaven in a fiery chariot drawn by fiery chargers would parallel Mars with his two moons, streaming fiery hair as he leaves Earth after his last close approach.

Firdausi's translators, the Warners, discuss the similarity of the Indian and Iranian epic, but feel they must be referred to a common origin of great antiquity. The 7th and 8th centuries B.C. are not so long ago, however.

Though Santillana and Dechend do not subscribe to historical catastrophism, they illuminate many things to that end, and are not above using the language of catastrophism. They well understand that an age is coming to an end here and a new one beginning, even addressing the birth of Kai Khusrau (in the Shahnama)[23] and the death of Krishna (in the Mahabharata)[24] as chronological turning points. Their notion of catastrophic beginnings and endings is encompassed, however, in the slow Precession of the Equinoxes over its 26,000-year circle of the zodiac.

The Chinese epic, "Creation of the Gods," or "Popular Account of the Promotion to Divinity," is a special case, a fey, fanciful treatment of events that is difficult to extract hard data from. Of catastrophic events there is a plethora, one unbeatable hero or weapon appearing after another to join the battle. The Mars wolf is here a vixen, a female fox.

[20] Mars is tilted at close to the angle of Earth.

[21] O'Neill, *Night of the Gods*, p. 887. Also see pp. 55, 165, 289, 593 n. 3, 625, 977.

[22] Santillana and Dechend, *Hamlet's Mill*, pp. 77–78.

[23] Santillana, *Hamlet's Mill*, pp. 38–39.

[24] Ibid., pp. 78–79.

It is the Shang Dynasty going down to defeat during the Earth-Mars Wars. Santillana and Dechend, fingers on the catastrophic pulse, call it a "fantastic description of a major crisis between world ages." The 365 new gods that are appointed on the Terrace of Creation are elevated to that position by a character who is prominent through much of the tale and enjoys several names. One of these is Flying Bear. The change of 360 days in the year to 365 must have necessitated the "Promotion to Divinity." [25]

It is worthwhile to recall at this time a paragraph in *Worlds in Collision*. "When the year changed from 360 to 365¼ days, the Chinese added five and a quarter days to their year, calling this additional period Khe-ying; they also began to divide a sphere into 365¼ degrees, adopting the new year-length not only in the calendar, but also in celestial and terrestrial geometry." [26]

In a previous chapter we learned of Panzer's formulation of the Bearson folk tale and how Carpenter compared it to the adventures of Odysseus. An entire chapter of his *Folk Tale, Fiction and Saga in the Homeric Epics* is devoted to this enterprise, proving Odysseus a bearson to his ears. It draws to a close with the charming insight that we still celebrate some pale reflection of the immortality cult of the bear when we check to see if the bear's modern stand-in hibernator, Arctomys monax, sees his shadow on Groundhog Day.

Carpenter, of course, had no thought of catastrophism, was unable to recognize the Bearson as the constellation at the pole of a new age and its first folk hero. Still, he has encouraged us to examine the *Odyssey* for further evidence of catastrophism. Homer is such a fine storyteller that we tend to follow him without noticing it, but there is a carefully devised catastrophic substructure in the epic that needs pointing out. It comes into focus first in the myth of Sisyphus, the Bear to Odysseus' Bearson.

It seems to have been an open secret that Odysseus' father was the crafty Sisyphus,[27] that Laertes was only an adoptive father (echoing the Bearson folk tale). Graves has accused Sisyphus of substituting patrilineal for matrilineal laws in Corinth, so that he appears very much at the center of the change between ages. Without ever having attached a bear name, he figures as the central character in a myth of the tilt of the pole.

The myth of Sisyphus echoes the bear's hibernation, his apparent death merely a temporary disappearance, a descent to Hell in which he had a confrontation with Persephone. As an unburied person (having cautioned his wife against funeral rites), he feels he has no business there. "Let me return to the upper world, arrange for my burial, and avenge the neglect shown me," he argues. "My presence here is most irregular. I will be back in three days." Persephone grants his request, but Sisyphus repudiates his promise and remains in the upper world.[28]

In the myth, Sisyphus represents Ursa Major after the catastrophic contact with Mars that tilted the pole, invisible while the atmosphere is obscured and presumably in hell making his bargain with Persephone. He is believed dead but reappears (in three days?), removed from the pole but still visible in the sky, where he remains.

Sisyphus earns an unusual punishment from the gods; he is condemned to roll a huge

[25] *Feng Shen Yen 1, Creation of the Gods*, trans. Gu Zhizhong (Beijing: New World Press, 1992) p. 469.

[26] Velikovsky, *Worlds in Collision*, pp. 342–343.

[27] Graves, *Greek Myths*, 67 passim. Carpenter in *Folk Tale* cites one play by Aeschylus, two by Euripedes, three by Sophocles.

[28] Graves, *Greek Myths*, 105.7.

rock up a hill. Graves is clear on the stone having to be rolled up the vault of Heaven.[29] It is not just the sun,[30] however, as many have concluded, but a stone, the same stone that has sealed the tomb of a number of sleeping gods, in this case that of the bear. The cave, the connection with the other world at Ursa Major, has been sealed by a violent meeting of Earth with Mars. It will open again in a new place when the sky clears. The gods make this happen by having the stone which sits at the mouth of the cave be moved from one opening to the next.

Amazingly, the myth reflects knowledge that the stone has to be rolled "uphill," actually about 16 degrees toward the pole of the ecliptic. It is only in a larger sense, recognizing the ecliptic as the base line of the solar system, that the pole of the ecliptic can be understood as "up." But the ancients will not cease to amaze as we continue into the past.

Since Sisyphus has been revealed as the Bear, a parallel can now be drawn between the punishment meted out to him and a set of incidents in the adventures of Odysseus, his Bearson. As Sisyphus has to move the stone from the mouth of one cave to another, so Odysseus has to move from one cave to another. Thus, as Bearson, he will be at the correct spot in the sky, i.e., Ursa Minor.[31]

The lengthy visit of our hero at Ogygeia with the nymph Calypso must encourage examination of the cave they slept in for seven years. Calypso is the daughter of Thetis, perhaps by Atlas,[32] weaves on a loom with a golden shuttle,[33] has plenty of nectar and ambrosia, but it is the four streams of the cave that immediately give away its polar position.[34] Odysseus' stay with Calypso was not a careless idyll; he was unable to leave until Zeus himself gave the order.

The second cave is perhaps better known, the Cave of the Nymphs on Odysseus' home island of Ithaca. Porphyry treated it at length in "De Antro Nympharum." The nymphs, like Calypso and her companions, are naiads, associated with the flow of water, but there is only one spring there, though it "flows unceasingly." The nymphs weave on stone looms and there are stone amphoras where "bees lay up stores of honey." The honey connection reflects the first visit of Venus, after which manna, or madhu, fell to Earth.

". . . Porphyry considered Homer a visionary allegorist and the episodes in the *Odyssey* screens for transcendent meaning."[35] As Porphyry notes, "It is impossible that he should have successfully created the entire basis of the story without shaping that creation after some sort of truth."[36] Indeed, this style of creation after truth is that which lasts two and a half millennia. The connection of the Bearson Odysseus with two caves contributes to the power of the story. A modern might call it a subliminal message.

Once noted, the two caves can be seen to form parentheses around a catastrophic substructure of the *Odyssey*. Books V through XIII contain the exposition of how Odysseus spent this time between caves. Although 22 days are

[29] Ibid., 67.2.

[30] As a result of the tilt, the sun will be "higher" in the sky, not wandering as far north and south.

[31] In actuality, it will take two catastrophic interactions to move the pole from Ursa Major to Ursa Minor: the tilt to Alpha Draconis, then a precessional spin to Ursa Minor.

[32] Graves, *Greek Myths*, 170.v.

[33] From the Japanese Amaterasu to Spider Grandmother, weaving is done at the pole.

[34] See the four heavenly rivers in "The Saturn Configuration."

[35] Robert Lamberton, *Porphyry on the Cave of the Nymphs* (Barrytown, N.Y., Station Hill, 1983) p. 9.

[36] Ibid., p. 40.

involved, the bulk of them (19 by Homer's count) find our hero alone at sea. During only three days with the Phaeacians on Scherie is the rest of the action telescoped. These are a strange three days. If Odysseus is not reciting his adventures, he is either asleep, screened from sight by Athena, or wearing a cloak which he does not trouble to remove to throw the discus at the Phaeacian Games. We are reminded of the disappearance of Sisyphus and his promise to return in three days.

To recapitulate: In Book V, there is a description of Calypso's cave, symbol of the cosmos to the ancients. After Zeus gives permission, she sends our hero off, aided by Boreas the North Wind, with instructions to keep a certain constellation on his left hand while crossing the sea: "The Great Bear, nick-named the Wain, which always wheels round in the same place . . . the only one who never bathes in Ocean's Stream."

First day: After 17 days of sailing and two of hard swimming (after the steering oar "is wrenched out of his hands"),[37] Odysseus lands on the island of Scherie, home of the Phaeacians, immediately covers himself with leaves, and goes to sleep. Homer has been counting the days for us but now we are on our own. Odysseus wakes, has a brief meeting with Princess Nausicaa, then is wrapped in a mist by Athena to get him safely before King Alcinous and Queen Arete. This despite the Phaeacians being friendly folk with a reputation for returning lost sailors to their home. That night Odysseus recounts his seven-year stay with Calypso to his hosts, thus setting it off from the rest of his adventures.

Second day: Everyone sits down for a feast while the bard, Demodocus, sings of the "Quarrel of Odysseus and Achilles," mentioning "those days when almighty Zeus was conjuring up the great wave of disasters that would strike Trojans and Danaans alike." These would be the Earth-Mars Wars. The Phaeacian Games take place, but Odysseus participates in only one, winning the discus throw, "not even troubling to remove his cloak." Entertainment begins afterward and Demodocus sings at length (other lays are merely mentioned) of the "love of Ares and Aphrodite." Odysseus is then called upon to give a true account of his wanderings, and he responds with Books IX, X, XI, and XII. This "flashback" of his adventures corresponds to the Bearson folk tale.

Third day: Preparations are made for Odysseus to be taken home to Ithaca by Phaeacian sailors "whose ships know by instinct what their crews are thinking and propose to do." A bed is made for the hero on the deck of the boat and he sleeps all the way back to Ithaca. He is lifted ashore, deposited on the sand still asleep in a spot near the Cave of the Nymphs.

A description of the second cave, another symbol of the cosmos, closes the catastrophic centerpiece of the epic. At this point, readers familiar with the *Odyssey* will have figured out that the remainder of the story, wherein the hero returns to Penelope and punishes her suitors and faithless maids, corresponds to the end of the Bearson folk tale, wherein he deals with his false friends and marries the princess.

Homer does not always string his beads in the same order as Bearson, nor are they in the same style, but the parallels between the two stories are hard to overlook. The contrasts between the way the material is handled, the purity of Bearson versus the artistry of Homer in the *Odyssey,* may someday tell us much about the workings of the human mind.

Thus Homer has featured telling information about the Earth-Mars Wars between the two caves of the *Odyssey*. The somewhat strained sequence would be awkward in the hands of anyone other than the master, but the story never flags. Over the next 2,700 years,

[37] The "steering oar" would be the axis column.

the catastrophic substructure will sink into a few hardly noticeable threads in the warp of Homer's tapestry, so far beneath consciousness that when we think of Odysseus in the land of the Phaeacians, we first recall his innocent flirtation with Nausicaa.

Only in the lay of Demodocus do two major players in the sky story overhead make an appearance: Ares and Aphrodite, Mars and the Moon. But the image of the two lovers trapped in bed by Haephestus' bronze net, with all the gods looking on, making comments, does not encourage us to look at the sky. The position of the lay of Demodocus in the *Odyssey* indicates that this particular interaction of Mars and the Moon took place at the same time as the tilt of Earth. Certainly there was disorder in the cycles of the Moon at that time.

Both Homer and his audience were probably quite conscious of the events of the Earth-Mars Wars. Certainly the poet was operating under the continuing spell of bringing heavenly events to Earth.

The preoccupation of our present culture with time travel is ongoing. This can be perceived as another instance of bringing heavenly events to Earth. In the following chapter we will address two major disruptions in earthly time that occurred during the Earth-Mars Wars.

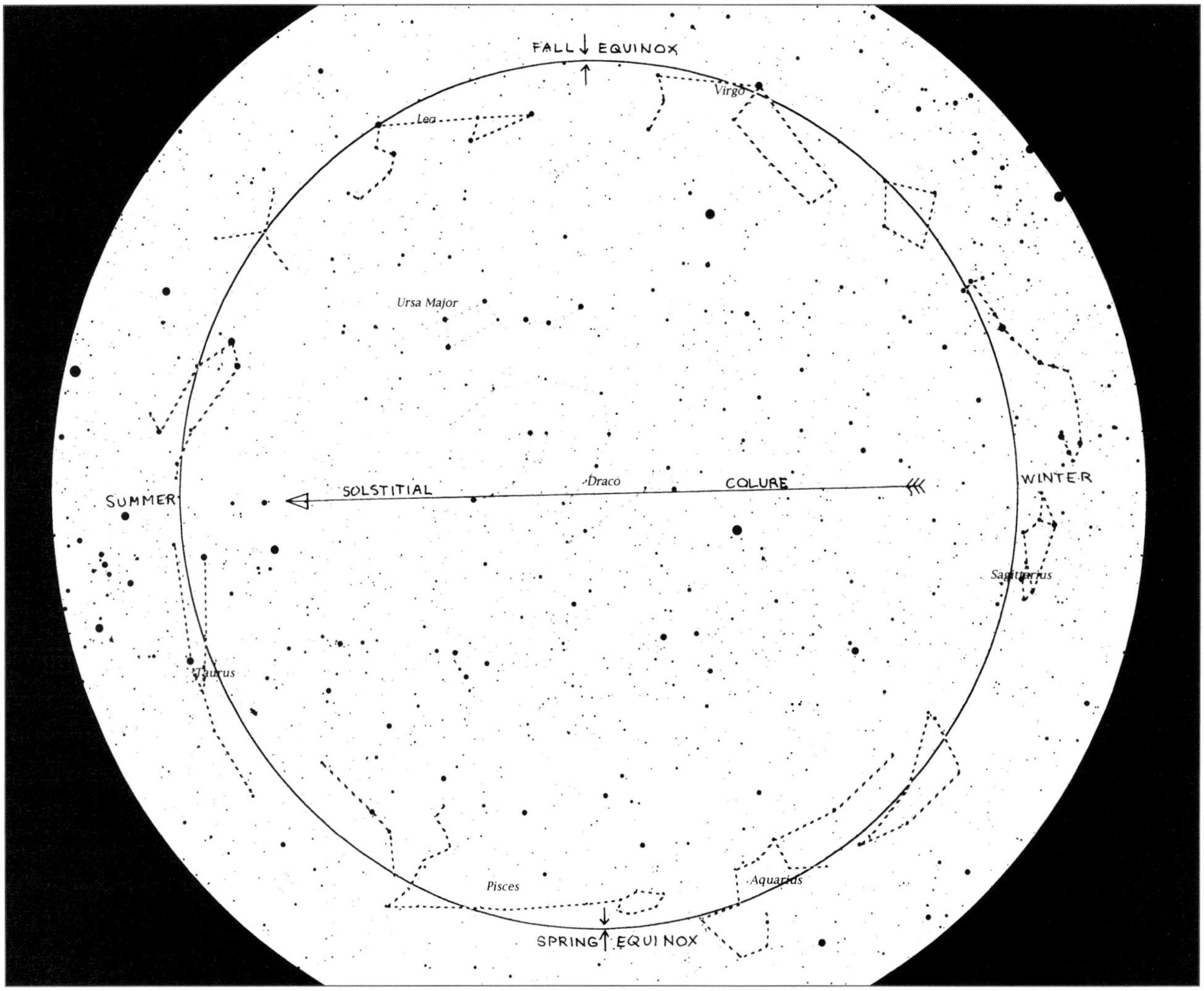

The precessional clock. There is only one indicating hand, the solstitial colure, the summer half presently running two degrees from the North Star, and pointing to the constellation and zodiac sign of Gemini. Equinoxes are marked on the circle of the ecliptic/zodiac, the Spring Equinox well along in Pisces. It will move into Aquarius in about 160 years.

10. Time Travel on the Precessional Clock

The stars began to fray, and time and earth washed hands in mischief.

– The Shahnama of Firdausi

After beginning work on the terrestrial tilt of the Earth-Mars Wars, I took advantage of a family visit to chat about it with my son-in-law, the physicist.[1] We sat down at the dining room table, talked, and moved models around as I explained the concept, bolstering it with various bits of research. We worked our way through the tilt, Dan nodding, occasionally questioning. Finally he got up, pushed his chair in to the table, leaned on it, and smiled.

"You're going to have to relate this to the galaxy, you know."

"No!" Mentally I dug in my heels. "I'm not ready for that." The words spilled out of my mouth as I continued to backpedal, making excuses. It began to seem as if someone else were talking. A few weeks later I had the following dream.

I am wearing a dark blue coverall, overly large, dirty and greasy, and walking into a huge garage with a high ceiling. A number of mechanics wearing the same type of garment are working on equipment. My coverall snaps aren't properly aligned, so I unsnap and fix them, faintly surprised that the nasty garment doesn't seem to bother me. The other mechanics look up briefly, then ignore me.

My feet carry me to the right toward a strange-looking machine dominated by what looks like a wide conveyor belt, just visible under a litter of dirt and unidentifiable fragments. Somehow I know this is a candy machine. Another mechanic is trying to deal with it, but without success. He gives up, moves to one side, and watches me. I sense he wants the same kind of candy bar that I do.

The dream leaves me standing in front of the machine, convinced that I can get candy out of it, mainly because there are two quarters in my pocket from a dream the previous night. Thus I have "change."

It is not until I wake up that I remember that my favorite candy bar, the one I would have bought, is Milky Way, the nickname for our galaxy. I had a good laugh and shed resistance to my new career as celestial mechanic.[2]

✦ ✦ ✦

When modern man wants to know what time it is, he lifts his hand and looks at his watch, often a digital one. Days and nights were divided into hours and minutes so long ago that most of us are hardly aware that time came and still comes out of the sky, and from the stars rather than the sun.[3] The Babylonian system of time is called sexagesimal because it rests on the base 60. It was developed under the orderly skies of the Age of Heroes, 1450 to 747 B.C., when there were 360 days in the year. The circle of the zodiac was divided into 12 equal parts of 30 degrees or 30 days. Sun time was easy to keep with twelve 30-day months. Thus sun and Moon could be seen as "married."

[1] Dan Willey gave me my copy of *Worlds in Collision*, the Book-of-the-Month Club edition.

[2] Pursuing the tilts and spins of Earth into the past will bring the connection with the Milky Way into focus in a later chapter.

[3] The Babylonians divided the day and night into 12 equal parts. *JNES*, Babylonian Astronomy, p. 24.

One of the most fascinating chapters in *Worlds in Collision* is the one in which Velikovsky assembled references revealing the existence of 360-day calendars in many parts of the world: India, Persia, Babylonia and Assyria, Palestine, Egypt, Rome at the time of its founding, Mesoamerica, and Peru. These calendars frequently agreed on another unusual phenomenon, months that were actual lunations or moons. The new Moon was always on the 1st of the month, the full Moon on the 15th. Later 5¼ days were added to these calendars under circumstances that caused them to be considered unlucky, i.e., the 60 years of the Earth-Mars Wars.[4]

Although we still use number systems worked out by the ancients, modern orthodox astronomy views their careful timekeeping and astronomical records with a jaundiced eye, as "... a rigid schematism unrelated to reality."[5]

The largest clock that mankind can look to for time is that of the precessional motion of the Earth, often compared to a spinning top as it tilts and winds down. Far from winding down, however, Earth is and has been inclined to tilt upward in recent millennia, both gradually and catastrophically.[6] The face of the precessional clock is the cosmos itself; the numbers are the signs of the zodiac in the horizontal plane of the ecliptic. There is only one indicating hand, the circle or colure of the solstice,[7] and a full sweep around the circle would take 26,000 years.

As the sun appears to move forward, circling through the zodiac in the course of a year, so the armature of the colures moves slowly backward with the Earth, thus "precessing." We have chosen the term "Precession of the Equinoxes" to describe the motion but, in fact, the entire Earth precesses. The precessional clock, a nuisance to timekeepers because it keeps jiggling sun time askew, cannot be overlooked. It is the bottom line on cosmic time for us here on Earth. As we shall see, the ancients used precessional time to record both world ages and the cosmic catastrophes that separated them.

For us here on Earth with the vault of Heaven in ceaseless apparent motion overhead, the "heart of Heaven"[8] seems to be the still, turning point at the center, where the pole of Earth, extended infinitely, pierces the sky. At present, at the beginning of the 21st century, it is within two degrees of the star at the end of the handle of the Little Dipper or bear, Ursa Minor. With knowledge of the precession, however, we realize that this "north star" is a temporal fiction, a center that does not hold. The real heart of Heaven is the pole of the ecliptic, a point in the constellation Draco (dragon) around which the pole of Earth traces a circle every 26,000 years.

Before we can proceed to describe another sudden motion of the Earth, it is necessary to tap into knowledge of the ancients so that we can grasp it properly. One can open any textbook on astronomy and read that the Precession of the Equinoxes is caused by the sun and Moon acting as a couple on the Earth. This is where Newton left us, but ancient astronomers would not agree with him.

The ancients understood the model of the precession thoroughly because they had seen it at work under both normal and catastrophic

4 Velikovsky, *Worlds in Collision,* Chapter 8 of Part II.

5 Otto Neugebauer, "The Astronomical Chapters of the Ethiopic Book of Enoch," Det Kongelige Danske Videnshabernes Selskab Matematisk-fysiske Meddeleser 40:10, Kobenhavn 1981, p. 13.

6 Throughout the present Iron Age the Earth has gradually tilted very slightly upward.

7 The equinoctial colure cannot be a hand on the precessional clock since it does not intersect the pole of precession, but its position can be indicated on the circle of the zodiac.

8 This is a Mesoamerican term.

conditions. We, on the other hand, have the good fortune to be able to reconstruct it rather than experience the catastrophic motions of Earth. Unfortunately, despite the work of Jung with mandalas and Eliade in the history of religion, Western Civilization has only the most tenuous grasp of the symbolism of the center that pervades ancient art, architecture, and religion. In order to move back in time, we must listen to those who knew something about the center.

The real heart of Heaven is not well marked. The true center lies at the pole of the ecliptic, the "empty place" in the sky mentioned in the book of Job. "He stretcheth out the north over the empty place, and hangeth the Earth upon nothing." The constellation Draco is draped around this empty place, and to the ancient Babylonians, Draco was "Yoke-Star of the Sky." It was their way of describing precession, seeing the North Pole of Earth as linked to the pole of the ecliptic, slowly circling around it. The view was comprehensive; the South Pole, too, had a mooring line on it. Fortunately, the southern hemisphere has a bright star near the south pole of the ecliptic. Thus the South Pole of Earth was tied to the "heavy-weighing" Canopus, the "Yoke Star of the Sea." The Arabs called Canopus the "Heart of the South."[9]

Classical Greeks had forgotten a good deal of their astronomy and had a very problematical view of the underworld. Hell or Tartarus was sometimes out of sight at the south pole of the ecliptic, but could be reached through the Earth by means of a rope, well, or whirlpool, sometimes by Ocean's Stream. There were rivers—some of us will remember Lethe or Styx—and a ferryman to take us across.

Earlier views were clearer. There was actually a conflux of four rivers, and it was way down, at Canopus. If one is going to have a Hell, it ought to be in the very pit, at the Yoke Star of the Sea, just as the heart of Heaven is at the Yoke Star of the Sky. Santillana and Dechend provide a fine model of the underworld in *Hamlet's Mill* as it was understood by the ancients, though it is not generally perceived to be an objective one. We will return to it as we proceed into the past. For now, all that is needed is this expanded view of the Precession of the Equinoxes, Earth slowly spinning like a top in the plane of the ecliptic, yoked to two points in the sky, directly above and below.

Adding the yoke stars makes a more interesting model, but one that must be articulated further in order to be understood. In a previous chapter we saw the Earth tilt, pole in line with the solstitial colure, noted that it is hinged at the equinoxes, saw the North Pole of Earth move up toward the pole of the ecliptic. It certainly looked like Mars somehow tilted the Earth. But now that we have seen the ancient model and understand that the pole of Earth is yoked to the pole of the ecliptic, perhaps that wasn't exactly what happened.

The precessional model begins to look like a balance of forces: "pull" of the yoke stars balanced by the precessional "spin" of Earth.[10] When Earth and Mars approach each other at –717 B.C., we see an upset in the balance of precessional forces. Suddenly "pull" kicks in at the solstice positions while "spin" holds steady at the equinoxes. This combination resulted in the tilt of Earth. In a later chapter, we shall see this particular set of forces in operation during one of the Venus catastrophes.

In the previous chapter we left the Earth just after the tilt, situated in a respectable position recorded by the Khufu pyramid as well as the Mithraic tauroctony and the Labors

[9] Sigmund Mowinckel, *Die Sternamen im Alten Testament,* Norsk Teologisk Tidsskrift 29, 1928, p. 12.

[10] These forces will eventually have to be analyzed in several frames of reference: rotation, revolution, and precession.

of Hercules. The tilt is right, somewhat more than the 23.4 degrees at present, and orthodoxy would accept the model if it were 3600 B.C. But it is not, and after the Hercules Equator has been in position for 15 years, it is 702 B.C. and Mars is approaching for its fourth encounter with Earth.[11]

The Earth is about to participate in a kind of time travel. How else to describe a sudden spin along its precessional path, the indicating hand of the solstitial colure moving suddenly along with it. For this is an entirely different motion from the tilt just dealt with. This time when the precessional forces are upset, the

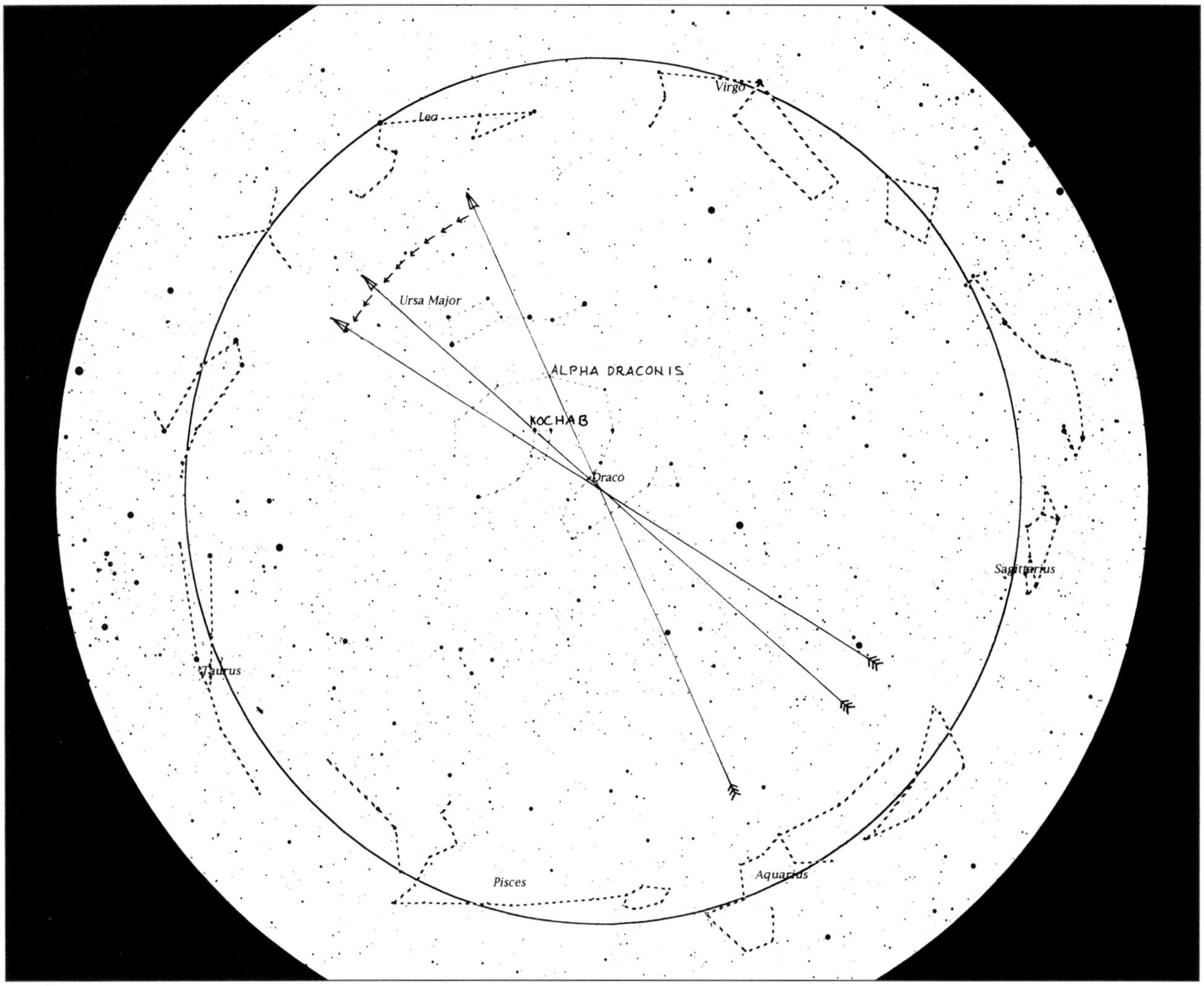

Time travel on the precessional clock.
In 717 B.C. the solstitial colure ran through the star Alpha Draconis, Ursa Major, and Leo on the ecliptic/zodiac.
In 702 B.C. it moved suddenly to run through the star Kochab (in Ursa Minor) and Cancer on the ecliptic/zodiac.
In 687 B.C. the hand of the solstititial colure moved suddenly again, to a position nine degrees further along in Cancer.

[11] We must always keep in mind that Mars was only a danger to the Earth every sixth time around, the period corresponding to its favorable opposition today, the time at which it is closest and best visible from Earth.

"spin" kicks in while the "pull" holds steady. Suddenly the Earth moves quickly along its regular precessional path; the gates of the Summer Solstice slide from Leo to Cancer, the Spring Equinox from the middle of Taurus to the middle of Aries. A full zodiac sign reads out at about 2,200 years on the precessional clock.

The precessional clock now reads 1400 B.C. (retrocalculating into the middle of Aries) but it is actually 702 B.C. The Babylonians, noticing the change in the heavens, now do something that we would rather expect: they record the new rising times of their "36" stars about 700 B.C. in a well-known text, Mul.apin.

The Babylonian text states, once on the first tablet, twice on the second, that the equinoxes are in the middle of Aries and Libra, the solstices in the middle of Cancer and Capricorn. Orthodoxy agrees that Mul.apin was inscribed in 700 B.C. The data on it does not fit the time frame, however, so it has been decided that it is a compilation of data from as far back as 1400 B.C. "The final check on this reconstruction [of Mul.apin] is that it agrees better with modern calculation than the list of dates given in the text."[12] Urdummheit raises its ugly head. The Babylonians couldn't keep track of the calendar or record what they saw.

It is to Egypt that we must look for confirmation of Mul.apin, which places the Spring Equinox in the middle of Aries. As we saw earlier, the Khufu pyramid is a celestial marker devoted in part to recording the Earth-Mars Wars, since a shaft in it is directed at the famous Alpha Draconis. This marks the end of the most recent terrestrial tilt from Ursa Major toward the pole of the ecliptic.

After the precessional spin of 702 B.C., the Babylonians recorded the new sky at the Spring Equinox. But this sort of motion can also be recorded elsewhere, a project that was taken up by the Egyptians. If one draws the summer solstitial colure of the period, 90 degrees from the spring equinoctial colure, it will intercept a star of Ursa Minor, Kochab. Although the star is above the present circle of precession, the point where they cross is easily deduced. This particular star was chosen to mark the solstitial colure of the period, and a shaft of the Khufu pyramid was directed at Kochab.[13] The block of time/motion indicated by the Egyptians is exactly the same as that indicated by the Babylonian Mul.apin.

The fifth and last encounter of Earth and Mars in 687 B.C. resulted in another precessional spin of the Earth, this time a shorter movement along its precessional path. Velikovsky can help us here since he twice noted the movement of the Spring Equinox by nine days: once in Jewish records, again in the records of the Babylonians.[14] This is the block of precessional time needed to get the Earth on schedule for Greco-Roman times.[15]

Sudden movements of the equinoctial and solstitial colures recorded by the ancients cannot be swept under the rug. We can now see that in massaging the data of Mul.apin back to 1400 B.C., orthodox scholarship was unwittingly dealing with the catastrophic precessional spin of 687 B.C. which moved the Spring Equinox by approximately nine degrees. It is only by checking the precessional clock that we can grasp the nature of the catastrophe. Fortunately the ancients made lasting records for us. The clay tablets of the

[12] B. L. Van der Waerden, "Babylonian Astronomy. II. The Thirty-Six Stars," *Journal of Near Eastern Studies*, Vol. VII, 1949, p. 6.

[13] Robert Bauval and Adrian Gilbert, *The Orion Mystery* (New York, Crown, 1994) p. 209. Santillana notes that Kochab (beta Ursae Minoris) has been called "mill peg," indicating a central position. *Hamlet's Mill*, p. 137.

[14] Velikovsky, *Worlds in Collision*, pp. 350, 354.

[15] That the Babylonian "first meridian" is located at eight degrees Arietis reflects the beginning of the Iron Age.

Babylonians withstand everything except storage in damp museum basements, and the Khufu pyramid is an enduring marker, difficult to overlook.

Readers who are having difficulty accepting one catastrophic Earth motion, much less two, should consider the gyroscopic force pair that is taught in physics. The spinning Earth acts like a gyroscope, resisting any attempt to change its axis of rotation. When force is applied to change it, the spinning body moves at right angles to the force.[16] Try holding the axle of a spinning bicycle wheel in your hands, and see what happens when you try to tilt it. Since events happen more slowly in the macrocosm, the sudden precession occurs considerably after the tilt (and seems to need the electromagnetic "trigger" of planetary interaction).

Both Mars and Hercules are given credit for moving the Spring Equinox from Taurus into Aries. Hercules killed the Cretan Bull in one of his labors. The case of Mars is more subtle as it is indicated in the tauroctony. Mithras is conceived in the style of immanence and transcendence, working behind the scenes as it were. In the tauroctony, the bull is killed by the dagger, a weapon of Mars. The image can therefore be seen as catastrophic.

Astrology preserves the catastrophic precessional motion of the Earth-Mars Wars by placing Venus in the house of Taurus, Mars in the house of Aries. This reflects the dominance of Venus in the Age of Heroes when the Spring Equinox was in Taurus, while Mars initiated the present Iron Age which began with the Spring Equinox in Aries. The present age, with its accent on war and violence, owes much of its character to the God of War.

In the following chapter, we review Hesiod's "Shield of Heracles," wherein the mythographer tells us of the five interactions of Ares and Hercules (Mars and Earth). For students of catastrophism, we also provide a chronology of the main events of the Earth-Mars Wars as a point of departure for further work.

[16] World Book Encyclopedia: gyroscope.

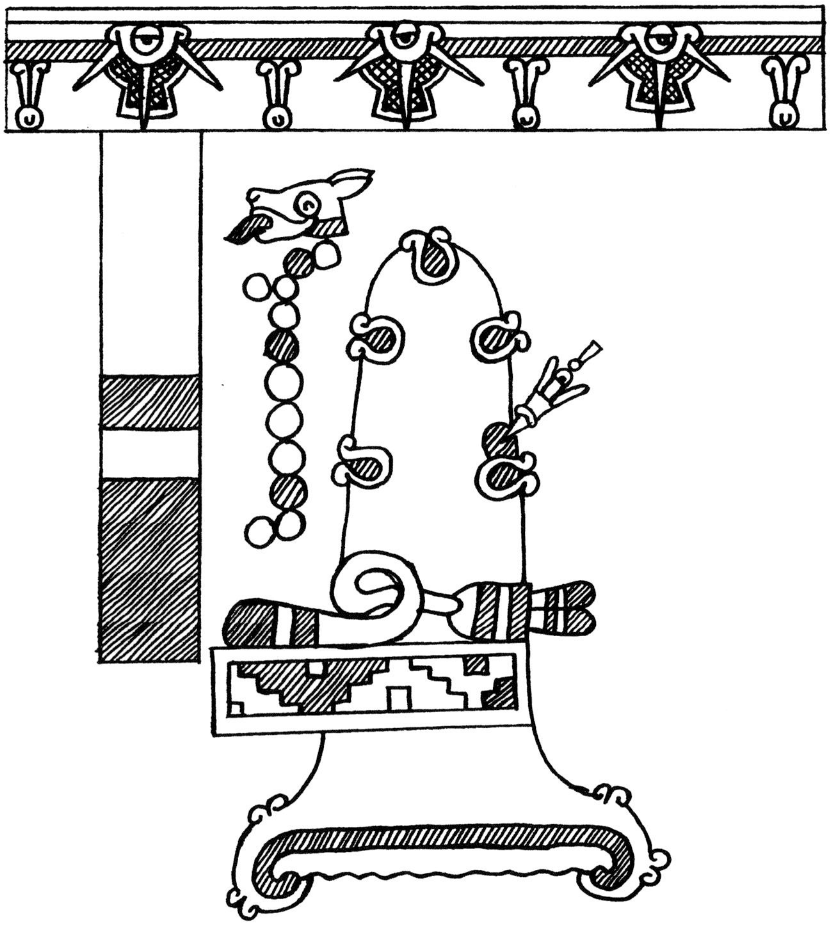

The five catastrophic episodes of the Earth-Mars Wars as indicated in this detail from the Codex Nuttall (omega symbols striking prominence) where it is repeated several times. See record in the Vienna Codex on page 214.

11. The Age of Heroes Goes Down

Ares, . . . sceptred King of manliness, who whirl your fiery sphere among the planets in their sevenfold courses through the aether wherein your blazing steeds ever bear you above the third firmament of heaven; . . .

– Hesiod, "Homeric Hymns"

The Greek mythographer Hesiod well knew the planetary character of the gods of Olympus, and his mythological lore reaches far into the past. He gave us knowledge of the five ages of man: gold, silver, bronze, Age of Heroes, iron, and genealogical tables of the gods that include and precede the Olympians. His record of the Earth-Mars Wars, "Shield of Heracles," was written down within a century after the events, lending them additional weight for Westerners.

Hesiod gives us perhaps the most lucid, easily read myth of the Earth-Mars Wars. It is a unique view of the powerful war god in that Mars is clearly outmatched and loses out in every encounter. This is exactly what would be expected given the red planet's smaller size and weight vis-à-vis Earth. Since the Labors of Hercules regularly involve the mighty hero in tracing out Earth's catastrophic changes, we are not surprised to see him, the Superman of his time, defending the home planet.[1] "Shield of Heracles" focuses on the final fifth encounter.

Fully one-third of the poem is taken up in describing the shield itself, decorated with imagery which relates to earlier mythological events that we will presently address. That Hercules is a twin aspect of Mars becomes clear when we find an image of Panic, one of Mars' horses, in the center of his shield. The "Shield of Heracles" is analogous to the shield of Achilles in the *Iliad* which also contains records of catastrophic changes. Like Homer, Hesiod knows whereof he speaks.

Hesiod tells briefly of the four earlier times that our hero threw the god of war to the ground, then proceeds with the fifth encounter. ". . . Herakles . . . stabbed with full force into the thigh left bare under the elaborate shield, and beat him to the ground between." The description is much the same as on the fourth fall. In both cases Mars was struck in the thigh and a hole torn there.[2]

As usual in these tales of Hercules, Athena is present, advising and encouraging. Since the Olympians are in power during this era, the outcome of the battle has already been decided by Zeus. Hesiod counts the five knockdowns, but both Hyginus and Apollodorus mention that a thunderbolt separated the contestants during their last bout.[3]

Hesiod's account underscores again the five planetary interactions between Earth and Mars. But we must recall that Velikovsky's Martian episodes have been largely written off by historical catastrophists. Here we have reopened the issue, and now construct a chronology in the hope that it will be of help to interested students. We begin with the biblical framework of Velikovsky and the

[1] Zeus chose Alcmene with a view to begetting a son powerful enough to protect Earth. Graves, *Greek Myths,* 118e.

[2] The injuries to Mars' thigh during the fourth and fifth bouts with Hercules may bear some relation to the precessional spins that occurred at those times, possibly planetary "scars" or "wounds" caused by electrical interaction.

[3] Hyginus, Book XXXI, Apollodorus, 2.7.7, 2.5.11.

story told by Olmec iconography, then add the significant substance of ancient media where it is indicated. References are given only when they are not treated elsewhere in the present volume.

747 B.C.: *Earth-Mars Interaction #1:* (The pole is in Ursa Major tilted about 37 degrees, Spring Equinox in the middle of Taurus, ratio of longest day to shortest night 2:1,[4] 360 days in year, 12 months of 30 days each.)

Earth changes: Commotion (raash) in days of Uzziah, Moon disturbed.

Afterward: Moon cycle is changed.

Mythical/historical reflections: The months were perturbed at the foundation of Rome in the time of Romulus. There were ten months in the year, the first named after Mars, the second after Venus.

The Jews begin to call the months by number.[5] New Moon feasts, Sabbaths, and other feasts are considered to be frowned on by Yahweh because the lunar cycle has changed and the weeks are disrupted.

A new calendar is instituted in the Middle East initiating a new astronomical era.

Commencement of the "Twilight" between the Dvapara and Kali Yuga during which the Mahabharata War was waged. Wars between the Shang and Chou dynasties, Iranians and Turanians, also took place during this period.

732 B.C.: *Earth-Mars Interaction #2:*[6] (The pole is in Ursa Major tilted about 37 degrees, Spring Equinox in the middle of Taurus, ratio of longest day to shortest night 2:1, 10 months in year.)

Earth changes: None recorded.

Mythical/historical reflections: Isaiah 8:22 and 9:1, previous to the "commotion" of 717 B.C., refers to two previous occasions on which the land was "afflicted."[7] These would be interactions #1 and #2.

The Olmecs record two 45-degree crossed bands on their serpent bas-relief, also two 45-degree crossed bands with an "opening of 15 between them." The angle of the crossed bands, again, is a close match with the tilt of the pole during this period (allowing for the difference between amplitude and declination).[8]

717 B.C.: *Earth-Mars Interaction #3:* (The pole is in Ursa Major tilted 37 degrees, Spring Equinox in the middle of Taurus, ratio of longest day to shortest night 2:1, 10 months in year.)

Earth changes: The pole tilts directly toward the pole of the ecliptic, stopping at Alpha Draconis.

Commotion (raash) occurs at the burial of Ahaz. The sunset is hastened as the sun goes forward 10 degrees on the sundial of Ahaz, the only recorded forward motion of the sun.[9]

[4] B. L. Van der Waerden, "Babylonian Astronomy. III. The Earliest Astronomical Computations," *JNES*, pp. 20–21. This is the *"older system* . . . a very rough and primitive approximation." But it perfectly reflects the 37-degree tilt of the pole.

[5] This is suggested by information in "The Three Calendars of Ancient Israel" by Julian Morgenstern, *Hebrew Union College Annual,* Vol. I, 1924.

[6] This date is my own interpolation, based on the timing between the series of catastrophes and the weight of material indicating five interactions between Earth and Mars.

[7] Velikovsky saw these occasions as 776 B.C. (first Olympic Games) and 747 B.C. I feel they are 747 and 732 B.C., the latter being an event that fits a sequence of 5 every 15 years over a period of 60 years.

[8] Amplitude is measured on the horizon and increasingly greater than declination as distance from the equator increases.

[9] In all other instances of apparent motion of the sun, it is seen to go backward, agreeing with catastrophic precessional motion. Here the forward motion would have been the result of an axial tilt.

Klimasturz (climate disturbance) and high water catastrophes occur in association with the commencement of Fimbul Winter.[10]

The Moon is disturbed.

Afterward: Terrestrial pole points toward Alpha Draconis, tilted at an angle of close to 24 degrees (slightly more than at present).

Ratio of longest day to shortest night changes to 3:2.[11]

Months in year (actual lunations) change from 10 to 12.

The Hercules Equator is now in place.

A protracted winter commences in far northern and southern latitudes with longer freezing and snowfall as sun no longer moves as far north and south. Men and animals begin to migrate closer to the equator.

Mythical/historical reflections: The bear, Ursa Major, distanced from the pole, now dips below the ocean (horizon) for the first time.

Tezcatlipoca, Ursa Major, jumps down and has his foot bitten off by the earth monster.

The jaguar, Ursa Major, jumps down, attacks men, fornicates with women to produce were-jaguar babies.

Atlas shifted the burden of the sky at the death of Romulus.[12]

The Hercules Equator is recorded in the Labors of Hercules and the Mithraic tauroctony.

The appearance of the 90-degree-long constellation of Hydra (part of the Hercules Equator) at the equator is recorded in both the Sibylline Phaethon myth and the Bible. "Now the serpent scraped with his writhing belly the equator far away from the bear . . ."[13] In speaking of the "great red dragon," the Bible says, "And his tail drew the third part of the stars of heaven, and did cast them to the earth."[14]

Fenris Wolf swallows the sun as Fimbul Winter begins.

The Anugita records a threefold day and night concluding a world age.[15]

The Sisyphus myth, echoed by Odysseus' fall into the sea, describes the movement of the pole upward. Both mention a three-day period of obscurity.

Afrasiyab has a dream of the changed path of the sun.[16]

The army of Kai Khusrau cry, "The sun has wandered from his way."

A shaft is constructed in the Khufu pyramid directed at Alpha Draconis.

Sargon of Akkad expresses great interest in Alpha Draconis, called Tir-An-Na, the Life of Heaven.[17]

Numa dealt with the increase in lunations by adding two months to the beginning of the year, January and February, the first month in honor of Janus,[18] the second suggesting purification.

702 B.C.: *Earth-Mars Interaction #4:* (The pole points to Alpha Draconis, tilted at a slightly greater angle than we see today, Spring Equinox is in the middle of Taurus, the ratio of longest day to shortest night is 3:2, 12 months in the year.)

---

[10] Velikovsky, *Earth in Upheaval,* Chapter XI.

[11] Van der Waerden, "Babylonian Astronomy," pp. 21–22. Now we have the "latter, better value" which reflects the present tilt of the pole.

[12] Ovid, *Fasti,* II ll. 489 ff.

[13] W.H.D. Rouse, quoted in *Kronos* II, 4, p. 28.

[14] Rev. 12:3, 4.

[15] K. T. Telang, trans. "The Anugita," in *The Sacred Books of the East,* 1882, Vol. VIII.

[16] Firdausi, *Shahnama,* Warner translation (London, Kegan Paul, Trench, Trubner & Co, Ltd., 1906) Vol. II, p. 232.

[17] Allen, *Star Names,* p. 206.

[18] J. G. Frazer, *Ovid Fasti* (London, 1951), p. 9. Ovid quotes Janus. ". . . none but me may rule the wheeling pole."

Earth changes: Earth takes a sudden precessional spin of 30 degrees.

Afterward: The Spring Equinox was moved from the middle of Taurus to the middle of Aries.

The terrestrial pole, always on the solstitial colure, moved with it from Alpha Draconis to a point near Kochab, a star in Ursa Minor, thus defining the new solstitial colure.

Hercules Equator is no longer in position.

Mythical/historical reflections: Mul.apin records the new rising times of the 36 stars.

Hezekiah doubled the month of Nisan to keep the seasons in order.[19]

The sun moves backward in the Thyestes myth while the Pleiades and the rest of the stars maintain position, mirroring the precessional spin.

The Ritual of Karneios Dromios, the "ram runner," is instituted as Aries, the ram, moves suddenly into the equinox position.[20]

687 B.C.: *Earth-Mars Interaction #5:* (Pole is near Kochab in Ursa Minor tilted about 23¾ degrees, Spring Equinox is in the middle of Aries (15 degrees Arietis), the ratio of longest day to shortest night is 3:2, 12 months in the year).

Earth changes: Earth takes a sudden precessional spin.

Afterward: There is a disturbance of the sun's movement when people begin to sing the Hallel prayer of the Passover, an equinoctial feast,[21] and the army of Sennacherib in his second campaign against Judah is destroyed by a "blast."

The sun goes back 10 degrees on the dial of Ahaz as the equinox is moved from 15 degrees Arietis to about 6 degrees Arietis, a point that agrees with the present position of the pole.

Sun moves backward in the Mahabharata War.

Two instances are recorded of a nine-degree change in the date of the Spring Equinox.[22]

Seven Greek temples are oriented to the star Alpha Arietis, at six degrees Arietis.[23]

There are 365¼ days in the year from this point on.

Mythical/historical reflections: Kai Khusrau ascends to Heaven followed by the death of five paladins.

Yudishthira ascends to Heaven followed by the death of four brothers and joint wife, Draupadi.

Flying Bear invested 365 new gods on the Terrace of Creation, marking the end of Shang, beginning of Chou Dynasty.

The considerable confusion surrounding the exact position of the Spring Equinox after the precessional spin of 687 is mirrored by the fact that there were no New Year's celebrations for 20 years afterward.[24] Although Hipparchus is given credit for discovering the Precession of the Equinoxes in 128 B.C., the positions of 15 degrees and 8 degrees Arietis ring down the centuries in scholarly works until medieval times, echoing the precessional spins of 702 and 687 B.C.[25]

That Earth and Mars could have five catastrophic meetings in space at intervals of

[19] This would have worked perfectly, but only if there were already 12 months in the year. Numa's adding two months to the calendar must have occurred after 717 B.C. where Homer suggests a disturbance of the Moon, the love affair of Ares and Aphrodite.

[20] Jane E. Harrison, *Prolegomena and Themis* (New Hyde Park, N.Y., Univ. Books) pp. 234–235.

[21] Velikovsky, *Worlds in Collision,* p. 232.

[22] Velikovsky, *Worlds in Collision,* p. 349.

[23] Lockyer, *Dawn of Astronomy*, p. 420. This would have been on the equinoctial colure of the period.

[24] Velikovsky, *Worlds in Collision,* p. 346.

[25] Neugebauer, *Ancient Mathematical Astronomy*, pp. 594–600, Vol. II.

15 years is initially hard to digest, probably a result of our billiard ball view of the universe rooted in the celestial mechanics of Newton. In the electromagnetic view illuminated by Velikovsky, the Earth-Mars interactions begin to look more like normal phenomena than a series of celestial accidents. The Bible mentions an arbitrary period of grace of 15 years, and Olmec iconography also indicates one. The reign of Kai Khusrau was 60 years, preserving the pattern of five catastrophes at 15-year intervals.

The five-element associated with the five visits of Mars is widely recorded in myth, legend, and iconography, and commonly associated with warfare and catastrophe.

- The Olmecs erect the altar of the five were-jaguars.
- The five-petalled Mexican Star or flower is a symbol of warfare.
- Five Aztec warriors perform in the ritual of Tlacaxipehualitzli.
- Kai Khusrau's five paladins fight in the Shahnama.
- Five Pandavas fight in the Mahabharata War.
- Five men of Turkish antecedents bore the title of Atsina-wolf.[26]
- Five spears of Aswatthaman are painted with red-hot iron.[27]
- The Dresden Codex depicts Quetzalcoatl throwing five spears.
- The Borgia Codex depicts five spear "piercings."[28]
- Mars is recorded as being thrown down five times in "Shield of Heracles."
- In Agnicayana, the "piling up (of bricks for the altar) of fire . . . ," a five-course fire altar is built, five heads walled up in the first course.[29]

Those who object to the regular synodic-like time schedule of this series of cyclical catastrophes must learn to draw comfort from the detectable order of the phenomena. Another instance of order in what initially appears to be chaos was noted in *Worlds in Collision*: planetary catastrophe appears to take place at the Spring Equinox, as in the case of the last interaction of Earth with Mars in 687 B.C. The disturbance in the motion of the sun was noticed at a ceremony of Passover, the Jewish feast which was observed at the Spring Equinox.

But this was not the only visit of Mars that can be linked with the Spring Equinox. The very first action of the Earth-Mars series was in 747 B.C., a date associated with the foundation of Rome. Now the Romans would no more overlook their legendary beginnings than citizens of the USA would overlook the Fourth of July. The Parilia or Palilia of Rome was connected with fire[30] and celebrated at the beginning of spring.[31]

[26] Petitot, Emil, *Traditions Indienne,* Les Litteratures, pp. 62–63. "The Feet of the Dog" is a Dene legend relating to the myth of the origin of the Turks from a she-wolf.

[27] John O'Neill, *Night of the Gods,* p. 86.

[28] Borgia Codex, Plates 53 and 54 of the Diaz and Rodgers restoration (New York, Dover, 1993).

[29] Mircea Eliade, *History of Religious Ideas,* pp. 217, 228.

[30] Dumezil, ARR.

[31] O'Neill, *Night of the Gods,* p. 45. The Romans appear to have moved the start of their year to the Winter Solstice as a result of the movement of the pole, perhaps because it follows the solstitial colure. Calendric confusion, however, was the order of the day, compounded by the addition of 5¼ days to the year. In classical Roman times the Parilia was celebrated about April 21.

At the precessional spin of 702 B.C. caused by the fourth visit of Mars, the equinox was moved from Taurus to Aries, a full 30 degrees. This pushed the seasonal calendar back a month so that Hezekiah had to double the month of Nisan to keep it in order. This one-time action solved the problem. Here we must understand that Jewish skywatchers didn't need years to figure out what was happening. They would have seen the rising of different stars at nightfall on the first clear night after the catastrophe, and quickly grasped what needed to be done. The problem was dealt with immediately in doubling the month of Nisan. Since this month marked the Spring Equinox, Hezekiah's action strongly suggests the catastrophic connection.

Across the Atlantic, the ancient Mexicans took a more comprehensive view, celebrating the Earth-Mars Wars as a whole. At the time of the Spanish Entrada, Tlacaxipehualitztli, the ritual wherein five warriors attacked a single victim, was celebrated on March 20.[32] The divinity Macuilxochitl, Five Flowers, was celebrated at Xochihuitl. That it was a moveable feast[33] suggests the effort to honor two different world orders.[34] On balance, it is difficult not to be struck by the regularity of this catastrophic series.

Students of myth will find that confusion between bear and wolf began during the Earth-Mars Wars. In the following chapter we will attempt to shed light on the powerful and evocative mythos of the werewolf.

The five catastrophic episodes of the Earth-Mars Wars as indicated by Ehecatl raising up the sky with its five Mexican star signs, the symbol which grew out of the Earth-Mars Wars. See Chapter 23.

[32] Diego Duran, *Book of the Gods and Rites* (Univ. of Okla. Press, 1975).

[33] Sahagun, *A History of Ancient Mexico,* trans. Bandelier (Nashville, Fisk, 1932) p. 35.

[34] See the discussion of moveable feasts in the chapter on the Olympic Games.

The werewolf.

12. Patterns: The Werewolf

Ritual imitation of the wolf is a specific characteristic of military initiations and hence of the Mannerbunde.

– Mircea Eliade, *Zalmoxis*

Memory is inherent in nature.

– Rupert Sheldrake, *Presence of the Past*

The myth of the werewolf offers a unique opportunity to follow a pattern over more than two and a half millennia: the beginning in the wolf-like apparition of Mars in the sky during the 8th and 7th centuries B.C., the end in case histories of people taking on the characteristics of the wolf. The conscious effort to reproduce patterns from the sky in the past mysteriously connects to the unconscious compulsion to repeat these same patterns in the present. The werewolf myth certainly carries enough psychic weight to be worth some attention as to its origins in historical catastrophe.

There have always been a small number of people, unbemused by the novel and fashionable, who understand that myth has something to do with literature. In fact, myths rise into consciousness, interactive with outer events, and are brought forward into literature repeatedly.[1] All of us are familiar with the modern myth of the war between the sexes: it is most irrational behavior, but undeniably compelling. From the conflict between Elizabeth Bennet and Fitzwilliam Darcy to that of Scarlett O'Hara and Rhett Butler, the "war" between the principals turns us on. Few recognize the fiery, spark-shooting Venus and Mars, but our archetypal buttons are pushed, and we react to the dynamic.

Opening the book of historical catastrophism illuminates the power that is associated with myth and accessed, consciously or unconsciously, in literature. It is in the literature of film that the werewolf is most familiar to us. The vision of hair growing thick on Lon Chaney's arm, his voice slurring to a snarl, still lives in the memory of those of us who recall Saturday afternoon matinees of the 1940s. More recent film studies have followed a degenerate story line, losing the power of the archetype.[2] One must return to earlier wolf-men to see them clearly.

In the classic treatment, our subject is a normal, likable fellow, usually involved in a serious love affair with a young woman. In the course of the tale, we learn that he has been bitten by a wolf and become subject to strange seizures. At the time of the full Moon, he loses his humanity, takes on wolf-like characteristics, and attacks and kills helpless victims. Eventually, he becomes a threat to his fiancee.

His tragic situation is grasped by a mana-figure, often a doctor, and a cure is sought. Knowledge of his illness, as with that of the vampire, is available in eastern Europe, in the mountains of the Carpathians in the Balkans. Success in this direction, sadly, is antithetical to the myth, and the story properly has an unhappy ending, certainly for the werewolf.

[1] Irving Wolfe, "Collective Amnesia," *Kronos* VII, 2, pp. 69–84.

[2] But see *Wolf*, which hews close to the myth. Columbia film directed by Mike Nichols, 1994.

Only death can release him from a bondage that to this day is not understood.

The two myths mentioned above, the war between the sexes and the werewolf, both derive from the series of planetary interactions in the 8th and 7th centuries B.C., which Velikovsky described in the second half of *Worlds in Collision*. These star wars have provided patterns for human behavior which deserve careful scrutiny. Here we will examine the two somewhat interconnected myths with an eye toward raising them into consciousness.

It is a hallmark of Velikovsky's work that he considers myth and legend to be a kind of historical record. Thus he was possibly the first modern to realize that Mars had appeared in the sky in the shape of a wolf, specifically in the battle between Fenris Wolf and the serpent Midgard in the Icelandic epic. Those of us on the planet today can only speculate on the appearance of such a phenomenon. Venus (Midgard) would have lent Mars some of her cometary substance so that both planets were trailing gases in a state of electrical excitation with occasional exchanges of energy. The spectacle must have had something of the eidetic quality of clouds or the grain of wood, lending the transformed planets the ability to be perceived in both natural and fantastic shapes. The Dacian standard was figured with a wolf with the body of a dragon. Mars' appearance as a wolf seems to have continued throughout the series of interactions with Earth.

The wolf in the sky was reasonably well recorded: Skoll of the North Germanic tribes, Vukadlak of the Slavs.[3] Huitzilopochtli of the Mexicans was said to have appeared as a jackal,[4] while a vixen or female fox was the Chinese vision of the sky animal.[5]

In the various northern myths the wolf is described as pursuing the sun, swallowing or devouring it. Dumezil feels that ". . . the crisis in which the sun is devoured by Fenrir is regarded as Fimbulvetr," the "Great Winter."[6] The Fimbul Winter, a protracted period in which snow falls all year round, would have been brought on in the north by the tilt of the pole. Farther south, the sun is reported to have been darkened for three days and nights, but wolf-like Mars is responsible in both cases.

A deep-seated contamination between wolf and bear myth begins here, not difficult to understand when we consider the catastrophic motion of the Great Bear and the fiery appearance of wolf-like Mars in the sky. One can imagine the confusion attending a sacrifice by the bear-folk, the Arkades, on having a fiery wolf appear in the sky. The well-known myth in which Artemis chases the bears with dogs fits well here. Hesiod in his *Astronomy* says of the bear, "But while she was in the mountain she was hunted by some goat herds and given up with her babe to Lycaon [wolf]."[7] At least he counts bear and wolf as separate, but not so succeeding generations. In such confusion does a bear myth attach the name "Beowulf."[8]

The early Romans were apparently in no doubt of what they saw in the sky, for Mars became the war god and the wolf, lupus martius or lupus martialis, became his chief animal symbol. The lupercal or wolf's cave moves smoothly from legend to reality. Virgil

3 Velikovsky, *Worlds in Collision,* p. 262.

4 Ibid., p. 261.

5 *Promotion to Divinity*, or *Creation of the Gods,* trans. Gu Zhizhong (Beijing, New World Press).

6 Dumezil, *Archaic Roman Religion,* p. 337.

7 Santillana reminds us that "lyk" means both wolf and light. "The planets are the dogs of Persephone." *Hamlet's Mill,* p. 380.

8 Friedrich Panzer, *Studien zur germanischen Sagengesschichte,* Vol. I, *Beowulf.*

relates that the cave was among the first things shown to Aeneas by Evander when he came to settle at Rome.[9] Later it was consecrated by Augustus for a festival of the Lupercalia.[10]

If we are to believe Homer, there were two celestial battles between Mars and Venus (Ares and Athena), one in the 5th, one in the 21st book of the *Iliad*. In both of these Mars was defeated. It was later, during the Earth-Mars series, that his warlike reputation was enhanced. The battles between Mars and Venus were seen as generative couplings by the early Romans. Thus Mars became the father of Romulus and Remus, the first kings of Rome. According to Plutarch, Romulus was conceived during the first year after the second Olympic Games.[11] An appropriate mother was found for the two boys, Rhea Sylvia from Alba Longa, who sent them floating down the Tiber in a little god boat.[12] The twins were eventually suckled by a she-wolf, imbibing with her milk the wolfish tendencies that would later come to be associated with the armies of Rome.

The legend of Aeneas comes into play again here. After the fall of Troy, he led off a band of the defeated Trojans to found Rome. Mars had fought on their side and both had lost. Some determination not to eat the bread of defeat again may have motivated them to continued association with their god, resuscitated in the Earth-Mars Wars. The fiery image of Mars regularly displayed in the sky provided a model of martial power that the early Romans found desirable and lived up to.

Romulus would throw the javelin, symbol of the cosmic axis,[13] to mark the site of Rome on the Palatine Hill,[14] and Roma Quadrata would be laid out, holy and powerful from the beginning because it was assimilated to the center.[15] Although Jupiter was the head of the pantheon of Roman gods, Mars was from the first the particular genius of Rome, the archetypal warlike image that they would draw on to conquer the known world. He was Mars Pater (father) and Mars Caecus, blind when enraged and likely to strike friend or foe.[16]

Taking a leaf from our study of the bear, we ask ourselves what the real wolf has to offer to the prosecution of war. He has certainly been a threat to domestic animals, since by definition they cannot defend themselves. Other than that, he is a model citizen of the animal world, a family man who mates for life. He prunes the population of everything from mice to caribou, and avoids man, but makes an affectionate pet when domesticated.

In drawing on the wolf as a pattern for ferocity, the Romans and other ancient peoples could only have had in mind the wolf in the sky, Mars, a mysterium tremendum, a fearful apparition known to be destructive. It is a case of mistaken identity in which the reputation of the four-footed wolf has suffered accordingly, attaching qualities that simply do not apply. The early Romans were not the only ones to bring the "big, bad wolf" in the sky to Earth. European folklore is full of tales in which the wolf plays an unpleasant role. The fear and hatred generated by this fairly inoffensive carnivore can be observed on the editorial pages of any newspaper in the western U.S. where there is talk of reintroducing wolves.

Mars and the wolf entered the Roman character over the next millennium. As Mars is the red planet, triumphant Roman generals had their face reddened in imitation of their war god.[17] A statue of Mars stood on the Appian Way between figures of of wolves.

---

[9] *Aeneid*, Book VIII, pp. 327–358.

[10] Liebischuetz, *Continuity and Change in Roman Religion*, p. 62.

[11] Plutarch, "Romulus," XII.

[12] Graves, *Greek Myths*, 68.3.

[13] O'Neill, *Night of the Gods*, pp. 31–39.

[14] Graves, *White Goddess*, p. 173.

[15] Eliade, *The Sacred and the Profane*, p. 47.

[16] Dumezil, *Archaic Roman Religion*, p. 229.

[17] Graves, *Greek Myths*, 28.12.

His sanctuaries up until the time of Augustus were never in the city, always on the threshold of the wilderness.[18]

The bulk of the Roman celebrations in honor of Mars are held in his name month, March, or close to it.[19] Another cluster of his feasts is held in October.[20] They are generally associated with warfare, explained as opening and closing the season for war, but are noticeably close to the equinoxes as well. The Romans never displayed the fixation on solstice and equinox of earlier civilizations, but for the Angeronalia on December 21,[21] shortened days that they were only concerned to get through.

The Lupercalia, carrying the name of the wolf, is known in one aspect as a fertility feast, one that we still celebrate, in a manner of speaking. The early Church found the rituals not to their taste and gave the date, February 15, over to a feast of the Virgin. But imposing new festivals on old ones is problematical; the original energy continued to shine through, in this instance on St. Valentine's Day, one day earlier and now the accepted day of celebration for lovers.

The combination of elements in the Lupercalia—wolf name, fertility rites, and purification—deserves scrutiny. Eisler identifies a hunting ritual commonly overlooked.

On the morning of the 15th, two teams of naked boys, representative of the wolf-cub twins, Romulus and Remus, appear on the approaches of the Palatine Hill. Proceeding to the Lupercal or wolf's cave, they take part in a ritual. Goats are slaughtered and their blood smeared on the forehead of each boy, then wiped off with wool and milk. At this point the boys are supposed to laugh. They then dress in the flayed goat skins and take part in a feast followed by a race around the Palatine Hill. The boys all carry goat skin thongs, using them to strike people along the way, particularly women, to ensure their fertility.[22]

The "blooding" is that of the novice in at the kill, a part of hunting ritual, while the cleansing with milk and wool is the removal of guilt with childish innocence. The young must laugh to show they are unafraid.[23] Goat skins are "februum," signifying purification or purgation.[24]

The ritual of Lupercalia is akin to myth, a kind of historical record. Purification with milk and wool after blooding had probably been practiced time out of mind in hunting ritual. One did not kill casually or without consideration. The fertility that had produced the twins, Romulus and Remus, came from the sky at the time of the Venus-Mars Wars. The "wild hunt," well known in teutonic myth, is another artifact of the Earth-Mars Wars. In Greece it was expressed in the myth of Artemis chasing the pregnant Callisto, dogs (wolves) chasing a bear who would presently give birth in the sky to the new constellation at the pole, Ursa Minor.

The Romans were deeply involved with purification rituals that related to the prosecution of war and its aftermath, not only of soldiers, but of trumpets and weapons as well

[18] Dumezil, *Archaic Roman Religion,* p. 206.

[19] These associations would have begun when Mars was the first month of the year, beginning at the vernal equinox.

[20] Dumezil, *Archaic Roman Religion,* p. 206.

[21] Ibid., p. 335.

[22] Plutarch, "Romulus," 21.10.

[23] Eisler, *Man Into Wolf,* Appendix II.

[24] Dumezil, *Archaic Roman Religion,* p. 348.

"after the necessary or superfluous violence of war." [25] Plutarch says that the Lupercalia was anciently called Febrata, signifying purification. In equating the hunt with warfare, the concept of a ritual cleansing after killing animals was extended to a purification after killing men. Wolf energy was accessed for war, but afterward it had to be dissipated. The Roman soldier had to return to the farm.

Mars himself had two aspects according to Servius. "When Mars breaks into rage, he is called Gradivus; when he is tranquil, Quirinus." [26] Whatever faults may be ascribed to them, the Romans dealt well with warlike energy, conquered the known world, and established the Pax Romana. To the north and east, however, were other people who had also been visited by the wolf in the sky and did not handle it well.

North of the Danube, their territory including the lion's share of the Carpathian Mountains, the Dacians from earliest times called themselves wolves. They borrowed the Phrygian word "daos," which did not distinguish them, however. The Thracian war god was part wolf, Kandaon; another wolf god was Daunus. In Moesia, directly across the Danube was Daousdava, literally "village of wolves." According to Strabo, certain Scythians to the east of the Caspian Sea were also called "daoi," wolves. South of the Caspian was Hyrcania, "Varcana," the "country of wolves." [27]

Mircea Eliade, on his home ground in the Balkans, devotes a chapter of *Zalmoxis* to the Dacians and wolves. Examining ways in which they may have attached the name, he considers the ability to change into a wolf through ritual. "Such a transformation may be connected with lycanthropy properly speaking—an extremely widespread phenomenon, but more especially documented in the Balkano-Carpathian region—or with a ritual imitation of the behavior and outward appearance of a wolf."

It is not a matter of the spread of an idea. Everyone has seen the wolf in the sky and many are drawn into the new vision, generally aggressive and violent. Eliade quotes previous work in describing the ritual military initiation. [28] "It was not solely a matter of courage, physical strength, or endurance, but of a magico-religious experience that radically changed the young warrior's mode of being. He had to transmute his humanity by an access of aggressive and terrifying fury that made him a raging carnivore.' " Such rituals were an Indo-European phenomenon, widely traceable.

The literature on werewolves is enormous and in many languages, reflecting mankind's fascination with the phenomenon. Persistent belief in werewolves is frequently connected with the widespread existence of these rituals in which uncounted young men, often fugitives and exiles, were assimilated to wolves. The pernicious effect of casting off humanity with its laws and codes hardly needs pointing out.

It has always been assumed that the four-footed wolf, canis lupis, provided a sufficiently horrific model for the above-mentioned rituals, but it will not wash. Another model is necessary. Eliade realizes this and formulated "the hypothesis that there was a ritual transformation into a wolf, implying mystical union with a war god capable of manifesting himself lycomorphically (like Mars, for example)." [29]

The ancestral home of the werewolf in the Carpathians is more easily understood since the inhabitants got a triple dose of the wolf: from Mars himself in the sky, from the Roman legions who conquered them under the genius of Mars, and from the invasion of Genghis Khan who claimed descent from a wolf. The phenomenon of "man into wolf" continues: in literature and film, in actual case histories.

---

[25] Ibid.

[26] Dumezil, *Archaic Roman Religion*, p. 262.

[27] Eliade, *Zalmoxis*, "The Dacians and Wolves."

[28] Eliade, *Naissance Mystique*, p. 181.

[29] Eliade, Zalmoxis, "The Dacians and Wolves."

A chilling study would be the assimilation of man into wolf associated with the Nazi party in Germany earlier in the present century. Eisler maintains that Hitler's title, "Der Fuhrer," denotes the leader of a hunting, lupine pack.[30]

Most moderns, indeed thoughtful Germans themselves, have difficulty understanding the power that Hitler appeared to exert over his followers.[31] But a people who have powerful ancient traditions such as the berserkir and ulfhedhnor (bear and wolf warriors), and the "wild hunt" of Wodan and his wolves, have singular vulnerabilities to manipulation. Like the Dacians, the Germans would have profited by rituals of cleansing and purification such as were used by the Romans.

In a later chapter, "Mankind in Amnesia," we will examine a new theory of how archetypal patterns such as the werewolf move through both time and space. Of course, not all warriors are evil; some fight on the side of right. In the following chapter, we learn more about the proto-superman, Hercules, whose feats and adventures are truly cosmic.

[30] Eisler, *Man Into Wolf,* note 128.

[31] He wanted "to see once more in the eyes of a pitiless youth the gleam of pride and independence of the beast of prey." Ibid., p. 35.

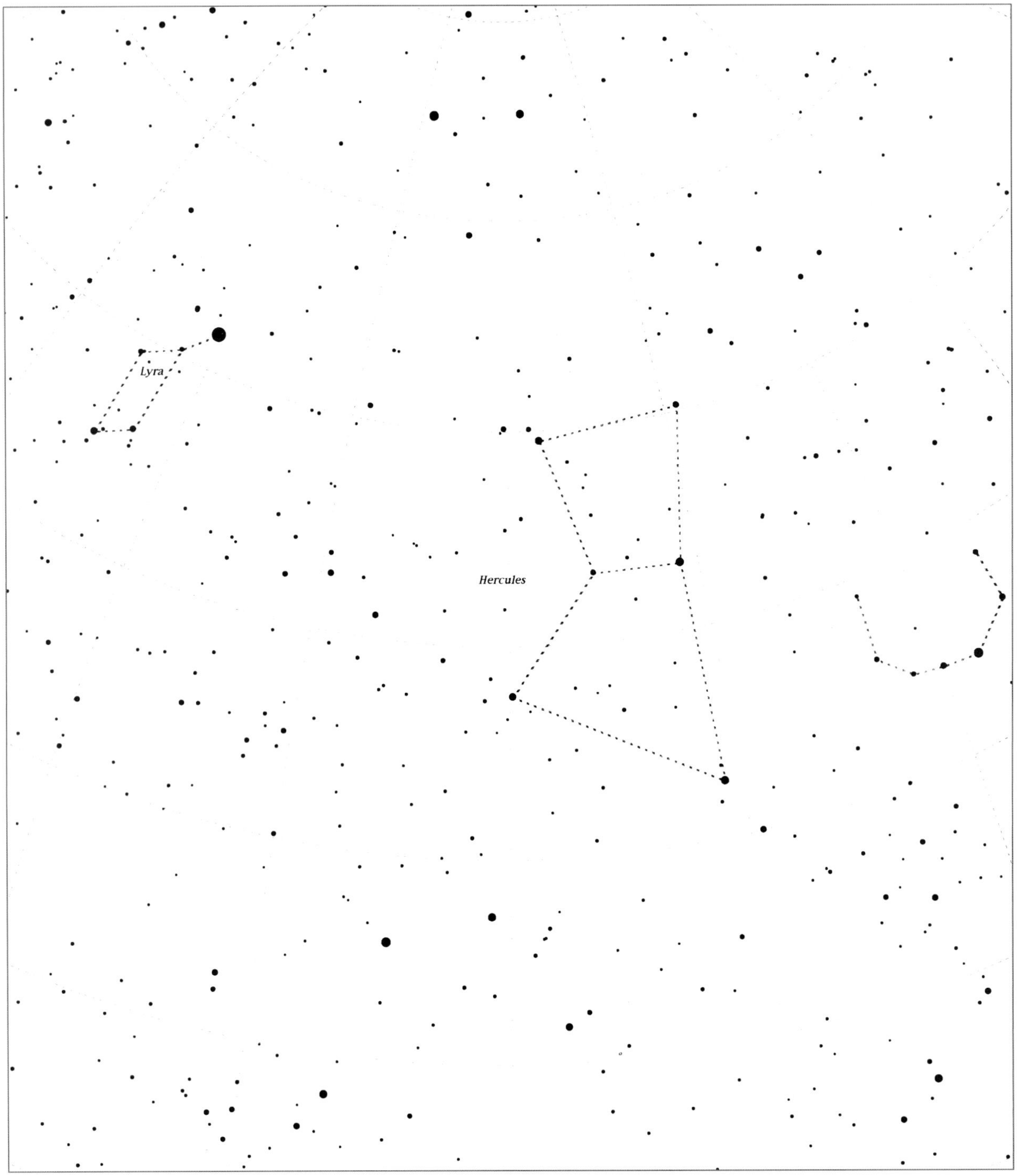

The constellation Hercules, a lopsided butterfly or bow knot.

13. Mighty Hercules

I will sing of Heracles, the son of Zeus and much the mightiest of men on earth.

– Hesiod, "Homeric Hymns"

Of all the gods and heroes of ancient Rome and Greece, there is only one whose name is still in common currency. Hercules,[1] mythical hero and strongman, is still involved in exciting adventure, his appeal greatest among the young to whom movies, cartoons, and comic books are addressed. Since the Earth-Mars Wars, hero and villain have become polarized and Hercules has assimilated everything good. His character as a protector is never violated, even in the case of comic book spinoffs such as Superman and certain film characters played by Arnold Schwarzenegger. It would not "play." He is on our side and will protect us, and we know this in our bones.

In the stories handed down to us in myth, Hercules is decidedly less than perfect, but his character as we know him now was fixed forever in the Earth-Mars Wars when he fought on our side and won. The Labors of Hercules have revealed themselves to be associated with a catastrophic event rather than a path through the zodiac as was formerly believed. He marked out the new equator at the most recent tilt of Earth, and myths of catastrophe are often filtered through him, a protector who will always take care of us. But where did this strongman come from?

Zeus claimed to be his father in the hero-god's most recent incarnation, but this is problematical since legends of Hercules predate this particular beginning. His identity as bull calf expresses his relationship to Saturn, Hercules as calf to Saturn's bull. Be that as it may, the thrust of the most recent myth was for Zeus to beget a son powerful enough to protect both gods and men against destruction.[2] So Zeus lay with Alcmene for a night which had the length of three, and nine months later boasted of fathering a son who was soon to be born. His wife Hera was incensed, the more so since Zeus planned to call the child Heracles, according to some, "Glory of Hera."

As it happened, the succession of the House of Perseus was at stake, as Hercules was in a race with his cousin Eurystheus to be born. The wily Hera, determined to cross her royal husband, hurried to the home of Alcmene and squatted at her door with her clothing tied in knots and her fingers locked together. Her actions were designed to delay the birth of Hercules so that his cousin Eurystheus, a seven-month's child, was born first. In such fashion did Eurystheus become king and head of the house of Perseus. Later he would set the labors for Hercules to perform.[3]

The myth sets up a pattern associating Hercules with knots, a pattern that will repeat, and a particularly strong one because it comes from the sky. For when we look at the constellation of stars that bears his name, we see a bow like that of a bow tie, a knot that will become one of his signatures.[4] Now a constellation will have different names in different cultures. That of Hercules is distinguished because "... no other sky group seems to have borne so many titles."[5] There was good reason for this, for it was prominently figured in the

---

[1] Hercules is the Roman form, Heracles the Greek.

[2] Graves, *Greek Myths,* 118 d.

[3] Graves, *Greek Myths,* 118 passim.

[4] See astronomical map on page 124.

[5] Allen, *Star Names,* Hercules.

heavens at a time when attention was focused strongly on the sky. We will consider some of the most evocative names.

To the Greek astronomer, Aratus, Hercules was the "Kneeling One"; to the Romans he was Saltator, the "Leaper." The Phoenicians worshipped the configuration of stars as Melkarth, "King of the City" in Tyre, and he was Melicertes in Corinth. The Chinese knew the constellation's brightest star as Ti Tso, "Emperor's Seat." Since some of my readers are probably ahead of me by now, I will take a little time to link up two myths: one took shape at the end of the Age of Heroes, the other, less familiar, was generated at the end of the age previous to it.

Several chapters ago, when we examined Odysseus' three-day visit to Phaeacea in the critical period when he (as the celestial bear) was between caves, we passed over an important incident which now comes into focus. Our hero had been at sea for 17 days and the island of Phaeacea was in sight. But his enemy, Poseidon, blew up a storm and stirred the sea to a froth with giant waves. The steering oar was torn from Odysseus' hands[6] and he was tossed out of the boat and into the water. Homer lingers over his point.[7]

"For a long time Odysseus was kept under water . . . he found it no easy matter to fight his way up against the downrush of that mighty wave. But at last he reached the air and spat out the bitter brine that kept streaming down his face." Homer does not merely wax poetic about the "bear" going beneath the sea as so many others have done; he half-drowns him in the Mediterranean!

Odysseus manages to get back in the boat, ". . . safe from immediate death." Homer tells the story best. "But there was a witness of Odysseus' plight. This was the daughter of Cadmus, Ino of the slim ankles, who once was a woman speaking like ourselves, but now lives in the salt depths of the sea, and, as Leucothoe, has been acknowledged by the gods." Ino advises Odysseus to strip off his clothes and swim for the shore, giving him a magical veil to wind around his waist which will give him divine protection. Then she slips back into the sea. Odysseus swims for two days and two nights while Athena, his patroness, beats down the waves in his path until he finally reaches shore.

The myth of Ino is not as well known as it should be. She has come to live in the salt depths of the sea since she jumped into it with her younger son Melicertes to save him from the wrath of his father.[8] She was made immortal and her son sent off on dolphin back to the Isthmus of Corinth. That Odysseus and Ino should encounter each other in the sea is not happenstance; the idiom of going under the sea is the same for both. Ino's jump into the sea with Melicertes is analogous to Odysseus falling out of the boat and into the water. A movement of the pole is involved in both instances. The old polar constellation goes down as the new one rises.

In case his audience hasn't picked up on Odysseus being dumped into the sea, Homer drives his point home by having Ino help our hero, thus drawing a parallel between two displacements of the pole: the present one from Ursa Major to Ursa Minor, the past one from Melicertes/Hercules to Ursa Major. Melicertes, "sweet power,"[9] or "honey-eater," was at the pole during the 52 years between the two major destructive visitations of Venus, providing the manna, madhu, or honey lash which sustained mankind during an extended period of darkness when nothing grew on Earth.

The power of bringing the honey-dew to Earth was attributed to whatever god a people

6 The steering oar represents the terrestrial pole.

7 Homer, *Odyssey*, E. V. Rieu trans., Book V.

8 Graves, *Greek Myths*, 70.h.

9 Graves, *Greek Myths*, Index.

happened to look up to, literally to the region of the north celestial pole. Western Civilization is most familiar with the manna from the Bible, but the phenomenon is known from the Edda, the Kalevala, and the Rig Veda and Atharva Veda of India. Velikovsky treated the heavenly food at some length in *Worlds in Collision* in the section on ambrosia in Chapter 6. In a London papyrus, a conflated set of gods, Zeus-Helios-Mithras-Sarapis, is referred to as "honey-creator, honey-sustainer, honey-destroyer."[10] An unusual set of ideas constellates around honey, bees, etc. in antiquity.

Homer does not fail to mention the ambrosia in the cave of Calypso, the honey in the cave of the nymphs, thereby relating the two. Porphyry also connects honey to nectar and ambrosia, the food of the gods, in his "De Antro," mentioning that "the theologians have used honey to symbolize many different things since it combines multiple powers, . . ." He waxes eloquent on the subject: its relation to purification, its prevention of decay, its symbolism of the sweetness of descent into the flesh, of life and death itself.[11] All this must be considered in relation to the cave of the cosmos where the bears were lured by honey when they were elevated to the pole.

In the fourth level of initiation into the Mithraic Mysteries, the grade of "lion," honey is placed on the tongue of the initiate to purify it of all sin.[12] And we may consider this seriously, but not the fall of manna related in the Bible, at least not within the present paradigm. It is a case where the qualities of the invisible heavenly manna have attached to the visible honey, much as the horns of Saturn and Venus have attached to the more visible Moon. Confusing at first glance, but a sound way to save the phenomena. We must find our way back to the fall of manna from Melicertes/Hercules to grasp the mythologic of a bear seeking honey at the pole during the Age of Heroes.

Between the two catastrophic visits of Venus in the 15th century B.C., then, Melicertes "sweet power" was at the pole, better known to the Phoenicians as Melkarth, to the Greeks as Heracles. Such was the importance of the polar position that even a short stay of 52 years was enough to immortalize a hero, a god, or a constellation.

Following Velikovsky, the first Venus event brought Earth-wide catastrophe and a lengthy period of darkness preserved in the myth and legend of many people, notably by the Jews in their years of wandering in the desert. It is likely that when the sky finally cleared, the Phoenicians would have been among the first to trace out the new constellation at the pole and name it. Their seagoing habits and knowledge of navigation required a practical familiarity with the night sky. The new god at the pole, Melkarth, "King of the City" at Tyre, gave direction to a race of sailors in a literal way. Perhaps that is why the Phoenicians continued to sacrifice children to him as they had to an older polar god. Infants were also sacrificed to Melicertes on the island of Tenedos, and probably at Corinth as well.[13]

Here we may gain some insight into ritual prostitution as it was practiced among the Phoenicians and others. Young women were not allowed to marry until they had made themselves available in the temple where any stranger could approach them and enjoy sexual congress. This ritual made allowance for the possible visitation of the god in the form of a stranger. A child born of this union

[10] Roger Beck, "Interpreting the Ponza Zodiac II," *Journal of Mithraic Studies*, Vol. II, 2, pp. 133–134.

[11] Robert Lamberton, *Porphyry on the Cave of the Nymphs* (Barrytown, N. Y., Station Hill, 1983) pp. 30–32.

[12] Ibid., p. 15.

[13] Graves, *White Goddess*, p. 44. Also Eliade, *Myth of the Eternal Return*, p. 109.

would naturally belong to the god and be earmarked for sacrifice. Although animal sacrifice is reported, child sacrifice was probably offered to Hercules as well.[14]

There is a common misconception that the Pillars of Hercules were the heights on either side of the Straits of Gibraltar. Strobe informs us, quoting Poseidonius,[15] that they were actually brazen pillars, eight cubits high, in a temple at Gades.[16] At one time the city stood on an island outside the straits, near the point where the Guadalquivir empties into the Atlantic. It provided both a jumping-off place for mariners and a sanctuary where they customarily offered thanks after the completion of a successful voyage.

The Pillars of Hercules were called by the Phoenicians Pillars of Melkarth. They worshipped similar objects in their home port of Tyre. It was a time of tree, pillar, and stone worship,[17] usually in a high place open to the sky, for God was perceived to be outside. Celestial sweetness attached to Melkarth as well as Melicertes, for his pillars were called ambrosial stones.[18] "But the later Semitic pillars are very frequently of hewn stone in the shape of a cone," and pictured on the coins of Tyre.[19]

Graves counts Hercules birth as having occurred at the Spring Equinox,[20] agreeing well with Velikovsky's scenario. Earth's first catastrophic contact with Venus tilted the terrestrial pole into the constellation of Hercules at the same time it precipitated the Exodus of the Jews which also occurred at the Spring Equinox. The relationship between Venus-Athena and Hercules is very close, the two actually knotted together in symbol. The two catastrophic visits of Venus, represented by omega signs, configure together as a square knot. And since Hercules stood at the pole between the two visits, and the constellation resembles a bow-knot, the symbol devolves on him as well.

Hercules with his lion skin tied around his neck in a square knot (a statuette in the Bibliotheque Nationale in Paris).

An examination of classical art will discover this square "herculean" knot in association with Hercules. The hero-god frequently wears his lion skin knotted thus around his neck. The knot in a bride's girdle, consecrated to

[14] Graves, *Greek Myths,* 145.k.

[15] Graves, *White Goddess,* p. 232.

[16] Modern Cadiz.

[17] Sir Arthur Evans, *Mycenaean Tree and Pillar Cult* (New York, London, Macmillan, 1901).

[18] Ibid., p. 35.

[19] Ibid., p. 35. Also see Stephen H. Langdon, *Mythology of All Races* (Boston, Marshall Jones, 1916) Vol. V, p. 51.

[20] Graves, *Greek Myths,* 118.6.

Juno, was known as a "herculean knot," while the girdle itself is a metaphor for the one around the waist of Mother Earth, the zodiac, which is loosened from time to time in a catastrophic generative event.

Another square-knotted lion skin on Hercules, a bronze statuette in Copenhagen (with Hera).

Venus not only established her protégé at the pole to begin with; on her return swing, she removed him as well. This brief world age is recorded in myth as the time when Heracles took over the support of the cosmos from Atlas.[21] According to one legend, Typhon killed Heracles,[22] Typhon having been identified by Velikovsky as a cometary Venus. In another, Hercules is swallowed by Tiamat,[23] another name for Venus. Not surprisingly, our hero survived these incidents and was eventually introduced to the company of the gods by Athena-Venus.[24] Seven hundred years would pass before Hercules returned to defend Earth against Mars and perform his labors; myth is scandalously careless of time sequence.

John O'Neill reads the name Heracles as "keystone of the heavens-vault," an interpretation that squares up solidly with the hero-god's former polar position. O'Neill's *Night of the Gods* is a study of northernism and the axis of the universe, material that has escaped the present scientific paradigm, but begins to make sense in the light of a comprehensive historical catastrophism.

Before we can describe the physical motions of Earth vis-à-vis Venus, it is necessary to identify the constellation at the terrestrial pole previous to the first contact. Just as there was a connective thread between Odysseus and Hercules through the myth of Ino, so another exists, a familial one, between Hercules and his grandfather Perseus. It is through his mother Alcmene and her father Electryon that Hercules enjoys a familial tie to the entity previously at the pole.[25] One of the Church fathers, Hippolytus, quoting an unknown exegete of Aratus, notes, "Perseus is the winged axis that pierces both poles through the center of the earth and rotates the cosmos."[26]

As a constellation, Perseus is poorly defined, not remotely resembling an important mythological hero. The configuration of stars is perhaps most distinguished by the second-brightest star, the binary and variable Algol, the "Demon Star." Its Chinese name, Tseih She, "the Piled-up Corpses," clearly suggests disaster if not catastrophe.[27]

Perseus is better connected catastrophically at the beginning of his stay at the pole, which we will examine presently. Here we are concerned with events toward the end. Perseus is best known for beheading the Gorgon Medusa,

[21] Ibid., 133.e.

[22] Graves, *Greek Myths*, 133.11.

[23] Ibid., 137.2.

[24] Ibid., 145.h.

[25] Graves, *Greek Myths*, 118.9.

[26] Hippolytus, *Refutation of All Heresies*, quoted in Ulansey, *Origin of the Mithraic Mysteries*, p. 94.

[27] Allen, *Star Names*, p. 332. Algol varies from 2.3 to 3.5.

Venus symbols decorate Perseus' Phrygian cap in the seventh grade of Mithraism.

and it is noteworthy that he eventually gave the head, the Gorgoneion, to Athena, who placed it on her aegis.[28] Medusa, as we shall see, has strong solar connections,[29] and the appearance of the Gorgoneion on Venus' aegis would have lent considerable light and heat to be subsequently displayed in the Earth-Venus Wars. The identification is so strong that we will eventually find the Gorgoneion's snaky hair tied in Venus knots.

Mithraism features Perseus, linking him with the Venus glyph. The harpe, with which Perseus decapitated the Gorgon, is included in a set of symbols which represent the fifth grade in Mithraism, called "Perses." Among the symbols of the seventh grade, "Father," is a Phrygian cap, often worn by Perseus and in this case decorated with an omega glyph, sign of the cometary Venus.[30] As celestial personages, Perseus and Mithras have much in common.

Perseus met his death at the hands of Medeus, whose mother Medea gave him a magic sword to accomplish the task.[31] Medeus is little known despite a name that places him in the middle of things. In the following chapter we shall see how he brought about the death of Perseus, a great hero in his own right and grandfather of mighty Hercules.

The herculean square knot, a symbol of Venus.

[28] Apollodorus, II, IV, pp. 3–4.

[29] Medusa's ties were to the ancient sun of Saturn.

[30] *Mysteria Mithrae,* ed. Ugo Bianchi (Leiden, E.J. Brill 1979) facing p. 60.

[31] O'Neill, *Night of the Gods,* p. 833.

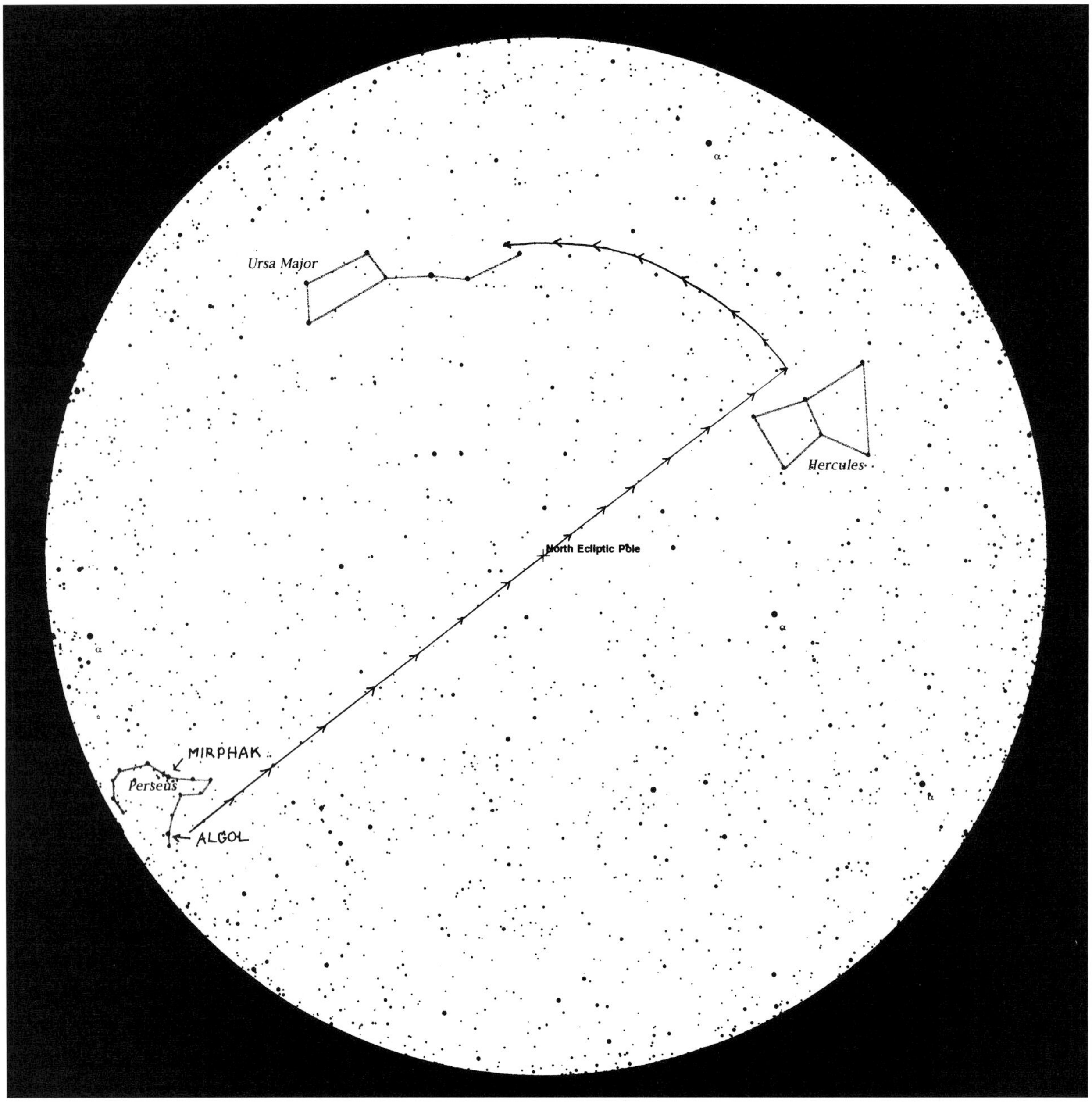

The precessional clock.
The solstitial colure marks out the tilt of the terrestrial pole at the first of the double visits of Venus. At the second visit of the comet/planet, Earth's pole moved in the curved line of sudden precessional movement as the solstitial colure moved with it.

14. The Comet Venus Strikes

I shall reverse the world. I shall bid the rivers flow upward; I shall cause the sea to gather itself into a huge towering wall which I shall hurl upon your wicked earth children, and thus destroy them and all life.

– Lapland Legends, Leonne de Cambrey

He gathereth the waters of the sea together as a heap.

– Psalms 33:7–8

The Earth is upside down, . . .

– Papyrus Ipuwer

The winter is come as summer, the months are reversed and the hours are disordered.

– Papyrus Anastasi IV

Ancient media has led us back to a world age when the constellation Perseus is at the pole. Now we will turn around and move forward in time through a most unhappy period of Earth history. The stories of destruction have been told repeatedly in myth and legend all over the world. It was left to Immanuel Velikovsky to correlate the information and identify a cometary Venus as the agent of Earth changes and worldwide destruction in the 15th century B.C.

Ancient witnesses speak throughout *Worlds in Collision* as in the quotations above which refer to the first visit of Venus. They tell us of things we might prefer not to know, like a rain of fire and red-hot stones that brought burning death to whole populations. This phenomenon was recorded in countries known for their oil deposits today: Mexico, the East Indies, Siberia, Egypt, and Iraq.[1] The near approach of Venus caused enormous stresses on our planet. Existing volcanoes erupted and new ones were created; mountains rose and plains were tilted; the Earth cracked open and lava flowed out; seas and rivers boiled; gigantic waves of translation swept over the land. Velikovsky treated the mythical and historical record in *Worlds in Collision*, the geological record in *Earth in Upheaval.*

In this volume we are concerned to plot the tilts and spins of Earth against the vault of Heaven, the zodiac and the frame of time we call the Precession of the Equinoxes. In the chapters on the Earth-Mars Wars, we developed a sequence of Earth motions, a tilt followed by two precessional spins. Now we will compare them to the Earth motions caused by Venus seven centuries earlier. Primeval media tells us that the terrestrial pole pointed successively at the constellations Perseus, Hercules, and Ursa Major. Let us examine the path through the sky described by the pole of Earth.

The first visit of Venus brought on the move of the pole from Perseus to Hercules. It is a tilt upward toward the pole of the ecliptic, exactly the same motion which occurred in 717 B.C. when Mars approached Earth on its third pass. But the difference in size and mass between Venus and Mars is

[1] Velikovsky, *Worlds in Collision*, p. 56.

considerable, Venus over five times the volume of the red planet. We should not be surprised to discover a correspondingly greater exertion of force on the Earth in the interaction with Venus.

Myth informs us that Perseus wore winged shoes to sally against the Gorgon.[2] He must still have been wearing them at the first visitation of Venus, for the constellation moved approximately 115 degrees in relation to Earth during the exchange of forces. At the same time, Hercules saltator, the leaper, jumped up to the pole. For us here on Earth, the old polar constellation goes "down," while the new one goes "up." Viewed from outer space, the terrestrial pole tilted up from the constellation Perseus, through the pole of the ecliptic, and continued on down to the constellation Hercules.

Tracing this tilt of Earth in the sky throws a flood of light on the mythical death of Perseus. Hyginus in his *Fabulae* XXVII tells how Medea and her son conspire to kill Perses (a doublet of Perseus). Medea actually departed from one myth, familiar to us from Greek drama, in a chariot drawn by dragons. She is an analog of Venus, her name indicating that she came from the center just as Venus spun off the Saturn Configuration. In both cases, the center is the real one at the pole of the precession.

"The central goddess Medea gave the magic sword to her divine son Medeus[3] and with it he slew Perseus." An unusual death, run through by ". . . this sword blade which I would identify with the universe axis." John O'Neill wrote these words in 1893 in his perceptive *Night of the Gods,* foreshadowing the importance of the universe axis (precessional axis) both in catastrophic theory and in a new vision of the mechanics of the cosmos. There are other elements in the fable, but essentially Medea-Venus arrives in her dragon chariot-comet and arranges to have Perseus "killed" by the magic sword/axis-pole of the ecliptic.

While the move of the pole from the constellation of Perseus to that of Hercules does not turn the Earth upside down, it certainly gives that appearance to that part of the sky that is between the two constellations. Only half of the celestial vault is theoretically visible from any point on Earth. At any time that both Perseus and Hercules appeared in the sky after the first catastrophic interaction between Earth and Venus, the largest part of it would be in a reversed orientation from the previous age.

The tomb of Senmut, architect of Queen Hatshepsut, has its ceiling ornamented with extensive astronomical data,[4] including precessional material which we will discuss below. Alexander Pogo commented, "A characteristic feature of the Senmut ceiling is the astronomically objectionable orientation of the southern panel." Sirius, for example, is on the wrong side of Orion. ". . . Orion, the most conspicuous constellation of the southern sky, appeared to be moving eastward, i.e., in the wrong direction." [5]

Velikovsky discusses the Senmut ceiling in *Worlds in Collision*, drawing attention to it as a record of major Earth change. In this case, he went too far, assuming a 180-degree turn of the planet, and that the sun thereby rose in the west. One might easily imagine that turning the Earth upside down would cause the sun to rise in the west, but such is not the case. Enough support for this strange phenomenon was introduced in *Worlds in Collision* that it

[2] Graves, *Greek Myths*, 73.g.

[3] Jason is the father of Medeus. Hesiod, "Theogony," 980–1001.

[4] O. Neugebauer and R. A. Parker, *Egyptian Astronomical Texts* (xxxxx) Plate I.

[5] *Isis*, Vol. 30, 1930, p. 306.

would be wise to stay open to the possibility. However, the Earth does not appear to have changed the direction of its rotation as a result of the Earth-Venus Wars.

What actually happened is that the 115-degree tilt of Earth moved the Orion-Sirius group from the northern to the southern hemisphere as viewed from Earth. Rather than facing north to view the constellation and star, one would have to face south. The left-right orientation of Orion and Sirius would be reversed, but not the order of rising.[6]

A consequence of this tilt of the terrestrial pole through the pole of the ecliptic was a change in the order of seasons, an immediate six-month displacement. For it is the tilt of the Earth that determines the seasons, causing the apparent motion of the sun as it sweeps back and forth between northern and southern hemisphere. "The winter is come as summer" complains the Papyrus Anastazi, a strange enough situation in the tropics, but what of the northern regions just coming out of winter and then heading back into it again?

As it happened, the northern hemisphere moved even more deeply into winter, for the tilt was such that the sun would no longer move as far north as previously. The tropic zone narrowed and became warmer as would happen again 700 years later after another tilt. Plutarch mentions Typhon (cometary Venus) as the agent who caused "abnormal seasons and temperatures . . ."[7]

The Exodus occurred at the Spring Equinox on the 14th day of the month of Aviv in the old Jewish calendar. As a result of the tilt of Earth, the Jewish New Year was moved suddenly to the fall where it is celebrated today. "This month shall be unto you the beginning of months; it shall be the first month of the year to you." The tilt occurred at the last Spring Equinox of the Bronze Age, immediately changing it to the fall equinox,[8] so that Earth was headed back into winter, the legendary Fimbulvetr when snow fell year round.

Uninstructed by Yahweh, Assyria, and later Babylon, would celebrate their New Year, Akitu, at the new Spring Equinox, after the Fimbul Winter. The Phoenicians would honor both old and new age, beginning their ritual year in the month of Nisan at the new Spring Equinox, their civil year in the month of Tishri six months later.[9]

Graves addresses the problem in his *Hebrew Myths*. "Irreconcilable views as to the season of Creation were held by rival Jewish schools from the first century A.D. onward."[10] He notes that there are different seasons suggested in the two relations of Genesis: the first suggests spring, the second, autumn. Graves connects these with the Babylonian Exile of the Jews, claiming that the New Year was celebrated on the First of Tishri (autumn) before the exile, on the first of Nisan (spring) afterward,[11] but Velikovsky correctly placed the Earth and calendar change many centuries earlier. The deeper issue is at which equinox a particular World Age began, admittedly a somewhat mechanical problem in retrospect since spring follows winter as autumn follows summer, even if winter and summer are exchanged. Jews and Christians see only one Creation.

---

6 This view was suggested in a letter to *Kronos* by R. M. Hendrickson, Jr., Vol. VIII, 3, p. 84.

7 Plutarch, *Isis and Osiris*, 49.

8 The equinoctial colure is an imaginary circle around the Earth, one half marking the Spring Equinox, the other half the fall equinox.

9 *Epigraphic Society Occasional Papers*, No. 17, pp. 284–286.

10 Graves and Patai, *Hebrew Myths* (New York, Doubleday, 1964) p. 26.

11 The old Hebrew months (Aviv, etc.) were kept up until the time of the Earth-Mars Wars, months by number during that twilight period, and the names of the Babylonian months were used during and after their captivity.

In the cyclic view, Earth is created anew at the beginning of each World Age.

The first visit of the cometary Venus plunged the Earth and its immediately reduced population into a lengthy period of misery. In *Worlds in Collision,* Velikovsky correlated the ten plagues of the Bible and the Papyrus Ipuwer to demonstrate that they both described the same period in history. The proximity of the incandescent Venus brought the rain of fire while the tilting of Earth caused disruption of the seas and atmosphere, waves of translation that swept back and forth over continents, winds beyond hurricane force. Stresses on the crust of Earth were enormous, extending the pattern of faults that we are more familiar with under the subject "plate tectonics." Topography and climate were suddenly and violently changed.

Tidal forces must have been raised by orders of magnitude by the close approach of another body almost as large as Earth. But as for the waters of the Sea of Passage rising up on either side for the Jews to escape Pharaoh, then suddenly falling down, only an unknown and uncalculated electromagnetism could be imagined to produce such an effect. Velikovsky was concerned with this speculative branch of knowledge from the beginning, for planetary interaction seemed to extend the physics of the microcosm to the celestial realm.

Venus left the scene on its cometary orbit, not to return for 50 years. The atmosphere of Earth was completely obscured, in "the Valley of the Shadow of Death," its air heavily polluted by its own vulcanism, hurricane winds, and the unwelcome contributions of the cometary train of Venus. One would imagine that the precipitations of this strange atmosphere would be equally strange, not the rain, snow, and hail that customarily fall from the sky. Of all the unusual substances that fell to Earth from its interaction with Venus, none was stranger than manna.

Manna, honey-dew, corn of Heaven. Just to name it makes us uneasy. We have struggled mightily to "educate" primitive peoples, to distance them from their myth and legend. The degree of success achieved can best be measured by the distance we have put between ourselves and our own roots. Although warned against it, we have done unto others and it has happened to us. The modern religion of science has so alienated us from our past that the manna that kept our ancestors alive is now a subject for ridicule. The degree of alienation may be fairly judged by the amusement evoked.

Manna fell in grains the size of coriander seed, had an oily taste like honey, and could be ground and baked into cakes. It would sublimate back into the atmosphere unless kept in a covered container.[12]

At length the sky cleared and the Earth became productive again. When the stars could be seen, a very different heaven appeared, with a vast portion in reversed orientation, rotating around a new constellation at the pole, Melicertes of sweet power, or Hercules Melkarth. Whether or not the Jews wandered exactly 40 years in the desert, there were precious few years to view the order of the heavens during the brief Age of Hercules.

As we proceed into the past, the solstices seem to gain in importance, particularly the Summer Solstice because of its nearness to the sacred center of the pole of precession during an earlier age. On a celestial map centered on this pole, the solstices are easily found by extending the line of the 115-degree tilt to

[12] Velikovsky, *Worlds in Collision,* p. 133. Velikovsky, trained as a physician, was often ridiculed for not knowing the difference between carbohydrates and hydrocarbons. In fact, Troperina, an animal feed, has been developed from hydrocarbons. *Velikovsky Reconsidered,* p. 83.

intercept the zodiac. The points will be found in Taurus and Scorpio in this case. The excessively long tilt of the terrestrial pole exchanged the two constellations at the solstices: Taurus at the Summer Solstice was replaced by Scorpio, and this unusual turn of events is supported by the Tauroctony of the Mithraic Mysteries.

Of more than 500 tauroctonies that have been discovered, the vast majority depict the scorpion with claws firmly clasped around the dying bull's testicles. It is a visual fix seldom discussed but difficult to overlook. Clearly the scorpion is going for the generative area of the bull. Once it is understood that the bull is at the Summer Solstice before the 115-degree tilt caused by Venus, the icon becomes clear. In fact, the tauroctony indicates that scorpion and bull are about to change places. Yet again the bull must die, but always catastrophically.

Another brighter constellation's trip to the southern hemisphere has already been marked. Orion, the hunter, was thought to have been killed by the bite of a scorpion.[13] Once we move the scenario to the sky, it begins to read out.

With Scorpius at the Summer Solstice during the Age of Hercules, the closest constellation on the ecliptic to the North Pole, an episode in the story of Gilgamesh, is clarified. On his epic journey, opening a new way, Gilgamesh arrives at the mountains of Mashu,[14] twin peaks where the sun both rises and sets." Since our present sun does not rise and set in the same place, this must be a reference to the sun of Saturn, which never moved but exhibited light and dark phases in its station at the pole.

Now Gilgamesh could not pass by the gates of the Mountains of Mashu because they were guarded by a scorpion-man and his mate, a reflection of Scorpius at the Summer Solstice, guarding the heavenly gates near the pole. The scorpion-couple questioned Gilgamesh and allowed him to pass, calling him "flesh of the gods."[15] As guardians of the gates of Heaven at the Summer Solstice, they will presently be succeeded by the lion gate guardians of the Age of Heroes. Scorpion-men also make appearance on Babylonian boundary stones as guardians of the limits of property.

Following the mythological model in the previous chapter, our second meeting with Venus moved the terrestrial pole from the constellation of Hercules to that of the bear, Ursa Major. This was a long precessional spin through three signs of the zodiac, a 90-degree move measured at the equator, three times as great as the precessional spin occasioned by our meeting with Mars in 702 B.C. It was at this time that Joshua ben Nun was said to have stopped the sun in the sky, as related in the Bible, to conclude a battle successfully.

Obviously a catastrophic event has been embroidered upon due to excessive religious zeal. A simple calculation reveals that a movement of the sun 90 degrees back along the zodiac during the course of a day will keep it in the sky an additional six hours. The Bible overstates the case when it claims that the sun "hasted not to go down about a whole day." Attributing the power to stop the sun to Joshua makes about as much sense as believing Moses could part the Sea of Passage with his rod. In any event, it was not the sun that moved, but the Earth, suddenly and along its natural precessional path.

As a result of the second visit of Venus, the Spring Equinox was moved from Leo to Taurus, where it would remain for seven centuries throughout the Age of Heroes, thus

[13] Kerenyi, *Dionysos,* p. 42. Also O'Neill, *Night of the Gods,* p. 837.

[14] Mashu means twins, and these twin peaks are analogous to the horns of the Cretan altar.

[15] Like Hercules, Gilgamesh is part god, part man.

continuing focus on the bull. Leo moved to the important Summer Solstice, the gates that opened to Heaven, located nearest the pole. It is at this point that lions begin to appear guarding earthly gates: the lion gate at Mycenae, Hatussas on the Anatolian plateau, Sincirli and Sakcegozu,[16] Puri in India.

The warrior-goddess Athena. The spinning whorls of swastikas on her clothing indicate her power to put Earth into a precessional spin. The head of a water bird looking out under her shield will be discussed in Chapter 25.

This second and final interaction between Earth and Venus imparted a sudden lateral spin to Earth while it caused Venus to move laterally as well, to settle into a planetary orbit, becoming our familiar morning and evening star. We recognize the transformation of Quetzalcoatl in Mesoamerican legend when he rose as the evening star. Thus the 52-year cycle of Venus is the older, cometary one, while the eight- and four-year cycles begin with the birth of the Age of Heroes in the 15th century B.C.

Now another pattern has appeared: tilt followed by precessional spin. In the Earth-Mars Wars, we saw a tilt followed by two precessional spins; in the Earth-Venus Wars, a tilt followed by one precessional spin. The power to put the Earth into a spin was primarily attributed to Venus, so that Schliemann was probably correct in assuming that the thousands of "spinning" whorls found at Hissarlik were offered to Athe or Athene Ergane.[17] Spinning symbols will be treated in a later chapter.

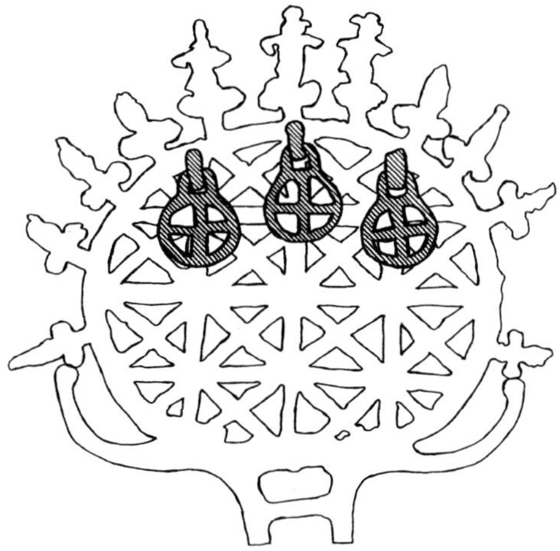

Ritual standard of an early middle eastern civilization, three symbols, probably representing months or zodiac signs, have been taken out of the pattern. The decorations at the edge are considered to be birds and flowers.

[16] These lions may be viewed in the archaeological museum at Istanbul.

[17] O'Neill, *Night of the Gods,* p. 880.

The 90-degree precessional spin of Earth circa 1450 B.C., as measured on the precessional clock, was equivalent to three months of yearly sun time. As in the precessional spin of 702 B.C., it was a phenomenon to be dealt with once, followed by approximately 700 years of regular calendar seasons. Although there is no written record of adding three additional months to the calendar after the second Venus catastrophe, as in the case of Hezekiah doubling the month of Nisan in 702 B.C., several evocative ritual standards have been recovered from what is deemed to be the Hittite civilization. One even features what is possibly the best known spinning symbol to be associated with Athena-Venus, the swastika.

Next, we press back into the cometary period of Venus, the Bronze Age of Hesiod, when the constellation Perseus was at the pole.

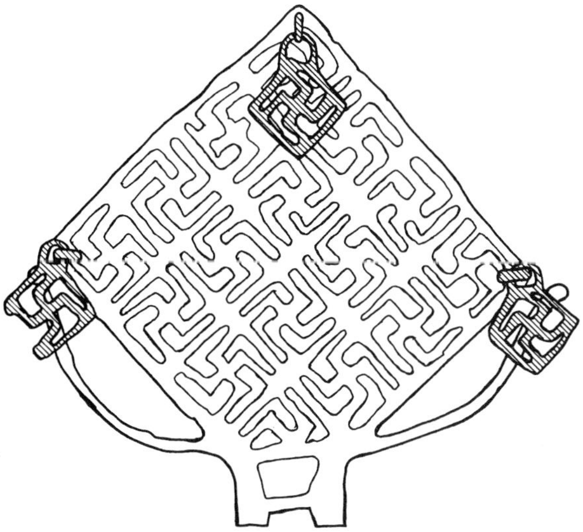

Ritual standard with swastikas, a symbol associated with Athena. From Alaca Huyuk, Ankara.

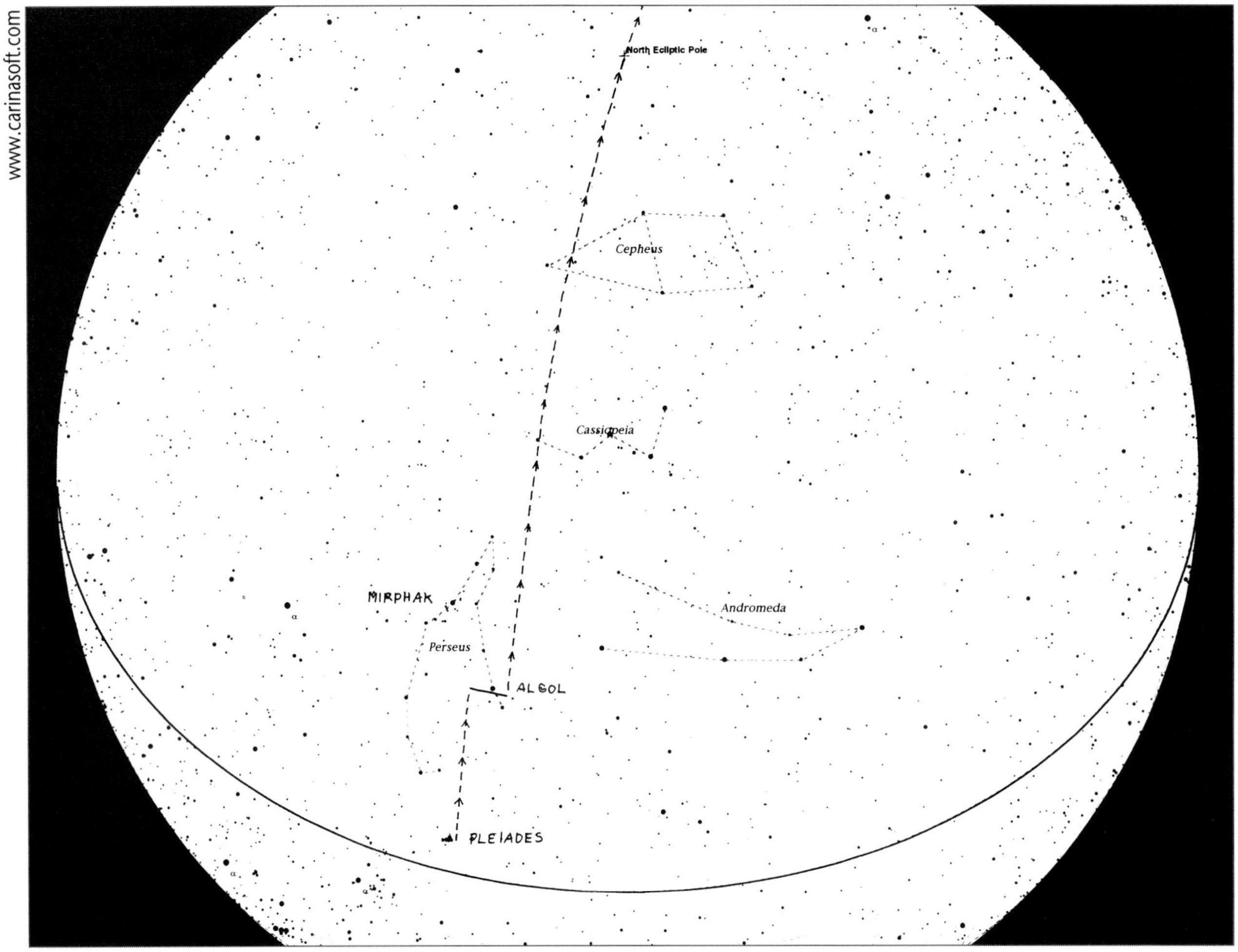

The Bronze Age when the constellation Perseus was at the pole. The solid line indicates over 400 years of regular precessional motion of the terrestrial pole; dotted lines indicate catastrophic motions of the terrestrial pole both before and after the Bronze Age.

15. The Bronze Age of Perseus

Perseus is the winged axis which pierces both poles through the center of the Earth and rotates the cosmos.

– Hippolytus, *Refutation of All Heresies*

Like his grandson Hercules who succeeded him at the pole, Perseus was fathered by Zeus. Also like Hercules, his mythical existence is not limited to his stay at the pole but splashes out elsewhere. Myth, like dream, outrages our sense of time, but the myth of Perseus is an outrage from beginning to end. Following Graves,[1] we string the events of his mythical existence in order, although ancient mythographers tend to give it in bits and pieces. Hyginus, for instance, relates the death of Perses in *Fabulae*, and the Andromeda episode in *Poetica Astronomica.* Relatively modern, Hyginus does not have the clout of a Homer or a Hesiod, but perhaps he knows something.

"Zeus came upon her [Danae] in a shower of gold and she bore him a son named Perseus."[2] Mother and son were locked up by her father in an ark and set adrift on the sea. Washed up on the island of Seriphos, the two were given shelter there and Perseus grew to manhood. In the course of defending his mother against a forced marriage, Perseus was led into making a promise to gain the head of the Gorgon, Medusa.

Graves continues, "Now Medusa had serpents for hair, huge teeth, protruding tongue, and altogether so ugly a face that all who gazed on it were petrified with fright."[3] Apollodorus says they were turned to stone[4] but Graves has grasped the truth. Medusa's horrible appearance was a punishment by Athena for lying with Poseidon in one of her temples. Ovid says that formerly Medusa had been a beautiful woman (else why would Poseidon have desired to make love to her).

To guard Perseus against the sight of Medusa's face, Athena gave him a highly polished shield so he could see the Gorgon's reflection in it. Hermes gave Perseus a weapon to cut off the Gorgon's head, and by stealing the single eye and tooth of the three Graeae, Perseus forced them to tell the whereabouts of the Stygian Nymphs, from whom he obtained the following: a pair of winged sandals, a magic wallet to contain the cut-off head, and Hades' dark helmet of invisibility.

Perseus found Medusa in the land of the Hyperboreans,[5] "fixed his eyes on the reflection in the shield, Athena guided his hand, and he cut off Medusa's head with one stroke of the sickle, . . ."[6] Astonishingly, the winged horse Pegasus and the warrior Chrysaor sprang from the open wound, both having been engendered by Poseidon.

The Gorgon's head, the Gorgoneion, was at length attached to Athena's aegis, but not before Perseus had used it a number of times against his enemies. On one occasion he turned the Titan, Atlas, charged with holding up the heavens, into a mountain. In a more romantic exploit, while flying through the air, he spotted the beautiful Andromeda chained to a rock and instantly fell in love with her.

Now Andromeda's mother, Queen Cassiopeia, had boasted of her beauty and that of

1 Graves, *Greek Myths,* 73.passim.

2 Ibid., 73.c.

3 Ibid., 73.e.

4 Apollodorus, *Library,* Book II, line 41.

5 These people live in the far north.

6 Graves, *Greek Myths,* 73.h.

her daughter's being greater than that of the Neireids, who complained to their father, Poseidon. He reacted by sending a flood and a sea monster against the unhappy family and their country. King Cepheus consulted an oracle which advised him to sacrifice his daughter to the monster. Enter Perseus who saved the day, beheading the sea monster and rescuing and marrying Andromeda despite the ill faith and devious machinations of her parents.

The myth of the axis goddess Medea contriving Perseus' death was related in the previous chapter, but the information is not as well attested as one might wish. That Medea in her dragon chariot is analogous to the comet Venus, that Perses is Perseus, can more easily be seen when the two myths are examined side by side: the killing of Perses by the central axis goddess, Medea, and the rescue of Andromeda by Perseus. At the first visit of the planet Venus, the terrestrial pole moved from the constellation Perseus up through the pole of the ecliptic. Looking up at the sky, or consulting an astronomical map, we can see the constellation Andromeda immediately next to that of Perseus, while the terrestrial pole moved through two constellations before reaching the pole of the ecliptic: Cassiopeia and Cepheus. Obviously we are dealing with the same hero and the same catastrophic event.

As the Greek myth of Odysseus is reflected in the folk tale of the Bearson, so is the myth of Perseus reflected in the same genre. Catastrophic events cannot be suppressed, but rise again and again to confront us. The themes of the European folk tale will be recognized by many readers: the "supernatural birth of the hero, his abandonment and persecution along with his mother, the theft of the single eye of the Phorcides, the overcoming of Medusa in spite of her power of turning people to stone through her glance, and finally the winning of the princess Andromeda as a prize for defeating the sea monster to whom she was to be sacrificed."[7]

Having found our way back through several world ages, we begin to see that the five ages of Hesiod dovetail neatly with the catastrophes related by Velikovsky. The Earth-Mars Wars separate the Age of Iron from the Age of Heroes; the Earth-Venus Wars divide the Age of Heroes from the Bronze Age. We are in Hesiod's Bronze Age now with the constellation Perseus at the pole, looking backward to define the time when he was "born" into that position.

As it happens, the engendering of our hero looks very much like a celestial event, Zeus-Jupiter descending upon Danae, the axis goddess imprisoned in a tower,[8] as a golden shower of rain. This agrees with Hesiod who relates that the Silver Age was brought to an end by Zeus, though he does not elaborate. It is to Hyginus and Apollodorus that we owe the celestial catastrophic vision of a golden shower that must have occurred just prior to, or perhaps accompanying, a sudden movement of the terrestrial pole.

The mythology surrounding Perseus tells us little else about the catastrophe that brought him to the celestial pole and the Silver Age to a close. The movement of the pole is expressed by Perseus and his mother Danae being set adrift in a boat to reach the place where he grew to manhood, i.e., became a constellation. Like Odysseus and Hercules, Perseus was abandoned to the sea. Students of ancient art are accustomed to see gods and heroes traveling across the sky in a god boat. The pattern is

7 Thompson, *The Folktale* (New York, Holt, Rinehart, 1946) p. 279. I have omitted Thompson's distracting "type" numerals from the quote.

8 O'Neill, *Night of the Gods*, p. 809.

brought to Earth to include infants such as Moses and Romulus and Remus.

To unravel the catastrophe in which Zeus-Jupiter sent a golden rain to Earth, engendering Perseus at the pole, we must return to that cornucopia of catastrophic information, the tauroctony of the Mithraic Mysteries. An entirely new aspect of Perseus can be seen when he is compared to Mithras as bull-slayer. Connections between the two (as well as with the country of Persia) are a lively issue among Mithraic scholars. Ulansey has pointed out some similarities in the artistic representation of Mithras and Perseus.[9] Both figures wield a weapon: Perseus a curved sword, Mithras a dagger. Mithras wears a Phrygian cap with the peak bent over; Perseus sometimes does. Both look away from their victim: Mithras from the bull, Perseus from the Gorgon. And when we look up in the sky at the bull Taurus, the figure looming over him is none other than the constellation of Perseus.

In examining the tauroctony, it is difficult not to be struck by the unlikely position of Mithras atop the bull. He looks extremely insecure with both his legs on one side of the bull, hardly a natural position, let alone one from which to make a kill. Yet this improbable attitude was repeated in tauroctonies all over the known world. And what of the spot the dagger appears to be entering? Anyone with a nodding acquaintance with bovine anatomy must wonder at this choice, for a solid sheath of bone and muscle lies between the shoulder of the bull and its heart. Mithras could scratch away there for a long time without risk of mortal injury to the bull. Perhaps the tauroctony is trying to tell us something.

Ulansey has noticed that Mithras' left knee points to the spot on the bull where the dagger enters; he quotes Aratus on the left knee of Perseus being near the Pleiades. (The graphic figures superimposed on old star maps do not always agree.) Only the forehand of the bull was drawn to begin with, and the Pleiades were in approximately the same position where the dagger of Mithras is seen to enter in the tauroctony, at the bull's shoulder. It is at this point that a catastrophe is indicated, moving the terrestrial pole from the Pleiades in Taurus to the constellation Perseus as Mithras-Perseus "kills" the bull.

Returning to our astronomical model, a line drawn from the Pleiades toward Perseus is decidedly headed north. It is here that Alpha Persei or Mirphak, the brightest star in the constellation, comes into focus. Astronomers may occasionally have scratched their heads over an Arab name for Alpha Persei, Muliammu al Thurayya, the "Concealer of the Pleiades." But it appears that a sudden movement of the pole from the Pleiades in the direction of Alpha Persei suggested the name, particularly since a line can easily be drawn in that direction to intersect the pole of the ecliptic.

It should be made clear that the tilt from the Pleiades was only in the direction of Alpha Persei, and probably stopped some degrees short of it. Thus the Earth would then be in position to approach Algol, "the Piled-Up Corpses," on its regular precessional path. We now count three tilts toward the pole of the ecliptic.

The bull will die once more as we continue into the past, but let us stop for a moment to examine this continuing model of the death or sacrifice of the bull. To begin with, it is an example of man's never-ending efforts to repeat on Earth what occurs in the sky, to bring Heaven to Earth. The bull has been a favorite cult animal and victim throughout history, and before we click our tongues at such primitive practice, we must recall that the bull is still sacrificed today in the bull ring. Bullfight is a misnomer since the dice are heavily loaded against the animal, and vanishingly few matadors die in comparison to their

[9] Ulansey, *Origin of the Mithraic Mysteries*, p. 25.

victims. Though outlawed in many countries, no one can deny that the bullfight has been romanticized by writers in the English language, notably Hemingway but also by James Michener.

The "death" of Taurus by a "blow" to the Pleiades was carefully noted by the ancients. Another element would be added to the bull sacrifice as it was performed on Earth. In Babylonian ritual, a bull worthy of sacrifice must be black and marked with "seven white tufts of hair like stars, . . ." Following Allbright, this was the mark of the Pleiades.[10] According to Thureau-Dangin, the black bull represented Taurus in heaven.[11] After sacrifice of the animal, the hide was stretched over a bronze drum for ceremonial use. Santillana has caught the sense of the ritual. "The striking of the drum covered with that specific bull hide was meant as a contact with heaven at a most significant point, . . ."[12] in this instance at a previous position of the pole. In such fashion did man stay in contact with the gods.

Returning to Mithras and Perseus, yet another parallel can be drawn between the two: they are both Aquarians, having the Spring Equinox in Aquarius. In the case of Mithras, this is known from iconography. There are a number of representations of the so-called rock birth of Mithras. A cursory examination of Housestead's Rock Birth reveals something closer to birth from an egg broken in two. The zodiac, frequently depicted in Mithraic art, forms an incomplete circle around the figure, but it is highly unusual. The zodiac begins at lower left with the sign we would expect to begin the year. To the consternation of Mithraic scholars, it begins—not with the Aries of classical Greece and Rome—but with Aquarius. Thus when Mithras was born, the Spring Equinox was in Aquarius.

The case of Perseus appears complex, but is actually simple. When Venus caused a 115-degree tilt of the terrestrial pole, exchanging winter and Summer Solstices while moving the pole from Perseus to Hercules, spring and fall were very naturally affected as well. Spring, after all, must follow winter as fall follows summer. Thus the points in the Earth's revolutionary cycle where spring and fall occur would be exchanged as well. The Spring Equinox in Aquarius was exchanged for Spring Equinox in Leo.[13] Thus when Perseus was at the pole, before the catastrophic tilt, the Spring Equinox was in Aquarius. Consulting an astronomical map, we see that the Spring Equinox was in Aquarius throughout both Silver and Bronze Ages.

Measured on the precessional clock, the Bronze Age of Perseus was the shortest of the Five Ages of Hesiod. It was a time when Venus was making cyclical but relatively harmless cometary visits about twice a century. The Aztecs possessed a double 52-year cycle based on the movements of Venus.[14] "In the *Codex Vaticanus* the world ages are reckoned in multiples of 52 years with a changing number of years as an addition to these figures . . . Four ages of 105 years are referred to by Censorinus as having taken place, according to the belief of the Etruscans, between world catastrophes presaged by celestial portents."[15] Thus, it appears that the Bronze Age of Hesiod when

[10] W. F. Allbright and P. E. Dumont, "A Parallel between Indic and Babylonian Sacrificial Ritual," pp. 118–122.

[11] Thureau-Dangin, *Rituels Accadiens,* 1921, p. 2.

[12] Santillana and Dechend, *Hamlet's Mill,* p. 125.

[13] Catastrophic motions of Earth can be seen clearly on an astronomical chart with the pole of the ecliptic or precession in the center of the projection. Conventional projections centered on the celestial pole of Earth are of no use.

[14] Thompson, *Mexico Before Cortez,* p. 203.

[15] Velikovsky, *Worlds in Collision,* p. 152, note l.

Perseus was at the pole lasted for approximately 400 and some odd years.

Because the length of the day, month, and year often changes from one world age to another, it is difficult to establish information of this sort, particularly working from the present paradigm. Velikovsky gives us help here, noting that the Palermo stone represented the length of the year during the Old Kingdom of Egypt as having 320 days.[16] Following his reconstruction of history would place this before the catastrophes involving Venus, perhaps during the Bronze Age when Perseus was at the pole.

We will meet again with Perseus on our journey into the past, but not before examining that spot on the shoulder of the sky-bull Taurus, the Pleiades, with some care.

[16] Velikovsky, "Calendric Changes and Sothic Chronology," *Kronos* VI, 1, p. 74. Also see Roger Ashton, "The Cycle of 320 Days," *Aeon*, Vol. V: No. 6, pp. 95–100.

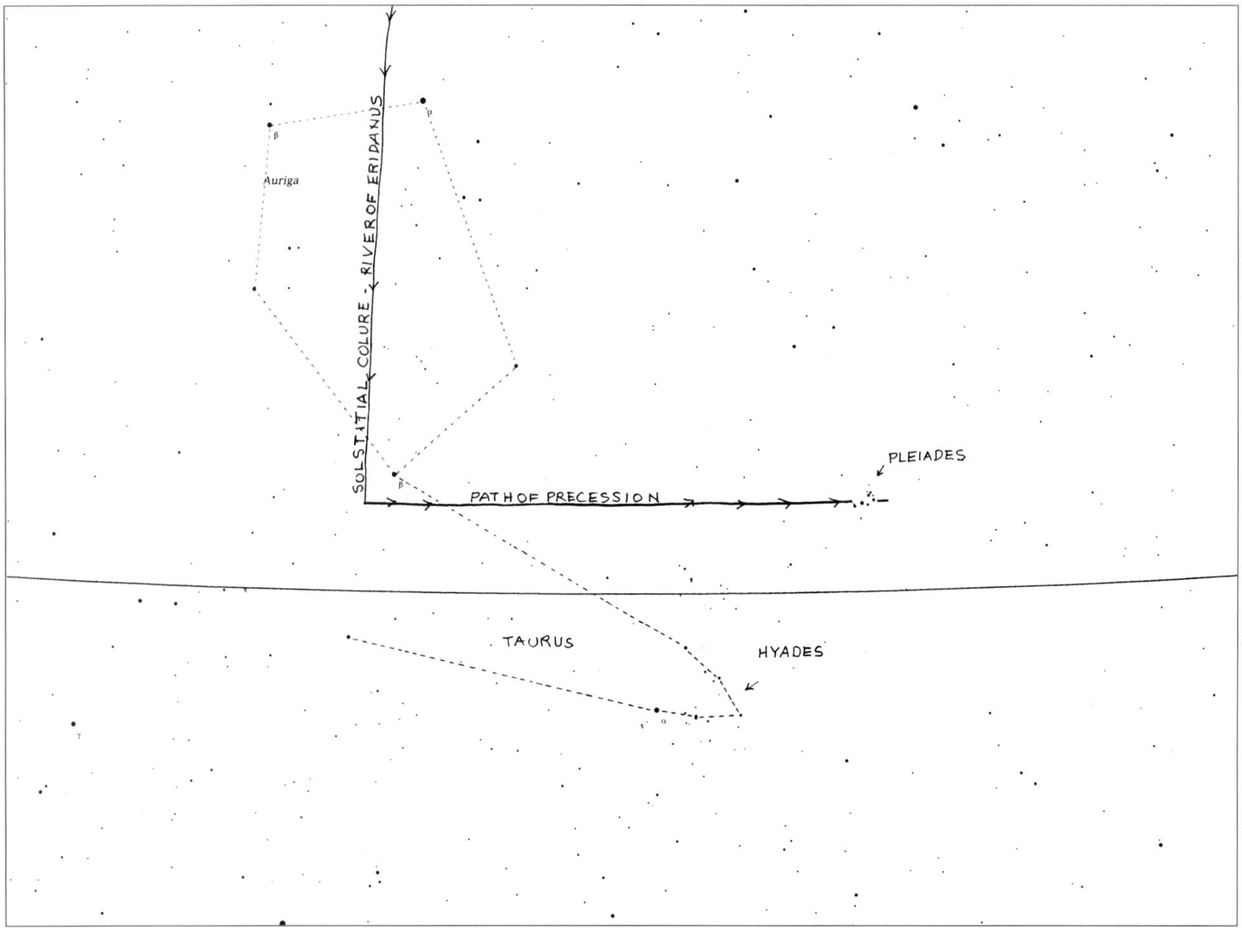

The path of regular precession of the north terrestrial pole when first Hyades, then Pleiades, marked the north celestial pole of Earth.

16. Taurus at the Pole: Hyades and Pleiades

Memorial festivals in honor of the dead . . . throughout a large portion of the ancient world . . . seem always to have been associated with the Pleiades.

– O. E. Scott, *The Stars in Myth and Fact*

The catastrophic episode that ended the Silver Age was represented in the tauroctony by Mithras plunging a dagger into the shoulder (Pleiades) of the bull Taurus. But when was the bull Taurus first recognized in the sky? What was the nature of the event that brought the star-bull into focus? We will answer the first question here, the other in subsequent chapters.

From the Pleiades we must work our way back into the past along a celestial meridian, the regular precessional path of the North Pole of Earth. Following this path back across Taurus, we are led to a point between the horns of the bull and a time indicated in many creation myths as a "Fall" or "descent of lordship." That we can cross almost an entire sign of the zodiac in a uniform movement without tilts or precessional spins gives the impression of a long period of stability, but as it happens, this was only relatively true. It does, however, lend support to the lengthy period in ancient history when the bull was celebrated in text, art, and iconography.[1]

The point between the horns of the bull that the terrestrial pole fell into locates the Babylonian constellation, The Chariot of Enmesharra.[2] This small configuration of stars comprises beta and zeta Tauri, well out on the horns of the bull. It is distinguished not only by lying athwart that enduring circle in the sky, the ecliptic, but by being within that other, more visible celestial circle, the Milky Way, not far from the galactic equator. Surely this position at the crossroads of two expressions of force is not due to happenstance or what some consider the vagaries of catastrophic motion. At any rate, I was spared the trouble of having to connect catastrophe to the galaxy as suggested by my son-in-law; they very easily connected themselves at the Fall.

Linking the Milky Way with Heaven at the terrestrial pole makes eminent sense, for this starry highway in the sky has always been regarded as a spiritual path to other worlds by many cultures, including the neopythagoreans at our own roots. Porphyry mentions the Milky Way thus in his Cave of the Nymphs.

It is more than a little troubling to imagine this Silver Age: Earth on its side in the ecliptic, the North Pole pointing to Taurus, sun and planets traveling a path that leads close to both poles. There is much to consider in such an order of the heavens. We will begin with the star clusters in Taurus: Hyades and Pleiades. Locating them in the sky is not difficult.

When the constellation Orion is in the sky on a clear moonless night, it will draw every eye with the pattern of its brilliant stars. His belt in the center, three-in-a-row, Tres Marias of the Mexicans, will lead the eye down to the dog star, brightest Sirius. Following the line of the belt in the other direction (upward, closer to the pole), we encounter two small star clusters, easily overlooked next to the mighty hunter: first the V of the Hyades, and

[1] Orthodoxy has always considered this to have been a result of the vernal equinox having been in Taurus from roughly 4200 to 2000 B.C. In fact, this was true only during the Age of Heroes, 1500 to 747 B.C.

[2] Santillana, *Hamlet's Mill*, pp. 266–267; also Appendix 22.

farther away, that mere smudge of stars, the Pleiades. These star clusters mark the constellation of Taurus the bull. In searching the north polar sky for the lost Bull of Heaven, Saturn, the ancients had to settle for the insignificant Taurus, probably one of the first constellations of the zodiac to receive a name in the present cycle of world ages.

In antiquity only the forehand of the bull was represented: head and shoulders, with the head and horns the most characteristic part of the animal. The V of the Hyades was seen as the bull's head, the first-magnitude star Aldebaran marking the bull's eye on our left, Epsilon Tauri the right eye. Drawing an imaginary line between the two and connecting it to the stars of the bull's nose gives us a triangle, brought to Earth as the white blaze on the head of the Apis or Hapi bull of Egypt.[3] And Hapi was "the second life of Ptah" and sometimes "the son of Ptah."[4]

Indic culture saw the new constellation of stars in the polar position differently. For them it was the head of a horse. "Vishnu assumed an equine head of great effulgence, . . . The firmament with all its luminaries and constellations, became the crown of his head . . ."[5]

In contradistinction to the clearly triangular Hyades, the Pleiades are simply a group of small stars close together, a fine sight in binoculars, but hardly noticeable against other marvels in the sky. Yet they are the star of stars, mul mul of the Babylonians. "This constellation is mentioned first in the Babylonian lists of the constellations belonging to the Babylonian proto-zodiac, consisting of groups of stars in the neighborhood of the ecliptic, in the zone called Harran Sin by the cuneiform astronomical texts."[6]

The Pleiades figure prominently in ancient cosmology and cosmogony, in mythology and folklore, in primitive tradition around the world. Temples are oriented to their rising in the present world order, and they are represented artistically in primitive art as well as mentioned by Homer on the Shield of Achilles. A historical catastrophist would begin to suspect that this undue prominence of an insignificant star group reflects another world age with a different sky. Thus we find them dominating the pole in the Silver Age along with their sisters, the Hyades. Both Pleiades and Hyades were daughters of Atlas, war lord of the Titans. On the losing side in the war against Kronos, he was charged by Zeus with holding up the heavens and later changed to a mountain by Perseus.

The other, even less noticeable star group in Taurus is the Chariot of Enmesharra, easily located by the extending the V of the bull's head, the horns, to the next two stars, beta and zeta Tauri. The Babylonians thought it important to mark the spot in the sky where the Elamitic chariot carried the corpse of Enmesharra.

There are many myths that refer to the "descent of lordship," to Tartarus, Eridu or Hell. Here we only mention two that have to do with catastrophes occurring to sun chariots. Most of us are familiar with the name of Phaethon, who drove his father's sun chariot and made a bad job of it because he was unable to stay on the proper path. In the passing of successive world ages, the myth has become a composite one referring to a changed path of the sun, but containing elements from different occasions on which such a change has occurred.

---

[3] O'Neill, *Night of the Gods*, p. 67.

[4] Ibid., p. 791. Ptah is another name for Saturn.

[5] Mahabharata, Roy trans., 12.348, Vol. 10, p. 605.

[6] Allbright, "Parallel Between Indic and Babylonian Sacrificial Ritual," *Journal of the American Oriental Society*, 1934, p. 124.

An early version is recorded by Hyginus.[7] It fits here because the sun was thought by some to be Saturn, who was "struck by a thunderbolt by Jove (Jupiter/Zeus), and fell into the river Eridanus." Santillana puts it literally. ". . . Zeus threw his father from the chariot and 'entartarosed' him."[8] There is need for a useful verb in the case because so many gods and heroes are "entartarosed," sent to Hell. In fact, here begins the obligatory visit to Hell emulated by gods and heroes ever since. The Pythagoreans say "that the Milky Way is the path taken by one of the stars at the time of the legendary fall of Phaethon."[9]

After the fall of Phaethon, Jupiter set him among the stars as Auriga, the charioteer.[10] This is a helpful clue, but cuneiform texts are more exacting, telling us where the corpse is, in the Chariot of Enmesharra, and we know that Enmesharra = Enki-Ea = Saturn.[11] This locates the point in the sky which appeared at the pole after Saturn was deposed and entartarosed. We will deal presently with a wealth of myth on this event.

Greek myth and Babylonian text fit remarkable well when we look at the sky. Just above the Chariot of Enmesharra between the horns of the bull, we see the spacious, showy pentagram of Auriga. In fact, just as the Square of Pegasus owes one star to Andromeda, so does Auriga, the charioteer, owe one star to the Chariot of Enmesharra.[12] They are meant to go together, charioteer and chariot. All that is lacking is the horse. But the horse is already there, for the V of the Hyades was seen as the head of a bull and of a horse (and many other animals) as well.[13]

The chariot action of the Silver Age can begin,[14] but since the chariot is at the pole, it can move little, only wheel in a circle as the V of the horse's head, the Hyades, revolves around it. Although disconcerting to moderns, the stationary situation of the Chariot of Enmesharra with its burden of a sun corpse would have been eminently satisfying to the ancients, accustomed as they were to the immobile Saturn at the pole throughout the Golden Age.

As we move back in time, tracing the precessional path of the pole from the Pleiades back along a celestial meridian, it leaves the Hyades considerably out of the way. We can only grasp its tight relationship to the pole when we imagine the apparent rotation of the star cluster around the pole at the beginning of the Silver Age. The well-proportioned V of the bull's head would have been a perfect mark to spot the pole, in the same way as we now use the two "pointers" at the mouth of the Big Dipper, Ursa Major, to spot the dim pole star today. Though at a slight distance from the pole, the association between it and the Hyades would have been very tight.

As the centuries passed, the pole would move from between the horns of the bull to points closer to both star clusters, eventually between them. We must understand that the ancients were always alert to an appearance of the Magic Twins in the sky. At that time the two star clusters could be seen as twin horses, wheeling like a good team (one going backward)

7 Hyginus, *Poetica Astronomica,* II, 42.

8 Santillana, *Hamlet's Mill,* p. 266.

9 A certain congruence between the river Eridanus and the Milky Way will be discussed in a later chapter.

10 Nonnos, "Dionysiaca," 38, 424–431.

11 Santillana, *Hamlet's Mill,* p. 267.

12 Beta Tauri.

13 Allbright, "Indic and Babylonian Ritual," passim.

14 Chariots were at the pole during both the Iron Age and the Age of Heroes. Both Ursa Major and Ursa Minor were Αμαξ, a chariot. Later they were a wain or cart.

around the pole. Here we can see the pattern for both the Roman October Horse sacrifice[15] and the Vedic Asvamedha,[16] for in both instances it is the horse pulling on the right of a team of two that is immolated. The Vedic horse must be "without a peer under the right end of the yoke . . ."[17]

Just as at present, there have always been two conventions for representing the patterns in the sky: from Earth and from space. The skywatchers of the Silver Age represented the chariot configuration as seen from Earth. As we view the chariot, charioteer, and horses, it is the "horse" of the Pleiades that is on our right.

The case of the Vedic horse sacrifice is also analogous to the Babylonian bull sacrifice, for both horse and bull had to bear the mark of the Pleiades on their bodies.[18] And it was when the Pleiades were at the pole that the bull, Taurus, met his death as a polar image.

It is worth remembering at this point that the mythical horse Pegasus, born from the decapitation of Medusa, drew the thunder car for Zeus. Since vehicle and horse were perma-

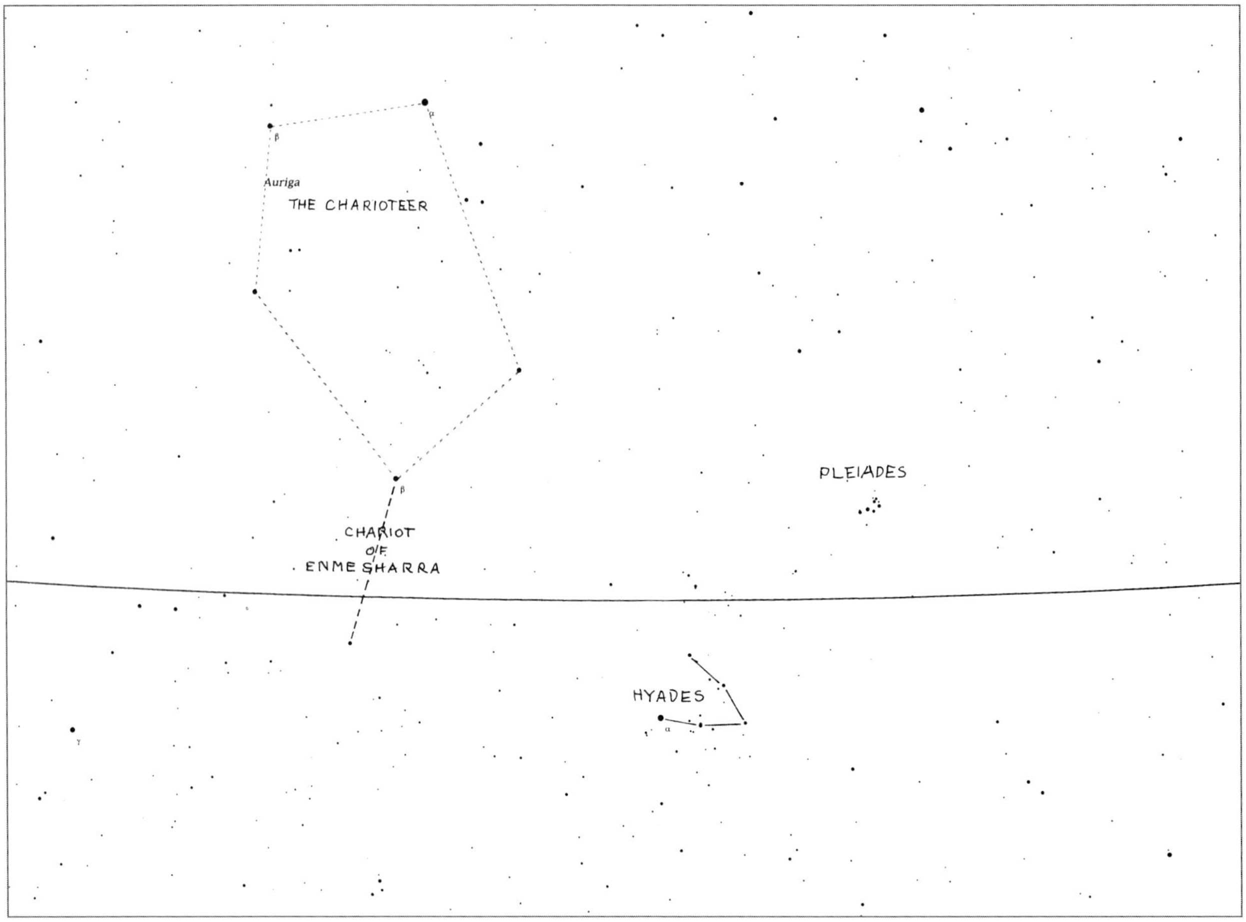

The horse (Hyades), chariot (of Enmeshara), and charioteer (Auriga) of the Silver Age. When the North Pole moved closer to the Pleiades, it was seen as another horse.

[15] Dumezil, *Archaic Roman Religion*, p. 225.

[16] Traces of the Asvamedha can be found among Germans, Iranians, Greeks, Romans, Armenians, Dalmatians. Mircea Eliade, *History of Religious Ideas* (Chicago, 1978) p. 218.

[17] Allbright, "Indic and Babylonian Sacrificial Ritual," p. 110.

[18] Ibid., pp. 120–123.

nently stationed at the pole, the thunder car could only be operational when Jupiter-Zeus was at the pole, and it was probably at this time that Jupiter gained his reputation for hurling thunderbolts. During the Silver Age, all the planets would have been circling through both polar regions, occasionally "splitting the wain of Rohini," wain being a four-wheeled wagon.

Indic astronomy found their celestial vehicle in the Hyades (head of the bull) rather than between the horns of the bull. Astrologically, splitting the wain of Rohini was very bad news, particularly if the planet should be Saturn or Mars. One is inclined to agree with Dechend. "The bad experiences which Saturn had with Auriga's vehicle—whether beta zeta Tauri or the Hyades—seem to have left a trace in the memory of Indian astrologers."[19]

Once the Hyades are realized as a horse's head, another horse comes into focus at the pole. Odin's (Saturn's) horse, Yggdrasil, grazes there beneath the world tree.[20] Though tree and horse were there during different ages, it is of no significance. Myth is epichronic, its evidence tending to be cumulative. Thus the cosmic pillar has been described as a "golden horse post,"[21] and the crescent of the Golden Age of the Saturn Configuration was suspended on the chest of the four-footed horse by Moslem horsemen.[22]

The horse's appearance at the pole has lent the animal considerable symbolic power, echoing the mythical flying horse Pegasus pulling the thunder car of Zeus and supported by the poetry of the book of Job. "Hast thou given the horse strength? Hast thou clothed his neck with thunder?"[23]

We begin to understand horse-goddesses. If a polar goddess was lost at the end of the Golden Age, she reappeared with a horse's head like mare-headed Demeter, who assumed the role of Nicippe, conquering mare.[24] Epona was a horse-goddess of the Celts.[25]

Graves calls the Hyades piglets,[26] but he does not make the obvious connection to the sacred boar. Zeus throwing Kronos out of the chariot is the beginning of ritual regicide in many cultures. The king never dies a natural death but comes to a bad end and is often, sorry to say, eaten by his devotees. Graves lingers over these ritual murders, but he is a self-confessed champion of the goddess. It is one of the more offensive aspects of her cult. We must assume that the Hyades were also seen as a boar's head since the tanist (twin and next king) wore boar disguise when killing the sacred king.[27]

An association with the boar was still preserved in Britain at the Winter Solstice festivities once ruled by Saturn. A roast boar was traditionally served with an apple in its mouth. Though Eve eventually confessed to plucking and eating the apple (of the Saturnian bull's eye), no effort was made to hide the boar's culpability. Interestingly, Egyptian taboos on boar's flesh were lifted at midwinter.[28]

[19] Santillana, *Hamlet's Mill,* App. 22, p. 399.

[20] Ask Yggrdrasill, "the ash tree that is the horse of Eggier." Graves, *White Goddess,* p. 57. Also Santillana, *Hamlet's Mill,* p. 236.

[21] No Holmberg, *Mythology of All Races* (New York, 1964) Vol. IV, *Siberian Mythology,* p. 337.

[22] O'Neill, *Night of the Gods,* p. 659.

[23] Job 39:19.

[24] Graves, *Greek Myths,* 24.b.

[25] *Mythology of All Races*, Vol. III, p. 9, 129 M.O.

[26] Graves, *Greek Myths,* 27.2.

[27] Ibid., 126.1.

[28] Ibid., 126.1.

Now it is a fact that both Hyades and Pleiades are associated with traditions of rain and flooding, the word Hyades actually indicating rain. Since they are well separated on the precessional path of the pole, we will look for two different watery events to assign to them. First in the sequence of time, the Hyades will be considered first. Now that we understand that they were seen as a horse's head, we see why there would be rainmaking rites of the sacred horse cult.[29] To the Babylonians the Hyades were mul is li-e, the jaw of the bull, and in the well-known story of Samson in the Bible, God brought water from the jawbone of an ass for Samson to drink.[30]

The tradition of a universal flood has always been a sensitive subject for Biblical scholars, geologists, and now historical catastrophists. Flooding is invariably associated with the tilting Earth motions described in these pages as water is carried out of its natural boundaries and swept over the land. But information connected with the Hyades suggests only rain. No catastrophic motion need be involved, nor does myth or legend indicate one here. We will thus posit one flood due to heavy rain at the time the Hyades became prominent at the pole at the beginning of the Silver Age.[31]

Throughout the Silver Age, both the terrestrial pole and the Summer Solstice remained within the sign of the bull. The Vernal Equinox was in Aquarius, mul gula of the Babylonians. Perhaps it is significant that Aquarius is the water bearer since such heavy flooding occurred at that time.

Toward the end of the Silver Age, the Pleiades were close to the pole. The star Alkyone in the center of the star cluster was apparently at the terrestrial pole at some point, as its name means center, pivot, or turning point, and "the ancients in very remote ages undoubtedly believed that it was the center of the universe, the primeval home of our race and the abode of the Deity and of the spirits of the dead, was in the Pleiades."[32] *Legends of the Jews* relates an incident where Istehar learned the fabled name [of God] from the angel Shemhazi, information that allowed her to raise herself up to Heaven where God placed her in the Pleiades.[33]

The Flood of Noah, the Deluge, took place when the Pleiades were at the pole. "The Flood was produced by a union of the male waters, which are above the firmament, and the female waters issuing from the earth. The upper waters rushed through the space left when God removed two stars out of the constellation of the Pleiades."[34] The sense of waters above and below is repeated in Genesis although the Pleiades are not mentioned. ". . . the fountains of the great deep [were] broken up and the windows of heaven were opened."[35] The rain began and continued for 40 days, but it was a year before the waters subsided.

In a lecture given by Velikovksky at the University of the New World, Valais, Switzerland, he suggested that the Flood may have been caused by Saturn, citing the Biblical account in Genesis. "And it came to pass after seven days that the waters of the Flood were on the earth." Speculating on what occurred

[29] Ibid., 33.4.

[30] Judges 15:17–19.

[31] The Hyades were singled out by the Egyptians. The pyramids of Dasher are said to correspond to the "eyes" of the bull Taurus. See Bauval, *The Orion Mystery,* pp. 140–143.

[32] *Nature,* Dec. 1, 1881, pp. 100–101.

[33] Ginsberg, *Legends of the Jews,* pp. 68–69.

[34] Ibid., p. 76.

[35] Genesis 8:11.

during the seven days, Velikovsky reminded his listeners "... that Isaiah refers to the same seven days in his description of the messianic age to come, when '... the light of the moon shall be as the light of the sun, and the light of the sun shall be sevenfold, as the light of the seven days ...' "

Velikovsky suggested that such an intense light indicated a nova within the solar system, and he believed it to have been Saturn. This means that the waters of the Flood were of cosmic origin, which might explain several things: "the water was warm, ... the waters ... were increased greatly on the Earth."[36] Mars would also have been a recipient of cosmic water, which would go far to explain what is sometimes thought to be evidence of water erosion on that planet.[37]

Although the event was both horrendous and memorable, accounting for a flood in legends the world over, it lacks the marks of catastrophe we are accustomed to see. The dates do not correspond to either equinox; the 40-day period given is from the 17th of Heshwan to the 20th of Kislew.[38] And it is clearly stated in *Legends of the Jews* that "... the earth assumed its old form at the end of the year of punishment, ..."[39] Although there was great loss of life, it does not appear that the Noachian Flood, the Deluge, brought the Silver Age to a close. We can assume that the Pleiades were still close to the pole when the rain ceased, no new Heaven and Earth being indicated.

What did occur was a flood connected with the Pleiades and associated with great loss of life. In focusing on the Pleiades, we must keep in mind uniform precessional motion as well as catastrophism. The star cluster remained close to the pole on its precessional path for a period of time after the Flood and became associated with additional loss of life at the catastrophic end of the Silver Age. Thus a feast of the dead is frequently associated with the Pleiades in traditions around the world, as well as a primitive calendar governed by their appearance.[40]

The Maya had a curious name for the Pleiades, tzab ek, "rattles of the rattlesnake."[41] It is an echo of the tail of Draco, representing the cometary Venus unwinding from the Saturn Configuration. When the terrestrial pole fell from Draco into Taurus, the rattle at the end of the tail, in approximately the same position as the Pleiades, must have seemed the only part of the cosmic serpent left in the heavens.

The end of the Silver Age in a rain of fire will be dealt with in a later chapter. But first we must address the most difficult concept recorded by primeval media, a vision of the sky so alien that it is hard for us to credit, the Saturn Myth.

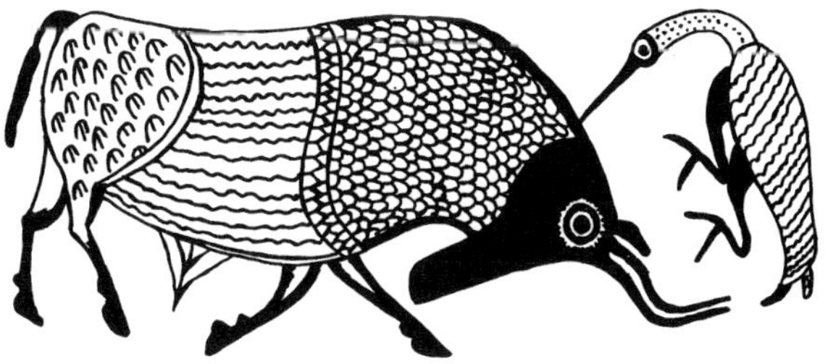

Bull and bird.

[36] Genesis 7:19. The Atlantic Ocean was called by the ancients "The Sea of Kronos."

[37] But see Wal Thornhill's CD Rom, *The Electric Universe,* for evidence indicating electrical scarring, available at holoscience.com.

[38] Ginsberg, *Legends of the Jews,* p. 77.

[39] Ibid., p. 78.

[40] *Nature* 9, Dec. 1881, "Primitive Traditions."

[41] Victor W. Von Hagen, *The World of the Maya* (New American Library, N.Y.) p. 178.

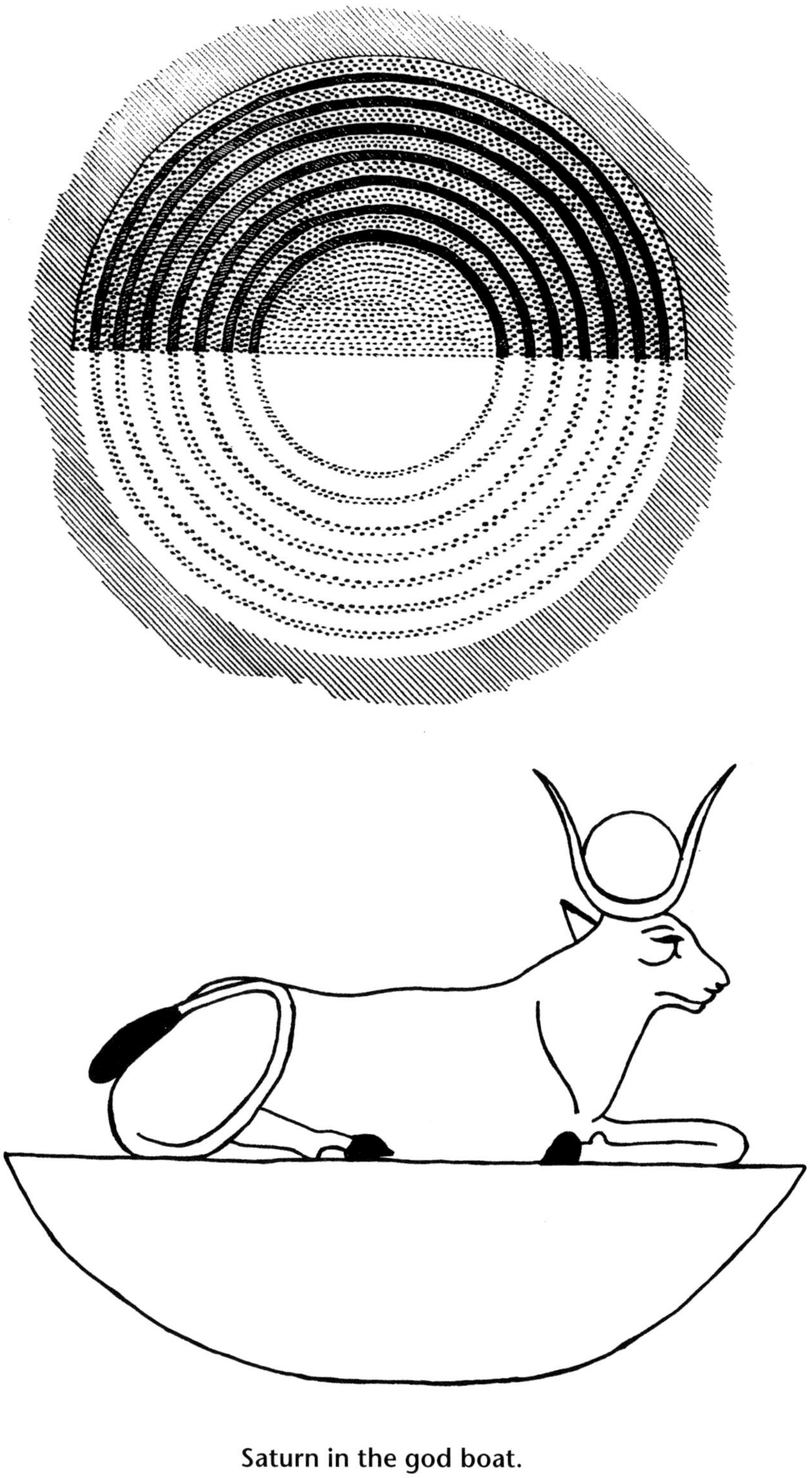

Saturn in the god boat.

17. The Saturn Myth

Anyone who has entered the labyrinth of an archaic culture's mythical compendia (The Pyramid Texts, the Vedas, the Theogony), can testify to a desperate suspicion that there is no thread of objective reality.

– William Mullen, "The Center Holds"

How can anyone hide from that which never sets?

– Heraclitus

Holy Cow!

– Anonymous

As secure in their world order as we are in ours, the ancient Egyptians freely described their visible god, looming in the sky overhead, in the following terms:

> Horus-Re, Mighty Bull—the sun with sharp horns who comes out of his Aten.[1]
> I am the Bull, the Old One of Kenzet [Kenzet, the horned pillar], . . . I support the sky with my horns.[2]
> The Great Bull of Osiris circles around![3]
> I am the sharp-horned Bull, who regulateth the sky, the Lord of the risings in heaven, the great Giver of Light, who issueth from Flame.[4]

In *The Saturn Myth,* David Talbott uses the above quotations to analyze the horns of Saturn as they appear in old Egyptian texts.[5] He is able to make sense of the imagery because he takes it as descriptive of a different order of the heavens in an earlier age. It is Tep Zepi, the First Time of the Egyptians, when Saturn stood above the North Pole, relatively near at hand, surrounded by the imposing ring that can be seen at present only with the aid of a telescope. Moreover, Saturn and Earth were connected by a line or lines of visible force, a cosmic axis commemorated in primeval media as the world tree, sacred pillar or mooring post.

Saturn was the sun god of the Golden Age or Tep Zepi, but an old and dying one. He gave light as the old vegetation god,[6] but was still capable of being illuminated by "our" sun as the two revolved around each other. Such double star systems are common in the skies around us. Our younger sun, with its more powerful light, illuminated Saturn on the side closest to it, just as it now illuminates the Moon. Seen from Earth, this illuminated portion of the planet and band would appear as a crescent. And it is this illuminated crescent, seen as a pair of horns, that gave Saturn

[1] Alexander Piankoff, *The Shrines of Tut-Ankh-Amon* (New York, 1955) p. 6.

[2] R. O. Faulkner, *The Ancient Egyptian Pyramid Texts* (Oxford, 1969) p. 148.

[3] Piankoff, *Shrines,* p. 84.

[4] P. Le Page Renouf, *Egyptian Book of the Dead* (London, 1904) p. 107.

[5] Talbott, *The Saturn Myth,* pp. 260–269.

[6] When the light, and likely heat, of Saturn was added to that of the sun, plants would have grown more rapidly, produced more bountifully. Thus Saturn became a god of vegetation.

his old name, "Bull of Heaven." Then, as the Saturn Configuration and our sun revolved around each other,[7] the illuminated crescent horns would appear to circle around Saturn and its band.

Readers are already familiar with the concept of the god boat in which gods and heroes are wont to ride. Saturn set the style for this mode of travel, the first god in the current series of world ages to ride in the god boat. Talbott analyzes the phenomenon at length in *The Saturn Myth*.[8] Once we understand that the illuminated crescent was also seen as a god boat traveling around the Saturnian band, other Egyptian texts begin to clear.

> Hail to you who sails in his Khu, who navigates a circle within his Aten.[9]
> I revolve round heaven like Re, I sail round heaven like Thoth.[10]
> Hail, Only One! Behold thou are in the Sektet boat as it goeth round about the Mount of Glory.[11]

The axis bond between Saturn and Earth has been seen variously as a rope, ladder, or bridge, world tree or mountain, the skirt of a goddess or the single leg of a god. Although Kronos is well known to be Saturn, ruler of the Golden Age, a parallel Greek legend claims that their earliest ruler was Cecrops, "eye with a tail," a similar Saturnian image. The faceless stick image of the Phoenician goddess Tanit is yet another, comparable to the Aztec goddess Chalchuitlicue, she of the jade skirt who connected Heaven and Earth.[12]

The image of the primeval Saturn was seen as male and/or female. Viewed as a giant figure of a god or goddess, the head could be seen as surrounded by the halo of the Saturnian band, eventually becoming the nimbus of Christianity. Indic culture saw lingam and yoni, the phallus of the axis bond entering the vagina of the Saturnian band. And however much the image may offend in some quarters, it can lead us to a more balanced view of our ancestors.

When the prototype of symbols discussed here is recognized to be the image of primeval Saturn, scholars will no longer be able "to build an entire interpretation of ancient ritual around the theme [of phallicism]," wherein "every pillar and every related symbol becomes an expression of a primitive preoccupation with human reproduction—and nothing else."[13] Phallicism must fall into perspective with a host of other symbols that derive from the primary image of Saturn connected to Earth.

We are all familiar with a few world mountains: the Biblical Zion, Mt. Olympus of the Greek gods, Mt. Meru of India. Zion has also been described as a pillar. "There is an upper and a lower Paradise. And between them, upright, is fixed a pillar; and by this they are joined together; and it is called 'The Strength of the Hill of Sion.' "[14] The culture of India is even more explicit. "Meru, in Sanskrit, signifies an *axis* or *pivot*."[15] Kwen-lun, the "Pearl Mountain," is the world mountain of China,[16]

---

7 This would be the same sort of motion that accounts for the variable stars in the heavens around us.

8 David Talbott, *The Saturn Myth*, pp. 269–282.

9 Ibid., *The Saturn Myth*, p. 272.

10 *Pyramid Texts*, trans. R. O. Faulkner (London, Oxford, 1969) p. 39, Utt. 210.

11 Talbott, *The Saturn Myth*, p. 279. Compare to Budge, *Book of the Dead*, CXLVI, p. 63.

12 Thompson, *Mexico Before Cortez* (New York, Scribner's Sons, 1940) p. 230.

13 Talbott, *The Saturn Myth*, pp. 206–207.

14 William F. Warren, *Paradise Found* (Boston, Houghton Mifflin, 1885) p. 232.

15 Wilford, *Asiatic Researches* (London, 1808) Vol. VIII, p. 285.

16 Warren, *Paradise Found*, p. 128.

Hara-berezaiti is that of Iran, "the seat of the genii: around it revolve sun, moon, and stars."[17] In Akkad is "the mountain on whose apex the heaven of the fixed stars is pivoted."[18] The glass mountain of fairy tale fits in here as well, but there are many others, all described in slightly otherworldly terms.

In Indic culture, "... a sea (of milk) turns seven times round Meru before giving origin to the four rivers."[19] The stars revolve around the world mountain, or the sun both rises and sets there as in the Gilgamesh Epic. The four rivers of Heaven, common to many cultures, find their source there. The present earthly mountains can best be understood as analogs of the cosmic axis that appeared to support Saturn.[20] When it disappeared at the end of the Golden Age, various local mountains were chosen to replace the original, but they would never live up to their reputation in ancient texts.

World trees are somewhat less familiar. That of the Germans was Irmensul, "the universal column which sustains all things."[21] Grimm relates it to the world tree of the Northmen.[22] The Egyptian tree housed the sarcophagus of Osiris, and Bennu the sun-bird sat in its branches.[23] The Persian tree has an aspect, Haoma, which is credited with supplying the world with tens of thousands of different seeds, surely useful after a world catastrophe.[24]

The Maya have depicted the world tree many times. In the Group of the Cross at Palenque, the three temples each contain a world tree,[25] while one decorates Pacal's sarcophagus in the Temple of Inscriptions. A bird sits atop each of these and on top of all four trees in the Fejervary Codex.[26] The Celestial Bird of Maya cosmology "resides at the top of the *axis mundi*, the world tree of Maya cosmology."[27]

A bird with outstretched wings is another image that derives from the illuminated crescent of primeval Saturn. The thunderbird is the well-known symbol of Native Americans, located as a constellation in Draco at the pole of precession.[28] As eagle or falcon, the bird has been chosen as a symbol by the United States of America and Mexico as well as many European countries. The bird is most often portrayed in an unusual position, wings outstretched, talons and tail beneath it. Probably only a chick in the nest watching one of its parents land sees this particular view on a regular basis.[29]

The world tree can bear fruit (usually red, an apple or pomegranate), or support a bird

[17] Spiegel, *Eranische Alterthumskunde* (Leipzig, 1871) Bd. i., S. 463.

[18] Lenormant, *Babylone et la Chaldee* (Paris, 1875) p. 46, cited by Warren.

[19] O'Neill, *Night of the Gods*, p. 888.

[20] The possibility of a lithic bulge will be examined presently.

[21] O'Neill, *Night of the Gods*, p. 292.

[22] J. Grimm, *Deutsch Mythologie*, p. 759, cited by Warren in *Paradise Found.*

[23] Warren, *Paradise Found*, pp. 265–266.

[24] *Bundahisn*, xxvii. 4., xxix, 5.

[25] Linda Schele and David Freidel, *Forest of Kings* (N.Y., Wm. Morrow, 1990) pp. 240–242.

[26] Fejervary-Mayer Codex, aka Codex Vaticanus B. The bird of the Saturn Configuration sat atop the four supports of heaven. All face counterclockwise.

[27] Linda Schele, *Blood of Kings*, p. 55.

[28] Ronald Goodman, *Lakota Star Knowledge* (Rosebud, Sinte Gleska, 1992) star map on inside back cover.

[29] The bird's head would be the "eye" of Saturn, as in the redheaded woodpecker of the Romans, discussed in the chapter on "Mars, The Magic Twin." The tail would be a part of the cosmic axis.

or a star. If a bear is in the tree, then it has been brought up to date with more recent ages. But world trees are not truly of this earth. As described in myth and legend, they have a disconcerting habit of reaching their roots down through the Earth to the Underworld, indicating their affinity to the axis bond. Four rivers frequently have their source at the world tree. "From beneath it [tree of life] flows forth the water that irrigates the whole earth, parting thence into four streams, . . ."[30]

Following the Maya *Popol Vuh*, the bird on top of the world tree is Vucub Caquix. His brief story has been thought to be an interpolation, but it belongs right where it is, an image, though a jaded one, of the primeval Saturn. Vucub Caquix is Seven Macaws, reflecting the seven rings around Saturn. Here we recall that the illuminated crescent of the Saturnian band was actually seven separate crescents, hence seven birds, seven limbs on the world tree. The Maya were careful observers.

Vucub Caquix is accused of being arrogant because he sits in his silver throne in the ceiba tree "pretending to be the sun." He is not the sun, of course, "because he can only see as far as the horizon."[31] The stationary sun of Saturn would necessarily have this limitation. We will continue with the adventures of the arrogant V.C. in the chapter on "Killing the Bull," where he comes to a bad end. This is always the case with Saturn, one of the clues that helps us recognize him: Osiris cut into a thousand pieces, Tammuz-Adonis relegated to the Underworld, Kronos thrown out of the chariot.

From Zelia Nuttall we learn that the polar god is usually depicted resting or stationary, for the turning point of the pole is still. And Saturn was immobile to those on Earth. The circular turn of the crescent god boat depends on the revolving motion of a double star system. All the efforts of translators to make Saturn "rise" or "set" in relation to the "horizon" are vain. He "comes forth and goes in" on the "Mount of Glory."[32] As Talbott remarks, "The problem does not lie with the texts, but with the solar interpretation, which looks for imagery of a rising and setting sun where there is none."

In his introduction to *The Saturn Myth*, Talbott writes, "Nothing came as a greater surprise to me than the sheer quantity of material bearing on the Saturn tradition." I wholeheartedly agree. Once I grasped the model, information on it kept dropping into my lap unsought while I was researching other world ages and the catastrophes that separated them. Myth has a decidedly epichronic character. One grasps a pattern, then perceives a similar one beneath or on top of it.

One aspect of the primeval Saturn that kept surfacing again and again until I could no longer ignore it was that of color. There are indications of brighter and darker periods during the reign of Saturn, in addition to "daily" coming forth and going in. An early purple light is indicated, but the most persistent color to be associated with Saturn is a blue-green. It must have been very beautiful.

Exactly what the Egyptians meant by the "Great Green" has never been agreed upon. Scholars have suggested the Nile Delta, or the Mediterranean itself. But when we read that "Horus is often represented as pilot of the sun ship in its voyage across the ocean of space, the 'Great Green,' " the mist begins to clear.[33] In the *Egyptian Book of the Dead*, a ritual for

[30] Ginsberg, *Legends of the Jews*, p. 37.

[31] *Popol Vuh*, trans. Recinos, Morley and Goetz (London, Edinburgh, 1951) pp. 93–94.

[32] Talbott, *The Saturn Myth*, pp. 172–173.

[33] G. R. S. Mead, *Thrice Greatest Hermes*, Vol. I, p. 65, note. Horus as the Martian "eye" of the configuration will be discussed in the following chapter.

the deceased is to be "said over an ark of Re, colored in pure green."[34] Putting this information together with what we already know about the god boat gives us the color of Saturn as well as solving the the Great Green mystery. Due to the "watery" or "oceanic" character of the image, the term probably devolved on the Mediterranean, for when one image disappears, the associated information attaches to another visible one.

We have already traced Venus back to the feathered serpent, Xiuhcoatl; her omega-shaped symbol is often figured prominently in his tail. This "blue-green snake" can be seen in Mesoamerican art as an enclosure, either square or rectangular, the latter being the present "shape" of the world. Toward the end of the Golden Age, the dragon Xiuhcoatl appeared to circle Saturn as his blue-green band, the cometary Venus spinning off the cosmic axis between Earth and Saturn, slowly circling through the Saturnian band. That Venus had a sea-green or blue appearance at the time[35] would help it achieve an identity with the band and an aspect as mother goddess.

In the *Popol Vuh* the forefathers "were hidden under green and blue feathers," of the Saturnian band.[36] We can probably assume that the Maya were looking at Saturn when they recorded Ah Raxa Lac and Ah Raxa Tzel, lords of the "blue-green plate and bowl" in their legend.[37] Viewed from below, this would be a fair description of Saturn.

In the Gilgamesh Epic, the horns of the Bull of Heaven are of lapis lazuli.[38]

As translated by W. G. Lambert, "... the middle heavens are of saggelmut-stone, of the Igigi. Bel sat therein on the lofty dais in the chamber of lapis lazuli ..."[39]

In the vision of Ezekiel, sappur (sapphire) is the color of the throne of god.[40]

The Pyramid Texts refer to the king's "throne of malachite, his food is in the Fields of Offerings in the Lakes of Turquoise ..."[41]

A Nahuatl hymn sings of "... the old god distended in the navel of the earth, engaged in the enclosure of turquoise ... He who dwells in waters the color of the bluebird."[42]

The "Old, Old one," Xiuhtecutli, as described by Burland, was the "Central Fire" and Lord of the Pole Star. Xiuhtecutli is seated (resting) and holds on his head an enormous circular stone with four crosses on it. Xiuhtecutli means "Turquoise Lord."[43]

At the summit of Kwen-lun, the legendary mountain of the Chinese, is a "resplendent azure hall."[44]

"Enki [Saturn}, the lord who decrees the fates, built his house of silver and lapis lazuli."[45]

Among the Maya "the four cardinal

[34] Renouf, *Egyptian Book of the Dead* (London, 1904) p. 259.

[35] Ev Cochrane, "The Birth of Athena," *Aeon* vol. II, No. 3, p. 26. But Venus is Athena, the grey-eyed goddess. Possibly her spinning-off tail acquired color when seen against the blue-green of Saturn and his band. See Talbott, "Mythical History of the Comet Venus," *Aeon* II, 4, p. 35.

[36] Recinos, *Popol Vuh*, p. 81.

[37] Ibid., p. 78.

[38] *The Epic of Gilgamesh*, trans. N. K. Sandars (N.Y., Penguin Books) p. 88.

[39] W. G Lambert, KAR, the Cosmology of Sumer and Babylon,"Ancient Cosmologies," pp. 58, 307–308.

[40] Ezekiel: 1:26.

[41] *Pyramid Texts*, Faulkner, Utt. 627. Also see Utt 350.

[42] Leon-Portilla, *Pre-Columbian Literature of Mexico* (Norman, Univ. of Okla.) p. 63.

[43] Sahagun, *Historia*, chapter 13, Book I.

[44] Wm. Warren, *Paradise Found*, p. 244.

[45] Samuel Kramer, *Sumerian Mythology* (New York, Harper, 1961) p. 63.

directions were seen in relation to the center, which also had its color (blue-green), . . ."[46]

The blue-green color seems to have suffused the entire configuration, even the axis bond with Earth. Thus the Mandaeans find their Light King at the pole "over the mountain of turquoises."[47]

We have already mentioned Chalchiutlicue, the Mexican Goddess of the Jade Petticoat who provided the bridge to Heaven.[48]

Nordic tales describe Kullervo with the "bluest of blue stockings."[49]

There is a "stone-born hero, Soszryho, whose body is blue shining steel from head to foot . . ."[50]

Hawaiian legends refer to the blue mountain Pali-Uli, where Kane, Lono, and Ku created man.[51]

The huge mosaic jaguar masks of the Olmecs were made out of green serpentine and sometimes set in blue soil,[52] honoring both the jaguar (Oceotl or Ursa Major) at the pole in the previous age and the color of Saturn which once filled the sky there. To commemorate the turquoise band of Saturn, Xiuhcoatl, they named the present, insignificant constellation at the pole Xonocuilli or Blue Worm (Ursa Minor).

Some of the blue-green color was brought to Earth. Budge describes a cache of tets in the British Museum.[53] A tet is a pillar-like reproduction of the axis bond between Earth and Saturn. Talbott calls the tet "the pedestal of Osiris."[54] Although a few are of carnelian (red) and agate, the color of the tets leans to blue or green: glazed faience, lapis lazuli, and opaque blue glass.

"The large, pierced, circular disks of jade, 'pi,' were used by the emperor when performing sacrifices to heaven."[55]

The remarkable habit of some men to paint themselves blue is recorded among the Berbers and primitive Britons.[56] Many cultures are less shy about identifying with their god than Westerners, so that this behavior would be another effort to connect with what had been seen in the sky.

Could it be that the Oriental penchant for jade, the Mesoamerican love of blue and green stone, the popularity of lapis lazuli and Egyptian Blue throughout the ancient world represents an association with the Golden Age of Saturn, remembered with affection in all cosmogonies? "Strangely, the Mayans of Mexico had a form of Egyptian Blue virtually identical to the Egyptian type."[57]

In many cultures, we learn that the gods, sometimes one in particular, taught men the art of writing. What this means is that the images came from the sky. We have already learned that the last letter of the Greek alphabet, omega, represents the constellation Draco, and earlier the cometary Venus unwinding from the cosmic axis. The mountain, tree, or pillar

[46] Schele, "*Forest of Kings*" p. 66.

[47] Frothingham, "Ancient Orientation Unveiled," *American Journal of Archaeology*, Vol. 21, 1917, pp. 423–424.

[48] Burland, Nicholson & Osborne, *Mythology of the Americas* (New York, Hamlyn, 1970) p. 159.

[49] Santillana, *Hamlet's Mill*, p. 97.

[50] Ibid., p. 264, n. 2.

[51] Ibid., p. 201.

[52] George E. Stuart, "New Light on the Olmec," *National Geographic*, Nov. 1993, p. 106.

[53] E. A. Wallis Budge, *Egyptian Book of the Dead* (New York, Dover) p. 357, n. 2.

[54] Talbott, *The Saturn Myth*, p. 180.

[55] William Watson, *Early Civilization in China* (New York, McGraw Hill) plate 126.

[56] Graves, *White Goddess*, p. 241.

[57] Dayton, *Minerals, Metals*, p. 74.

of the axis column appears to have been brought to Earth to form the initial letter of both the Greek and Hebrew alphabets. In both languages, the letter carries the meaning of "bull": alpha in Greek, aleph in Hebrew.[58] And the Western alphabet also begins with A, thus preserving the appearance of the world mountain.

Cone-shaped (aniconic) stones called baetyls stood in the place of God in some ancient cultures.[59] The cone was attached to the head in Phoenician culture, a pointed hat also worn by the Etruscan haruspex, held to be skilled in interpreting the will of the gods.[60] He was expected to interpret lightning, the flight of birds, and read the liver of sacrificial animals. Lightning and birds can be interpreted as polar phenomena, but reading a liver is more problematical. However, a model liver has been found at Piacenza, indicating that parts of the liver correspond to different parts of the sky.[61] Association of the pointed hat with occult spirituality carried down to sorcerers and witches of later times, while the symbol has been parodied on the television program *Saturday Night Live.* Everyone is amused by the "conehead's" otherworldly connections, unaware of the symbol being grounded in an objective phenomenon.

Perhaps the most offensive and unnatural effort to impose the celestial vision on humanity was the binding of the heads of infants to produce the elongated shape of the cosmic axis. This was practiced by the Olmecs and others, presumably to put the deformed subject in better touch with the gods.

As mentioned in an earlier chapter on Venus, seven separate bands surrounded Saturn, recorded again and again in ancient culture: Jamshyd's "sev'n ringed cup" of the Rubaiyat, the seven walls around some ancient cities and temples, and ritual practice of walking seven times around an object as in the sevenfold circumambulation of the Kaaba or the Buddhist dagopas of Ceylon, etc.[62] The rings of Saturn provide the objective basis for forms of circular worship, of which there are many.[63]

Ancient mythos devolves on more modern religions. Thus the bishop makes three circuits of a church to consecrate it, then goes seven times around the new altar.[64] Evidence suggests that at one time there were three rings around Saturn, for some ancient temples have three walls around them, often those which were used for child sacrifice.[65] The spiral which connects Venus with the mother goddess is present in cultures all over the world, from coins of early civilizations to Native American rock painting and sculpture. Nuttall reports a long cable carried into a spiral shape in ritual practice of the Aztecs.[66]

With Saturn begins the symbolism of the center where every culture seeks to found itself, the primeval image defining the difference between sacred and profane space. All within the circle of the Great Green is sacred and assimilated to the center; everything outside

[58] Lewis Greenburg, "The Lord of Light," *Aeon,* Vol. III, No. 4.

[59] Sir Arthur Evans, "Mycenaean Tree & Pillar Cult," p. 14 [112].

[60] Edward Bacon, *Vanished Civilizations of the Ancient World* (New York, McGraw-Hill, 1963) p. 176.

[61] Ibid., p. 197.

[62] J. Huxley, *The Ancient World* (London, 1968) p. 96.

[63] O'Neill, *Night of the Gods,* pp. 692–699.

[64] Ibid., p. 698.

[65] Cardona, "The Rites of Moloch," *Kronos* IX, 3, pp. 35–36.

[66] Nuttall, *Fundamental Principles,* p. 145.

is profane, assimilated to chaos and the other. Hence the importance of sacralizing city, temple, fields to the center, sometimes erecting an axis mundi in imitation of the visible connection between Earth and Heaven. In Eliade's words, "*. . . cosmic symbolism is found in the very structure of the habitation.*"[67]

We leave the numinous image of Saturn to take up some of the more mechanical aspects of the configuration. If the heavens had an unusual appearance, we should remember that this was the Golden Age, enshrined in the collective psyche as the best of all possible "good old days," and remember Saturn when we put a star on our Christmas tree.

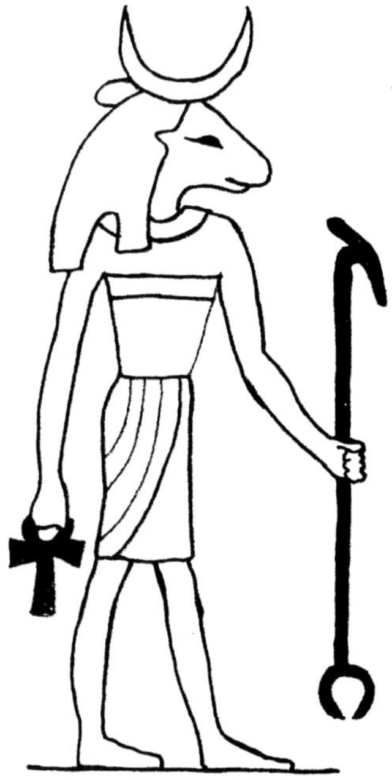

Although it is an outer planet and never displays a crescent, the Egyptian Saturn is depicted with one.

[67] Eliade, *The Sacred & The Profane*, p. 53.

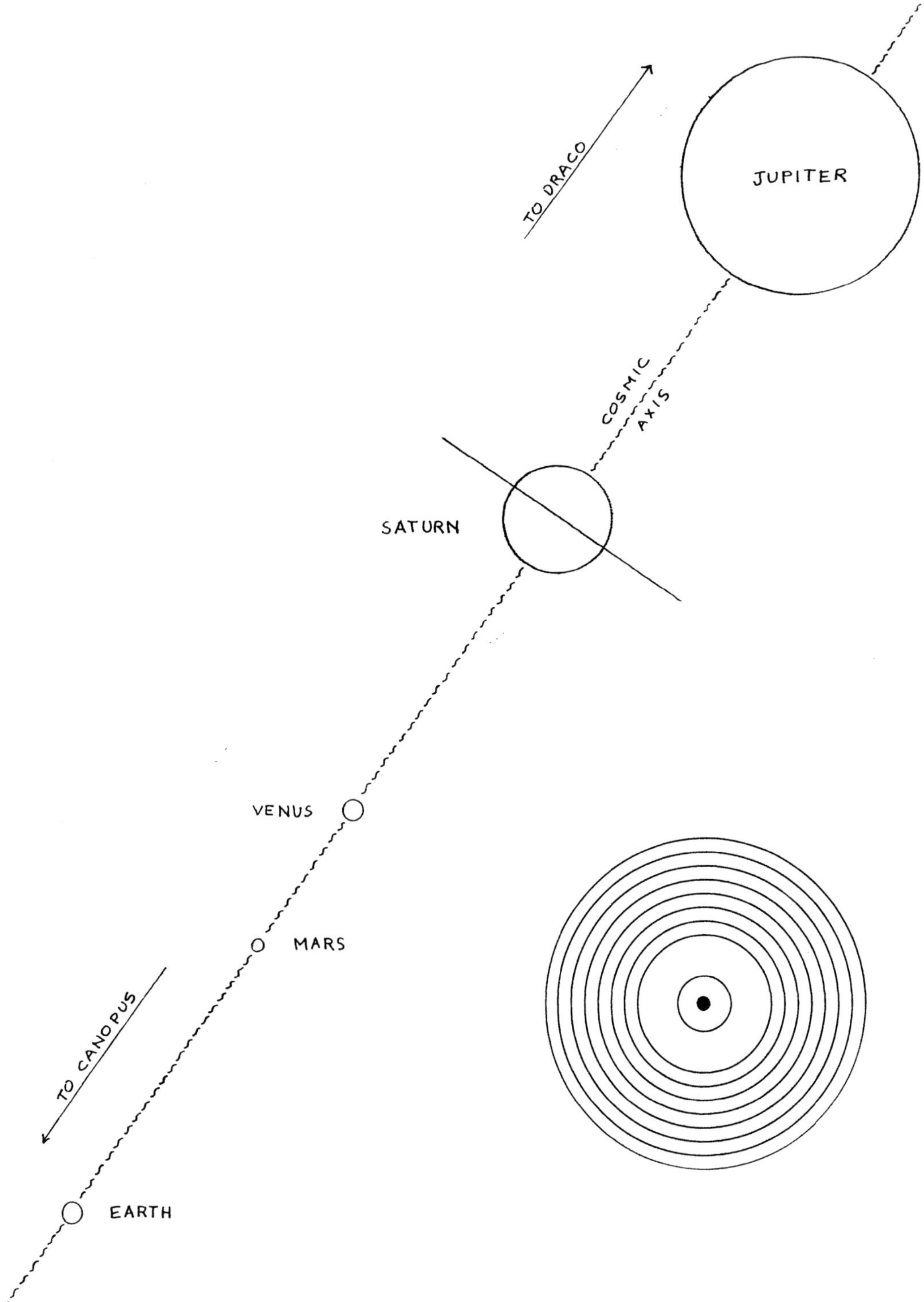

The Saturn Configuration upright between the poles of precession (not drawn to scale), and the Saturnian bull's eye as seen from Earth.

18. The Saturn Configuration

. . . a straight shaft of light, like a pillar, stretching from above throughout heaven and earth—and there, at the middle of the light, they saw stretching from heaven the extremities of its chains; for this light binds the heavens, holding together all the revolving firmament like the undergirths of a ship of war. And from the extremities stretched the Spindle of Necessity, by means of which all the circles revolve.

– Plato, *The Republic*

In *The Saturn Myth,* David Talbott developed a model of Earth connected to Saturn that faithfully reflects an enormous number of ancient texts as well as much ancient media. This is the same material that experts have long puzzled over in the mistaken assumption that it describes the present order of the cosmos. And if we accept Talbott's model? What then? We must immediately look for a more comprehensive, integrated view of these primeval heavens. Where were the other planets? What was the orientation of Earth in the heavens? What held it all together?

In the years since the publication of *The Saturn Myth* in 1980, Talbott and a number of interested scholars have carried the work forward, drawing on the legacy of the ancients to amplify the model of Earth and Saturn rotating on a common axis. The axis bond has been extended in space so that other planets can be imagined to share it, and a number of the present members of the solar system have been identified in the configuration.

Confronted with the reality of the present solar system, we accept the stability of a number of planetary bodies revolving around a central sun. To begin to understand the stability of the Saturn Configuration, we must imagine the same planets, but stacked up above each other on a common axis. Other smaller bodies would revolve around them much as planetary satellites such as the Moon do today. The same sun that shines on us today is present, and the sun and Saturn Configuration revolve around each other as double star systems are understood to do today.

Planets that have been identified in the Saturn Configuration include Jupiter, Mars, and Venus. The most commonly given account of the birth of Jupiter-Zeus suggests the identification. Kronos-Saturn's bad habit of swallowing his children was well known. Thus it was that when Zeus was born, Rhea hung his cradle on a tree so that Kronos could find him "neither in heaven, nor on earth, nor in the sea . . ."[1] The tree here is the world tree of the axis bond, and the god-planet Jupiter is on the other side of Saturn, invisible from Earth. The axis bond is also the golden chain by which Zeus can hang all the gods and Earth and sea to a peak of Olympus.[2]

When the planet Jupiter first appeared from behind Saturn, it would have first been seen in crescent, another crescent bird, the Jovian eagle with which he has long been associated. Considerable noise accompanied this event: the cymbals of the Korybantes, the clanging of the shields of the Kuretes.[3] We must expect a certain amount of noise when a world age

[1] Graves, *Greek Myths,* 7.c.

[2] *Iliad,* Book VIII.

[3] Graves, *Greek Myths,* 30.a.

comes to an end and another begins.[4] When we consider how the cymbals are played, with a sideways striking motion, we can see how the image arose, Jupiter sliding past Saturn to reveal himself. In some quarters, this was seen as a wrestling match.

At the beginning of the present series of world ages, all of the planets are said to have stood in line.[5] Astrology is obsessed with the concept of the conjunction of planets, while Kepler was particularly interested in the conjunction of Jupiter and Saturn. The thrice-repeated conjunction of Jupiter and Saturn in Pisces in the year 6 B.C. is thought by some to have been the star of Bethlehem. Virgil announced the return of the Golden Age,[6] but the model which has influenced all those which followed was actually the Saturn Configuration.

Both Mars and Venus were strung along the axis bond between Earth and Saturn, and it takes no great leap of imagination to visualize their appearance from Earth as a series of concentric circles. This is the origin of the target used in many games with the "bull's eye" of the Bull of Heaven in its center. The outer ring (or seven rings) is the Saturnian band, the next ring the circle of Saturn itself. Venus is within the circle of Saturn, the eye of the god,[7] while Mars within Venus plays the part of the pupil of the eye. This "eye" configuration answers all the eye and heart imagery embedded in primeval media, imagery that does not yield to the present order of the cosmos.[8] We will address this imagery in later chapters, only remarking here that when the illuminated crescent was above Saturn, it formed the eyelid of "... Balor, the one-eyed King of the old Gods ..." [9] recorded in the Irish Duanaire Finn.[10]

The model of the Saturn Configuration has been dubbed a shish-kebab by Fred Jueneman, and however irreverent, gives a clear idea of planets strung on a rod or spindle, expressive of a force of some kind. It is well to remember at this point that space exploration has established the existence of circling rings around the three other gas giants, Jupiter and Uranus and Pluto, as well as Saturn.[11] In visualizing the Saturn Configuration, planets with rings around them begin to exhibit a certain orderly appearance.

Any attempt to visualize the Saturn Configuration brings its most problematical aspect into focus. What is its orientation in the cosmos? We will assume that the forces we have identified as operational in cyclical catastrophe have led us in the right direction.

We tend to see the North Pole of Earth as "up." In the Golden Age, Saturn would have been up above us. But where does the cosmic axis truly point in this earliest of all recorded ages? The golden rope of the sun holding the Earth and planets in its present configuration gives us the flat plane of the ecliptic. And working our way through several catastrophic tilts has taught us that "up" is actually the pole of the precession, and surely enough force has been exhibited therein to suggest that it could at one time have held the planets in line.

---

4 The custom of greeting the sun with dances and the clash of instruments is worldwide. Jane Harrison, *Epigomena and Themis,* p. 201.

5 Alfred Jeremias, "Ages of the World," *Babylonian Ency. of Religion and Ethics,* Vol. I, p. 183.

6 Virgil, Fourth Eclogue.

7 Venus has always been the Great Eye in Mesoamerica.

8 But see Talbott, "The Mythical History of the Comet Venus," (part 1), *Aeon,* II, 4, p. 29.

9 Graves, *White Goddess,* p. 301.

10 See Augusta Gregory, *Gods and Fighting Men* (John Murray, 1904) p. 66. Also George Saul, *Traditional Irish Literature* (Lewisburg, Bucknell Univ. Press, 1953) p. 31.

11 "Rings Around Uranus," *Time,* Apr. 11, 1977, p. 55.

Therefore we will orient the Saturn Configuration with the pole of the axis bond held vertically between the poles of the ecliptic. But other primeval media also suggests the orientation.

The hippopotamus goddess Tau-ert, the constellation Draco.

Leading the parade of so-called northern constellations on the ceiling of Senmut's tomb and elsewhere is the hippopotamus goddess Tau-ert or Isis-Hippopotamus, she of the pendulous dugs and distended belly. Tau-ert is fated to have a long history of representation in Middle Eastern and Mediterranean cultures, usually with a crocodile on her back. She is the blue hippopotamus without which no collection of Egyptian art is complete. But this is the First Time of Egypt and the crocodile, a creature of the watery abyss, has not yet crawled up on her back. On Tau-ert's head in much Egyptian art, we find a symbol that is becoming familiar, the circle of Saturn resting in the crescent horns. This indicates the connection between the hippopotamus goddess and Saturn.

Tau-ert's identity has long been established; she represents the constellation Draco, our serpent or dragon. It is a large, spread-out constellation running between and around the two bears, actually an omega sign circling the pole of precession, located in that starless "empty place in the sky" mentioned in the book of Job. Thus we hang the axis of the Saturn Configuration from Draco much as God "hangeth the earth upon nothing." Lest this description seem too ephemeral, we must also recall that Isis-Hippopotamus also holds the "mooring post," or the chain that holds the mooring post.[12]

In Mesopotamia Draco bore the name "Yoke Star of the Sky," as the "heavy-weighing" Canopus was "Yoke Star of the Sea."[13] In the thought of the ancients, above the ecliptic (infrequently the equator) was the sky, below it was the watery abyss. Santillana also mentions "Yoke of Ea" for Canopus, useful since Ea has been identified as Saturn. That Canopus is very close to the southern pole of the precession is even more helpful. We thus make use of the yoke stars of the ancients to tie up the Saturn Configuration in vertical position to the poles of the precession. While we are at it, we will set the model rotating. It is an imposing image. One might be forgiven for taking it to be, as I believe it is, the turning mill of the gods.

Hamlet's Mill, as developed by Santillana and Dechend, is an elucidation of the turning mill of the cosmos, flawed as conventional Egyptian scholarship is flawed by seeing it against the present order of the heavens. Santillana dislikes catastrophism but has sufficient masculine intuition that he cannot help uncovering secrets. Dechend is cut from the same bolt of cloth, but more irascible, in one instance taking a bevy of Sumerian scholars to task for the poverty of their translations, in particular for ignoring Venus (Tiamat).[14] The work of

[12] See Appendix 32 of *Hamlet's Mill.*

[13] Santillana, *Hamlet's Mill,* p. 282.

[14] Santillana, *Hamlet's Mill,* Appendix 14.

Santillana and Dechend succeeds because there is relatively little premature closure.

Amlodi (Hamlet) owned a fabulous mill which ground out peace and plenty. Later, however, it ground out salt and finally broke down, tumbling into the sea where it continued to grind out sand and rock. The fable echoes the degeneration of subsequent world ages of Greek and Indic cultures. Santillana takes the mill for a model of the Precession of the Equinoxes. It is a pattern for orthodoxy, that the changes of world ages are merely the slow, barely noticeable precessional ones. There is a germ of truth here, but the frame of the precession in both time and space, so well modeled in *Hamlet's Mill,* provides only the background for the alternate uniform and catastrophic movements of Earth. Velikovsky's historical catastrophism as described in these pages articulates the model and makes it cohere.

Although the forces of the mill are still at work in the present, during the Golden Age the mill was the visible axis bond of the Saturn Configuration. It was the "host-cruel skerry quern," "spindle of necessity," male fire stick "pramantha," incomparably mighty churn (Amritamantha),[15] Sanskrit Skamba, and its derivative, the Sampo[16] with its three roots, "—one in heaven, one in the earth, the third in the water eddy."[17] All of these are examined with great care by Santillana and Dechend, considering that they are nowhere to be seen in the present world. In the Golden Age, the axis bond was visible, described by the ancients again and again.

Santillana traces the legend through Finland, Iran, and India, bouncing through time and space, Earth and sky as one must in the study of myth. He correctly equates the age when the mill ground out peace and plenty to the Golden Age of Saturn, marks changes as one world age comes to an end and another begins, but always attributes these to the Earth's slow precessional movement.

Where *Hamlet's Mill* shines is in its treatment of the underworld and the confluence of the rivers of Hell, almost grasping why a ferryman is needed to get there. The book locates Hell or Eridu at the southern pole of the precession, spotted close to Canopus which would put it exactly opposite to the primeval Saturn as viewed in our skies. "Canopus was the weight at the end of the plumb line, as befitted its important position as a heavy star at the south pole of the 'waters below.' "[18]

We should pause here and recall that the Egyptians were aware of "heavy-weighing" Canopus. In their concern for bringing Heaven to Earth, they made their country a model of the cosmos, even naming it after the eye of Saturn. Plutarch remarks, "Egypt, . . . they call by the same name as the black portion of the eye, 'Chemia,' and compare it to a heart."[19] Following this pattern, they gave the name of Canopus to a town at the southern border of Egypt. Canopic urns were used to hold the ashes of the dead. Similarly, a southern town in Mesopotamia was called Eridu. Following Plato's image of a ship of war, Canopus was the steersman of the ship Argo on the search for the Golden Fleece.

There was a quartering of the heavenly image of Saturn in which four "rivers" are described as flowing out from the center.[20]

[15] Ibid., p. 381.

[16] Ibid., p. 233.

[17] Ibid., p. 205.

[18] Ibid., p. 73.

[19] Plutarch, *Isis and Osiris,* Vol. V, Moralia, 83.

[20] It is characteristic of the poetic Mesoamericans that they saw the quartered Saturn as the four-petalled kin flower. It is the 20th day sign, "flower" in the Aztec series, "lord" in the Maya series. The image is common throughout the culture, even today, often reduced to rectangular shape because that is the present "shape" of the world.

These four heavenly rivers can be found in the myth and legend of many cultures, often confused with earthly rivers, into which they were believed to flow. Ginsberg, in *Legends of the Jews,* reports the Ganges, the Nile, the Tigris, and the Euphrates.[21]

The rivers of Heaven.

[21] Ginsberg, *Legends,* p. 37.

Examples are multiplied in *The Saturn Myth* to include the four directional rivers of Navajo legend as well as those of the Sioux, Aztecs, Maya, Polynesians, Slavs, Scandinavians, Iranians, Mandaeans of Iraq, Scandinavians, Chinese, and Hindus. Documenting the converging primeval media of the Saturn Configuration can be tedious. At the roots of our own culture, the Bible tells us: "And a river went out of Eden to water the garden; and from thence it was parted, and became into four heads."[22]

This image of the quartered Saturn is the beginning of cross symbolism around the world, the quartering of the heavens brought to Earth to pervade the culture of mankind as an archetype. It is an addition to the concept of the center, reflected in the quartering of cities and temples, the orientation of structures and fields, in mandalas and the drawings of small children.[23] Nuttall has traced the phenomenon around the world, perceiving it to have come from the sky, although she looks to the pole of Earth rather than to the pole of the ecliptic.[24]

The rivers of Heaven are clearly described in primeval media, flowing outward from the Saturn Configuration, but where do they flow to? The answer is that they continue on through the celestial vault toward Earth, eventually intersecting the horizon. At this point we see why they were imagined to be the source of important rivers on Earth.

Consider, then, these celestial rivers perpendicular to the horizon, surrounding us on all four sides. Now we can see them as the four supports of Heaven, celebrated in the primeval media of many cultures: the four Bacabs of the Maya; the four pillars of Hawaiian myth, Tane, Rono, Tanaroa, and Tu; the four sons of Horus who both support the heavens and "tie the rope ladder"[25] of the cosmic axis. In India, Vishnu is credited with a similar power, as "... with beams of light thou didst hold fast the earth on all sides."[26]

Returning to the ceiling of Senmut's tomb, we see these four sons of Horus following behind Tau-ert (Draco) and Isis in the lineup of "northern" gods: the flop-eared dog Hapy, the human Imsety, the jackal Duamutef, and the hawk-headed Kebsenuf. It begins to look as if these gods are rather polar than northern. At any rate, their position suggests close connection with this primeval order of the cosmos.

The four sons of Horus mirror Ixion spread-eagled on the cosmic wheel. "Your hands are Hapy and Duamutef ... Your feet are 'Imsety and Kebsenuf.' "[27] "... they are supporting themselves on their dcm-sceptres ..."[28] These would be the beams of light or rivers of Heaven.

The four sons of Horus (left to right): Hapy, Imsety, Duamutef, and Kebsenuf.

[22] Genesis 2:10.

[23] David Griffard, "Myth, Mandala and the Collective Unconscious," *Kronos* I, 1, pp. 27–32.

[24] Nuttall, *Fundamental Principles,* passim.

[25] Faulkner, *Pyramid Texts,* Utt 688.

[26] Eggeling, trans. Satapatha Brahmana (Delhi, 1963) 3 Vol. 5.3.14. Cited by Talbott.

[27] Faulkner, *Pyramid Texts,* ss 149, Utt 214.

[28] Wainwright, "Letopolis," *Journal of Egyptian Archaeology* XVIII, 1932, p. 171.

Imsety, Hapy, Duamutef, and Kebsenuf are echoed as the "four great beasts" of Chinese and Biblical mythos, and the concept appears to devolve on the four Evangelists as well.[29] The child's prayer:

Matthew, Mark, Luke and John,
Bless the bed that I lay on

raises up the "four poster" of the primeval cosmos.

Thus we have the rivers of Heaven and the rivers of Hell, the former clearly described, the latter somewhat less so. This netherworld and its environs are shadowy even to classical civilizations, known only from the poorly understood visits of gods and heroes. Socrates calls the rivers of Hell by name: Oceanus, Acheron, Pyriphlegethon, and Cocytus, in a relation marred by a confusing description of how they behave on Earth.[30] What is clear is that they are opposed to each other in pairs—Oceanus to Acheron, Pyriphlegethon to Cocytus—and meet in the ultimate pit at the confluence of the rivers, thus giving a quartered appearance to Hell. And having established a Heaven and a Hell quartered by celestial rivers, we will do the obvious and link up these rivers, allowing them to flow uninterrupted between the poles of the ecliptic.

Children of Western Civilization are well accustomed to dealing with imaginary lines. Using geometry and trigonometry, we divide up our world with them, some having a stronger basis in reality than others. Since we are familiar with the imaginary axis of the rotating Earth, it is easy for us to imagine another axis (hinge) running through the equinoxes (points on the globe where equator and ecliptic cross) and allowing the Earth to tilt on it. Having worked our way through a few tilts of the planet, we begin to feel that some kind of force is expressed there, something that holds the equinoctial hinges steady while the rest of the Earth tilts on it. In the case of a precessional spin/slide, the Earth moves rapidly, but only along its preordained path, thus exhibiting another kind of stable force. That these precessional hinges are in constant, uniform motion is a trifle disconcerting (timekeepers would say inconvenient), but thus is our world put together.

Astronomers and those entrusted with terrestrial navigation must keep more accurate time than the rest of us mortals if they hope to stay correctly oriented on the sea and in the cosmos. Astronomers do this by keeping clocks in their observatories regulated to sidereal (star) time, which is the real thing. If we were at the Spring Equinox and the Earth were to tilt, it would do so on the points on the equator where the equinoctial colure was registered.

To keep star time all over the world, astronomers calculate from the equinoctial and solstitial colures, imaginary lines, great circles that run north and south on the celestial vault of the planet Earth. The equinoctial colure passes through both equinoxes and both poles of Earth. The solstitial colure passes through both solstices, both poles of Earth, and both poles of the ecliptic. And having said that, the colures suddenly begin to seem more real, not imaginary at all in relation to the catastrophic events discussed within these pages. All the critical points we have been talking about are on the colures: the poles of the ecliptic, the poles of Earth, and equinox and solstice points. Far from just marking time, the colures begin to look like lines of force that determine how the Earth moves in catastrophic interactions.

Plato describes how the world is held together in the tenth book of the *Republic.* Efforts to relate this vision to the present order of the cosmos are vain, but let us superimpose

[29] O'Neill, *Night of the Gods,* pp. 184–188.

[30] Plato, *Phaedo,* 107D–115A.

the image on the Saturn Configuration with its armature of bright heavenly rivers.

"... a straight shaft of light, like a pillar, stretching from above throughout heaven and earth." Immediately we recognize the axis bond.

"... and there at the middle of the light [Saturn], they saw stretching from heaven the extremities of its chains [desmoi]; for this light binds the heavens, holding together the revolving firmament like the undergirths of a ship of war."

We pause to review ancient nautical design. Planks were laid from a keel, edge to edge or slightly overlapping, and fastened along their length and at stem and stern. The method produced a strong, integral hull, but nothing to give adequate support to a mast. Consequently supports for a mast ran under the hull, beneath the ship, probably ropes at first, later bands of metal. Ropes could then be attached to steady the mast which was stepped on the keel. Given the "mast" of the cosmic axis, the "ropes" of the colures, the "ship" of Earth, we can perceive a workmanlike metaphor for holding the firmament together.

The historian of astronomy J. L. E. Dreyer sees the "chains" as the colures. "The ligatures (desmoi) of the heavens are the solstitial and equinoctial colures intersecting with the poles, which points therefore may be called their extremities (akra)."[31] Dreyer naturally sees the colures as primarily intersecting the poles of Earth, as indeed they do today. But during the Golden Age, they also ran through the pole of the ecliptic. The solstitial colure still does, and the equinoctial colure has been working its way back up to the pole of the ecliptic during every one of the last three tilts of Earth.

Whatever force once held the Earth upright between the poles of the ecliptic appears to be still at work. Witness both the catastrophic tilts and the present, slight upward-tilting motion of the Earth's axis well known to astronomers.

"And from the extremities stretched the Spindle of Necessity, by means of which all the circles revolve." The "Spindle of Necessity" is identical with the "straight shaft of light" of the axis bond. The circles could be the bands surrounding the larger planets.

Plato's model is thus congruent to the Saturn Configuration: the "chains"/colures binding the heavens equal to the "rivers" of Heaven and Hell, the shaft of light or Spindle of Necessity equal to the axis bond. The armature of the colures, once it is grasped, easily lines up space and time in the series of cyclical catastrophes described herein.[32] En passant, since it is the hawk-headed Kebsenuf who is "spearing" the Bull of Heaven on the Senmut ceiling, we may conclude that this particular son of Horus represents the half of the solstitial colure that Saturn slid down, also known as the river Eridanus. In the background as the heavenly river approached Earth's horizon was the Milky Way.

Taken all together, Plato's description of how the world is held together fits the Golden Age of Saturn rather than the present one, although the same forces are still expressed in the Precession of the Equinoxes and the slight upward-tilting action of Earth's axis. We read in textbooks that the precession is an effect of the sun or the Moon and planets, but we begin to suspect otherwise. It appears that there is a very real magnetic force expressed at the poles of precession, above within the central curve of Draco, and below near the star Canopus. And this force is what holds the Earth in its precessional spin.

That the forces that hold the world together

[31] J. L. E. Dreyer, *A History of Astonomy from Thales to Kepler* (1953) p. 56ff.

[32] Another useful image of the armature of the colures attached to the axis bond is found in the *Pyramid Texts*, where the pole is referred to as "the foremost of its flagstaffs," with a note indicating "a pole with cords or stays attached." Trans. Faulkner, Utt. 519, note 13.

were visible during the Golden Age was probably due to a greater primeval force. Anyone who has seen lightning or witnessed the aurora borealis is aware that natural electromagnetic effects can be visible. The ancients certainly did their best to describe the appearance of the heavens.

The end of the Golden Age of Saturn, the Fall from the Garden of Paradise, has historically been considered to be the ultimate cataclysm in all quarters but those of scientific-minded modern man. In the following chapter we will review the evidence for exactly how the Bull of Heaven, Saturn, met his death.

Bronze amulets representing the quartered Saturn. (From the archaeological museum in Prague.)

The lion-bull combat: "... one of the oldest motifs and undoubtedly the most tenacious traceable in the history of Near Eastern art."

19. Killing the Bull

There was then a bull in Pohja,
Which had grown to size enormous,
And his sides were sleek and fattened,
And his sinews from the strongest;
Horns he had in length a fathom,
One half more his muzzles thickness,
So they led him from the meadow,
On the border of the ploughed field,
Up they plowed the roots of Sampo . . .

– Kalevala

The brilliantly conceived tauroctony of the Mithraic Mysteries has led us into the past, illuminating planetary catastrophes along the way. The short-lived equator of crow, cup, snake, and dog, bracketed by visits of Mars, illustrates the most recent "death" of the bull Taurus, replaced by Aries at the Spring Equinox. The scorpion seizing the bull's testicles represented another death of the bull as the two constellations exchanged solstice positions, winter for summer, at the first visit of Venus. And then there was Taurus struck in the shoulder (Pleiades) by Mithras-Perseus at the end of the Silver Age.

Each of these bull-slayings mirrors the original death of Saturn, the Bull of Heaven, the catastrophic event that initiated the series of "fallen" world ages. And this most important death is only mirrored in the tauroctony, for a mystery religion, after all, must keep its secrets. The deepest mysteries can only be hinted at and the tauroctony has both revealed and concealed for more than 2,000 years.

Other Mithraic icons are more explicit on the end of the Golden Age, though most have a remarkably static look about them. Mithras handing over the cosmic pole to the sun is a good example.[1] Who would guess that the initiation of a new planetary order lies behind this rather formal-looking transfer of a symbol? There would be no more Saturn Configuration. Our sun would now be at the center of the new order.

Other iconography treating this event are: Saturn handing over scepter and thunderbolts to Jupiter[2] as Silver Age follows Golden Age, the myth of Phaethon,[3] and Mithras shaking hands with the sun.[4] Investiture scenes mirror the beginning of a new age, as in CIMRM 1430 where Mithras has his left hand on the head of the sun while holding the bull's shoulder, Ursa Major, aloft with his right.

As a god, Mithras was lord of genesis, capable of catastrophic destruction while yet holding the cosmos together. He was known to have his seat of power both at the pole and the equinoxes where he exerted force to bring one age to a close and begin another. He is, therefore, the quintessential bull slayer.

Another intriguing Mithraic figure is the lion-headed god, frequently depicted standing on a globe. It can be seen as an artistic metaphor for the Saturn Configuration: the regal, upright human figure implying the axis bond, the monstrous head as the planet itself, the surrounding coils of a snake as the Saturnian band, the double set of wings that can move the axis in two different styles. Scepter and

[1] Maarten Vermaseren, *Corpus Inscriptionum et Monumentorum Religionis Mithriacae* (The Hague) 650.

[2] Ibid., 1283.

[3] Ibid., 1247.

[4] Ibid., 1430.

thunderbolts are frequently present, and what is often thought to be a key, but in some instances is clearly a square—Saturn's measuring instrument.[5] Cumont connected the figure with Zurvan, the Iranian god of time. Mithraism, following Cumont, has generally been assumed to grow out of the Iranian cult of Mitra.

The lion-headed god can be compared to Orphic Phanes and the Hellenistic god of time, Aeon.[6] All three have a similar iconography, often including the time context of the zodiac. The lion head is also reminiscent of the head of the Gorgon, Medusa, the curly lion mane similar to her lively snake hair. Ulansey has pointed out other similarities as well as explicit links between the two, noting "the remarkable existence of one statue of the lion-headed god wearing on its chest a Gorgon head."[7] We will presently examine the possibility that the hairy curls and snakes grew out of celestial phenomena.

When we pursue the derivation of the Gorgon's name, an aspect of Medusa much closer to the lordly lion-headed god reveals itself. The Greek verb μεδο means "to rule over or protect." Μεδων or Μεδωντοσ is "ruler or lord," of which the feminine is Μεδουσα or Medusa. Thus, despite her most familiar image, Medusa begins to appear other than a simple monster. The protruding tongue can be compared to that of the Egyptian dwarf Bes, or of the central deity on the Aztec calendar stone. The tongue of the Lord is a two-edged sword in Revelations. But the monstrous image known to us from the myth of Perseus is not the only sort of Gorgon depicted in ancient art.

The ugly Greek version of Medusa stands in opposition to the "vegetation Gorgoneion" whose tradition is traced by Frothingham in the urns, sarcophagi, and tombs of the Etruscans and Romans.[8] Cultures other than our own sometimes associate death with growth as indicated by the vegetation. Still, the other sort of Gorgoneion is a surprise, a pleasant round face, a few snakes, and much greenery. As a vegetation goddess, she can be admitted to the company of Adonis and Tammuz, alter egos of Saturn.

In his "Medusa, Apollo and the Great Mother," Frothingham came to the conclusion that "Medusa was not an evil demon or bogey, but primarily a nature goddess and earth spirit of prehistoric times identical with or cognate to the Great Mother, to Rhea, Cybele, Demeter, and the 'Mother' Artemis."[9] Through Talbott's work on the unity of the Great Mother as the primeval Saturnian band, we can now celebrate her with the intuitive Graves, who has always known the Goddess was one, despite her many names. And now even Medusa claims admission to that great feminine masonry.

Clark Hopkins, in "The Sunny Side of the Greek Gorgon,"[10] examines the solar aspects of Medusa, linking her with the Egyptian winged solar disc and accompanying uraeus (snake),[11] a symbol that was widely accepted in other Mediterranean cultures. "It is tempting to suggest that the averted countenance of Perseus represented originally the inability to look at the sun."[12] Hopkins has nearly got

[5] Ibid., 3122. See discussion in *Hamlet's Mill* of Saturn giving all the measures.

[6] Ulansey, *Mithraic Mysteries,* pp. 116–124.

[7] CIMRM, 1123.

[8] A. J. Frothingham, "Medusa II," *Am. Journal of Archaeology,* Vol. XIX (1915) No. 1.

[9] Frothingham, "Medusa II," p. 349.

[10] Clark Hopkins, "The Sunny Side of the Greek Gorgon," Berytus (Beirut, 1961) Vol. XIV, p. 25.

[11] Talbott, *The Saturn Myth.*

[12] Hopkins, "Sunny Side," p. 30.

it right, but it was not our sun but the sun of Saturn that became terrifying to look at.

Frothingham also sees Medusa as an emblem of the sun disk,[13] and goes on to connect her with the sun god Apollo through coins, art, and literature. He makes his case for Medusa's identity with other goddesses largely through their association with animals—birds, snakes, and horses—in a similar art and iconography. Medusa and Artemis share a particularly strong horse association. There was a cult of mare-headed Demeter (Demeter of Phigaleia) and Medusa is often a hippo (horse)-Gorgon. The heads of both goddesses can be flanked by horse's heads.

The Gorgon Medusa has three general identities then: a sun symbol, a vegetation goddess, and a horrible monster. She was all three, but at different times. A close reading of the Perseus myth will help us sort them out. Her attractive period as a vegetation goddess would have been when Poseidon courted and impregnated her in Athena's temple. Later, when Athena discovered this violation, she became angry and transformed Medusa into an ugly monster.

But Medusa was not just a sun symbol, but the sun of Saturn throughout: Saturn the beneficent ruler of the Golden Age, Saturn the vegetation god shining on growing things day and night, but later Saturn the changeful, exhibiting unusual frightening behavior immediately prior to his "death" at the break-up of the Saturn Configuration. In the behavior and appearance of Medusa and other figures in the myth of Perseus, we have clues as to what happened to Saturn at the end of the Golden Age.

Poseidon impregnating Medusa in the temple of Athena suggests an intrusion of some kind within the visible goddess, the Saturnian band. In myth, Poseidon becomes a bird to mate with Medusa,[14] apparently the bird of the illuminated crescent of the Saturnian band. Athena, then, expressed her anger by transforming Medusa into an ugly monster with a face that could turn the viewer to stone, or rather Saturn into a terrifying image that petrified the men of the Golden Age with fear. Their god/goddess had turned into a monster!

Athena's (Venus') anger was expressed in her spinning off the axis bond between Earth and Saturn, spiraling outward and leaving a visible trail behind her, obscuring and changing the face of Medusa-Saturn. As the spiral continued outward through the seven circles of the Saturnian band, it produced the labyrinth in which the Minotaur (the bull-headed man) was hidden.[15] Then the goddess "hung" for a time from the configuration just as the serpent "hung" from the wall surrounding Paradise.

The material trailing from Venus would not remain static, but move and change, become snake-like at the edge of the Saturnian band. In the center, the spiral would shift into the entrail-like face of Humbaba or Tlaloc. It was the beginning of the dissolution of the Saturn Configuration, frightening enough to petrify its viewers!

Eventually, of course, Earth would leave the old configuration as the axis bond dissolved and Heaven and Earth were separated.[16] The event entered myth as Perseus decapitating Medusa with his curved sword or harpe. This can be easily imagined as a movement of Saturn, apparent or real, with its illuminated

[13] Frothingham, "Medusa, Apollo, and the Great Mother," pp. 367–370.

[14] O'Neill, *Night of the Gods*, p. 144.

[15] Mars must have spun off the axis bond some time after Venus, as Ariadne (another hanging goddess) left a trail of thread to lead Theseus out of the labyrinth.

[16] We must assume that the Earth would rapidly follow the other smaller planets into an orbit around the present sun, the larger gas giants perhaps moving last.

crescent seeming to cut it off from the axis bond.[17] The harpe with its crescent-shaped sickle blade is the weapon of both Saturn and Perseus,[18] and was brought to Earth as a weapon although the shape is not particularly useful in combat. The Gorgon's head goes into Perseus' magic wallet (i.e., disappears in outer space), and, of course, Perseus has never actually been seen since he was wearing Hades' helmet of invisibility. And Earth falls over on its side into the ecliptic.

Now Pegasus and Chrysaor, engendered by Poseidon, sprang from the neck of Medusa (axis column). They are a part of the new order of the cosmos. Pegasus is the marker at the new pole, the horse's head of the Hyades. Following the myth, he was named after the springs of ocean (source of the four heavenly rivers), and the Hyades are slightly below the ecliptic, thus in the Watery Abyss, and only the horse Dadhyank drinks this water.[19] This myth agrees with the Vedic sacrificial horse being born from the "eye" of Prajapati,[20] an epithet of Brahma or Saturn.[21]

Both Frothingham and Hopkins feel that Chrysaor may be Apollo, ". . . not necessarily a new sun in the sky but a legitimate candidate if one were required."[22] Chrysaor has always been a shadowy mythological figure, but now bears further investigation as representative of our present sun. Indeed, Chrysaor is a Homeric epithet used for Apollo.[23]

That Perseus was never actually seen to decapitate Medusa is significant. He only appeared at the pole (as a constellation) in the beginning of the Bronze Age after "killing" the bull Taurus. Apparently he was then presumed to have also "killed" the bull Saturn. The involvement of Venus-Athena in the myth is critical. In some versions, it is she who kills the Gorgon. Medusa's head, the Gorgoneion, is fastened to Athena's aegis, suggesting that Venus took some of the brilliance of Saturn with her. Certainly this was the beginning of her long cometary period.

Returning to the lion with his curly mane as an analog of Saturn with his face obscured by the emanations of the spinning-off Venus, we recognize yet another bull-slayer. As it happens, ". . . one of the oldest motifs and undoubtedly the most tenacious traceable in the history of Near Eastern art,"[24] is the lion-bull combat (see page 174). Once we realize that the "combat" took place in the sky, it begins to read out, the lion-like apparition obscuring and "combating" the Bull of Heaven, Saturn.

To the list of bull-slayers headed by Mithras, we now add Perseus, the lion, and next, three animals, all represented by the Hyades. In the tradition of the Cuna, ". . . the tapir chopped down the 'Saltwater Tree' at the roots of which is God's whirlpool, and when the tree fell, saltwater gushed out to form the oceans of the world."[25] To a culture minus bull or horse, the tapir must serve. "The Arawak of Guyana call

[17] As the illuminated crescent of Saturn moves off, it is seen as the king flying away in the form of a bird. *Pyramid Texts,* Utt. 461–463, 890–891, 913, 1048.

[18] The "Flaming Sword" of god, hereb, is the harpe. O'Neill, *Night of the Gods,* pp. 687–688.

[19] Santillana, *Hamlet's Mill,* pp. 393–394.

[20] Allbright, "Indic and Babylonian Sacrificial Ritual," p. 110.

[21] Dwardu Cardona, "Child of Saturn," part II, *Kronos* VII, 2, p. 35.

[22] Frothingham, "Medusa, Apollo, and the Great Mother," p. 357. Also, Hopkins, "Sunny Side."

[23] Karl Kerenyi, *Apollo* (Dallas 1983) p. 52.

[24] Willy Hartner, "The Earliest History of the Constellations in the Near East and the Motif of the Lion-Bull Combat," *Journal of Near Eastern Studies,* Vol. XXIV, 1 & 2, p. 1.

[25] Santillana, *Hamlet's Mill,* p. 247.

our galaxy 'the Tapir's way.' "[26] And the Hyades marked the pole of the Silver Age which was within the Milky Way.

Chopping down the saltwater tree recalls the mill of the gods grinding out salt, as well as echoing the Mahabharata. "—It is for this curse that the waters of the Ocean to this day are saltish to the taste and are drunk by no one else than the Equine head."[27] The oceans of Earth are certainly salty, but here they are an analog of the plane of the ecliptic, surface of the Watery Abyss. Since the equine head of the Hyades is slightly below the surface of this space ocean, it can be understood to drink its waters.

It would appear that the salt referred to in so many myths was added to the waters of Earth at the time when the "rainy" Hyades were at the pole. As Velikovsky remarked, the enormous amount of salt in our oceans did not come from the rocks of Earth, since they are very poor in chlorine.[28] An extraterrestrial source is indicated.

In Finland's Kalevala, only the enormous "bull in Pohja" was able to plow up the roots of the Sampo or cosmic mill, with its roots ". . . firmly, in the depths nine fathoms under."[29] Note that the bull was led "from the meadow, on the border of the ploughed field, . . ." The "ploughed field" refers to the surface of the Watery Abyss; the border would be the circle of the zodiac with the Hyades of Taurus representing the bull. The bovine image and plow are actually repeated earlier in the Kalevala as a heifer with golden horns and a sun disc on her head who is replaced by a plow,[30] again the V of the Hyades. The metaphors are mixed, but the message is clear. The bull of Taurus killed, or replaced, Saturn, the Bull of Heaven. The connection between the two is clearly depicted in the Egyptian zodiac.

Taurus of the Egyptian zodiac carrying Saturn, the Bull of Heaven, on his back.

Adonis (Tammuz or Saturn) is generally held to have been slain by a wild boar, the "boar's head" of the Hyades. And this goes hand in glove with the sacred king being killed by his tanist (twin and successor) in boar disguise.[31] Nonnos introduces Mars into the equation: ". . . in the shape of a wild boar, Ares with jagged tusk and spitting deadly poison was destined to weave fate for Adonis in jealous madness."[32]

Mars' critical position in the Saturn Configuration, moving back and forth between Earth

[26] Ibid., p. 246.

[27] See discussion in Santillana, *Hamlet's Mill,* pp. 393–394. The quote is from the Mahabharata. Santillana gives Mbh. 12.343 (Roy trans.) Vol. 10, p. 583.

[28] Velikovsky, "On Saturn and the Flood," *Kronos* V, 1, p. 8.

[29] Santillana, *Hamlet's Mill,* pp. 105–106. Compare Lonrott, *Kalevala,* trans. Magoun (Cambridge, Harvard, 1963) Rune 42.

[30] Ibid., p. 101. Compare Lonrott, Rune 10.

[31] Graves, *The White Goddess* (New York, Farrar, Straus and Giroux) p. 210.

[32] Nonnos, "On Aphrodite," 41.208ff. Also *Movers,* Vol. I, p. 222ff.

and Saturn,[33] could easily have given him the appearance of responsibility for the downfall of the Golden Age, threatening the Saturnian orb each time he approached it. The myth of Vucub Caquix in the *Popol Vuh* casts light on this event.

In the study of myth one is occasionally struck by the shadow of human consciousness. The story of Vucub Caquix evokes this presence strongly. As he sits on his silver throne in the ceiba tree, the primeval Saturn, Vucub Caquix, is accused of arrogance, of "pretending to be the sun." His limitations are pointed out, for "he can only see as far as the horizon." This singular hubris puts us in mind of Aeschylus speaking through Prometheus who is suffering torments for bringing fire to man. "He is a monster . . . I care less than nothing for Zeus. Let him do as he likes." [34] But who would dare to put down the ruler of an age? Only one who had grasped the phenomena of the skies as being subject to discernible laws.

Vucub-Caquix receives his comeuppance at the hands of one of the Magic Twins; Hun-hunapu shoots him with a blow gun, knocking out his eyes and teeth. The inhalation and expression of breath can be heard in the pronunciation of hun "ah-pu," the Maya languages being somewhat onomatopoetic. The myth is depicted in the Ceramic Codex,[35] the bird Seven Macaws sitting in a tree, Hunhunapu crouching below with his blow gun. What is being recorded here is Mars in one of his aspects as Magic Twin, moving up through the axis column as the dart moves through the tube of the blow gun, shooting Vucub Caquix/Saturn in the "eye."

Mars' weapon is well known to be the spear, or occasionally the dagger. The myth of Vucub Caquix would be the only one associating him with the dart of a blowgun. Gods with a dart or arrow through the eye can occasionally be seen in the iconography of the New World,[36] although the motif, like all those connected with the Saturn Configuration, can degenerate into one of mere kingship. Such images should always be examined for other marks of Saturn.

Another bull-slaying we will consider is described in the Greater Bundahisn, thought to be a Zoroastrian astrological document, "the horoscope of the world," though it has more the flavor of a celestial catastrophe. In the Bundahisn, the Primal Ox (Saturn) is given a merciful death at the "onslaught of Ahriman." [37] This onslaught results in what appears to be a lively battle in the sky which we will consider as a description of objective phenomena. Although it would not have been possible, it is interesting that the event was said to have occurred at noon on the vernal equinox "when day and night were equal." [38]

We have been positing planets spinning off from the Saturn Configuration at its dissolution, taking up orbits around the present sun. This may be what is described in the Bundahisn at the onslaught of Ahriman ". . . when the aggressor rushed in . . . Those planets, when they rushed into the firmament in this manner, fell to striving with the fixed stars . . ." The planets were seen as evil, probably because

[33] The situation of Mars in the Saturn Configuration will be examined in "Mars, The Magic Twin."

[34] Campbell, *Myths to Live By,* p. 827.

[35] Francis Robicsek, *The Mayan Book of the Dead, Ceramic Codex* (Norman, Okla., 1981) p. 83, vessel 109.

[36] Epigraphic Society Occasional Papers, Vol. 21, pp. 165, 167.

[37] D. N. Mackenzie, "Zoroastrian Astrology in the Bundahisn," *Bulletin of Oriental and African Studies,* Vol. XXVII, part 3, p. 528. Mackenzie gives GB, IV, 20=43.11, Zadspram, ii, 9.

[38] Ibid., p. 513, V.A. [51.1] (1). Although Saturn dominated the sky in the Golden Age, the sun of today was also present, possibly used for marking time.

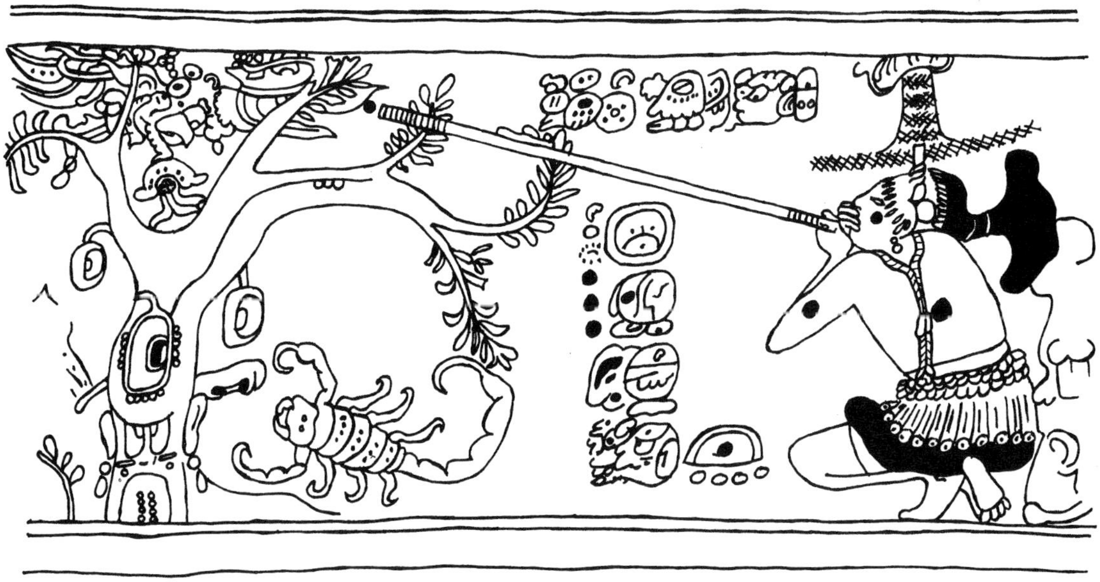

Hun-hunapu shoots Vucub-Caquix (Saturn) with a blowgun.

they appeared to upset the previous order. "All the arrangements of time, which is obvious to the eye, is connected with the fixed stars . . ." [39]

The account of the Bundahisn, seeing the planets as intruders on the cosmic stage, betrays an irritation and annoyance that would surely not have been directed against the planets had they always behaved thus. Two unusual images are described: the dragon Gocihr and the tailed Mush Peri. They are "opposed" to the sun, Moon, and stars. The commander-in-chief of the planets, Saturn, is "opposed" to the "peg in the middle of the sky."[40] This is generally supposed to be the pole star, but what we have learned about precessional forces suggests that it is the pole of the precession.

The opposition of the dragon Gocihr to sun, Moon, and stars appears to be that it follows a revolutionary path around the sun contrary to that of the planets. Gocihr is described thus: "The dragon Gocihr stood in the middle of the sky like a serpent, its head in the two images (Gemini) and its tail in the Centaur (Sagittarius), so that at all times are six constellations between its Head and Tail; and its running is retrograde (so that) every ten years the Tail reverts to where the head (was) and the Head to where the tail was." [41]

The description of Gocihr echoes that of Leviathan in the cabalistic *Book of Raziel.* Leviathan, well known from mention in the Bible, is described as having signs of the zodiac hanging from it. Apparently this sky monster was laid out in the sky just above the zodiac. But Leviathan is thought by some to have the sense of "twisted" or "wreathed," [42] which would recall the Saturnian band with spiral Venus trails overlaying it.

Here is a forceful image operating just above the plane of the ecliptic in a direction opposite to that in which everything else is moving, a difficult image to explain, but a possible solution for a handful of bodies in the solar system that move in retrograde fashion: moons of Jupiter, Saturn, and Neptune. The description of the dragon Gocihr is believed by orthodoxy to be an imaginative creation behind the discovery of the lunar nodes which

[39] Ibid., p. 516, V.A. [53.14] (9).

[40] Ibid., p. 513, V. [49.12] (4).

[41] Ibid., pp. 515-616, V.A. [52.12] (5).

[42] Oxford English Dictionary.

govern eclipses, a cycle of 19.5 years. But because of the historical catastrophes unveiled by Velikovsky, the present orbit of the Moon cannot be retrocalculated beyond the Earth-Mars Wars of the 8th and 7th centuries B.C.

There are a few other candidates for bull slayer. Eve did pick the apple (the red Mars) from the cosmic tree. In Mexico gods plucked the forbidden flower, the four-petalled kin flower at the pole, and Zeus threw his father Saturn out of the sun chariot. Another echo of this last bull-slaying is recorded in the Greek festival of the Bouphonia, literally "murder of the ox," which was celebrated at Athens. The ox was slain by an ax wielded by a priest of Zeus.[43]

In his groundbreaking *Saturn Myth*, David Talbott identifies the three-pronged Hindu trident as symbol of the cosmic column.[44] But the trident is also the weapon of Poseidon, impregnator of Medusa, and here the myth of the Gorgon and that of Vucub Caquix can be seen to reinforce each other.

Recalling that the axis bond and Saturn Configuration has been seen as an image of coition, Mars moving up the cosmic column is analogous to the dart traveling through the blowgun of the New World, and the impregnation of Medusa by Poseidon of the Old World (an ejaculation through penis to vagina). The Greek vision is a particularly positive one, a generative event with the promise of rebirth.

The vision of triplicity generated by the "trident" of Poseidon[45] relates strongly to the breakup of the Saturn Configuration. As we will see in the following chapter, gods, goddesses, the sky itself divides in thirds in the Silver Age. The vision was apparently compelling.

The death of the bull and the transition from the Saturn Configuration to the present order of the solar system has left a few other traces, both in the culture of the ancients and in our own. Since Taurus replaced Saturn at the pole, it is not surprising that Plutarch mentions the celebration at the 30-year sidereal cycle of Saturn when it enters the constellation of Taurus.[46] The Egyptian "sed" festival was performed 30 years after the enthronement of Pharaoh to renew energy,[47] and Persian shahs held their Jubilee after reigning 30 years.[48]

In reviewing the history of Earth in the cosmos, it begins to look as if today's astrology is a degraded astronomy, a half-remembering of very real planetary influence against the background of the signs of the zodiac or precessional clock. A particularly difficult astrological influence is held to be the one we must deal with at age 30; it is that of the "Saturn return."

[43] Mircea Eliade, *History of Religious Ideas* (Univ. of Chicago, 1978) Vol. 1, p. 450.

[44] Talbott, *The Saturn Myth,* pp. 244–245.

[45] The two outer prongs of the trident would have been the crescent horns of the Saturn Configuration.

[46] Plutarch, "De facie quae in orbe lunae apparet," 941.

[47] Henri Frankfort, *Kingship,* pp. 82ff.

[48] Santillana, *Hamlet's Mill,* p. 129.

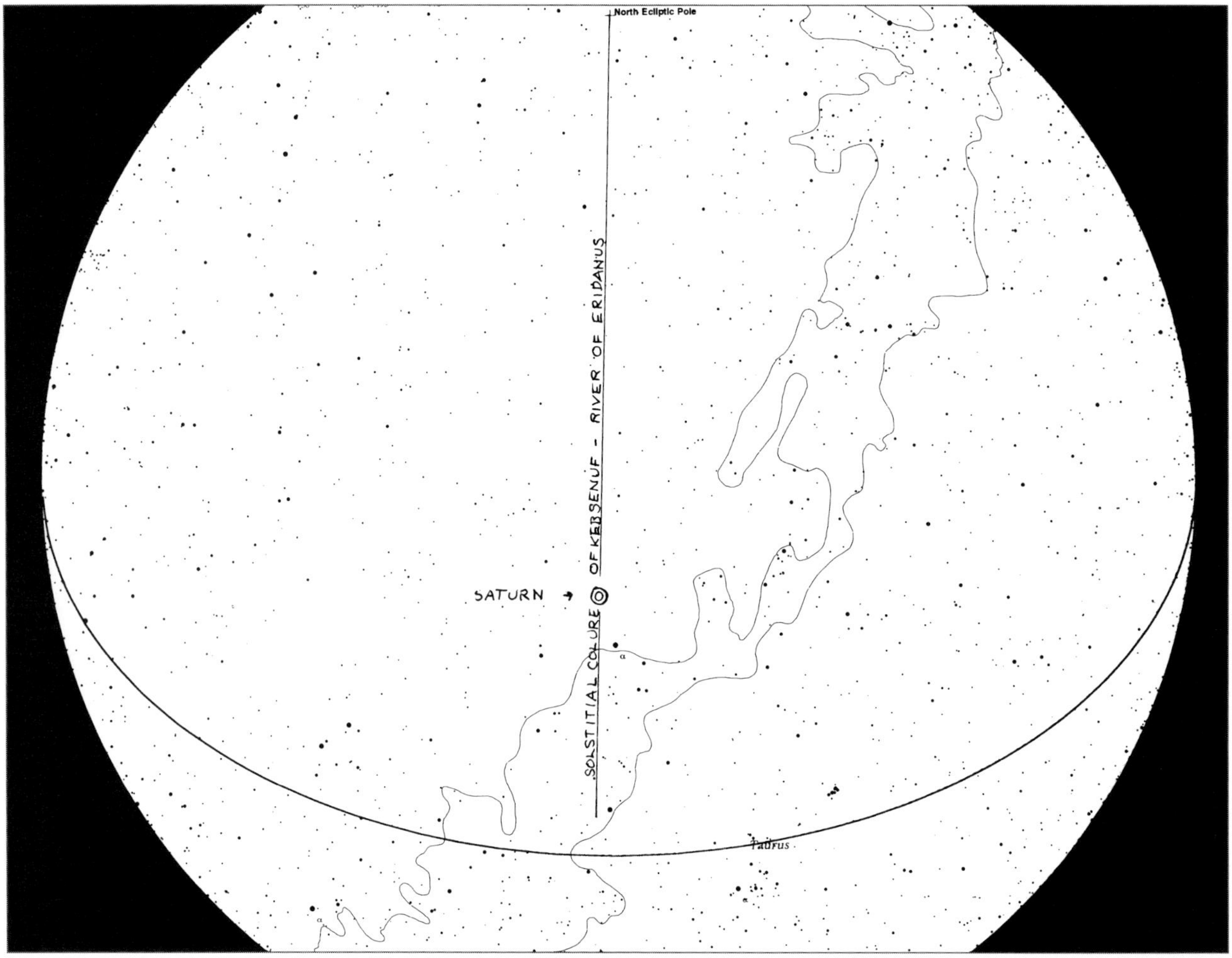

Saturn (Phaeton) slides down the solstitial colure of Kebsenuf (the river of Eridanus) through the Milky Way to a spot within the horns of Taurus, the Babylonian constellation Chariot of Enmesharra.

20. The Fall and Other Myths of Beginnings

Then kingship descended from heaven.

– Sumerian tablet from Nippur

Some say that the Milky Way is the path taken by one of the stars at the time of the legendary fall of Phaethon.

– Aristotle, Meteorologica

Orion, turn thy head that thou mayest see Osiris!

– Wainwright, "Orion and the Great Star"

. . . but the wide pathed earth yawned there in the plain of Nysa, and the lord, Host of Many, with his immortal horses sprang out upon her—the Son of Kronos, he who has many names . . . And Hades seized her and took her loudly crying in his chariot down to his realm of mist and gloom.

– Hesiod, Homeric Hymn to Demeter, "The Abduction of Persephone"

Therefore the Lord God sent him forth from the garden of Eden, to till the ground from which he was taken.

– Genesis, 4:23

With its dissolution signaled by movements of Jupiter, Venus, and Mars, the Saturn Configuration continued to fly apart. The pole of Earth tilted into the ecliptic, pointing to the Babylonian constellation Chariot of Enmesharra, while Saturn (Phaethon) reacted to the same force, appearing to slide down the heavenly river of Eridanus toward Hell and thence into outer space. Santillana suggests that Zeus established the equinoctial colure by hurling down Phaethon (Saturn),[1] but Homer knows better: ". . . the road to Hades lies near the gates of the sun."[2] And the gates of the sun are the solstices.

So the solstitial colure defined the line of force at the Fall and both Saturn and Earth followed the path. If Orion had turned his head, he could have watched Osiris (Saturn) fall[3] through the gates of the sun along the solstitial colure. Earth, meanwhile, tilted into the ecliptic, its North Pole following the same colure until it came to rest in the constellation Taurus.

From the Golden Age of Kronos, father of time, the Earth fell into time as we know it, into the months of the Moon, into the years and seasons of its own path around the sun, into the slow spin of the precessional motion. It was a time of beginnings, a rich, dense period of myth-making. Earth and Heaven were new, enmeshed in new, strange cycles, and everything had to be described, to be set

[1] Santillana is under the spell of Taurus at the Spring Equinox.

[2] Homer, *Odyssey*, 11–14.

[3] Orion's attention, as depicted artistically, appears to be determined by whether the constellation is in the northern or southern hemisphere.

Saturn, the Bull of Heaven, topples from the mooring post.

in order, to be spun into an intelligible matrix out of which life could proceed.

The Fall has long since passed into metaphor, but it was a very real one. Not in the gravitational sense of an apple falling from a tree perhaps, but a tilt of Earth to what must have seemed like a lower position. From being connected to a visible god above us, we fell into a state of dissociation. For some it was a fall from grace; for others, it was the gods who fell. The Golden Age was lost to us, to be enshrined in every culture as the best of all possible "good old days." In the Silver Age, we would have to catch our own rabbits, and under the most extreme conditions.

With the terrestrial axis almost parallel with the ecliptic, the seasons begin, but with a vengeance. The sun describes a course so far to the north and south that it passes close to the zenith of both poles. The seasons are eccentric in the extreme. To northerners—and most men were northerners—winter brings long darkness, summer endless light. From the northern hemisphere, one could watch the sun gradually spin down the vault of the sky over months until it was gradually circling the horizon, finally disappearing from view.

Now begins the great southerly migration of mankind. The age of hunter-gatherers commences as men follow the sun in its extreme path north and south. Only in regions relatively close to the equator would there have been a climate sufficiently stable for permanent settlement. But even there food would have to be stored against the extreme positions of the sun in winter and summer, as well as against the needs of the hunter-gatherers. Since "day" and "night" occurred but once a year, it is small wonder that they are so strongly emphasized, say, in the early Egyptian culture.

We are familiar with the birth of the seasons in the Greek myth of Persephone, who was abducted by her uncle Hades with the full knowledge of Zeus.[4] Hades carried her off to the "hidden places of Earth," rather than to Hell or Tartarus. That it is also a "realm of mist and gloom"[5] makes us suspect that some enterprising Greeks actually had a look at the southern hemisphere of Earth during the Golden Age, a region minus the comforting light and heat of Saturn.

Like Perseus, Persephone is a destroyer, ". . . she who brings destruction."[6] The known world, in the northern hemisphere, was in darkness, having fallen into a point in its new orbit where the sun was far to the south.

Helios the sun sees everything, and he it was who told Demeter of how her daughter had been abducted. Failing to find Persephone, Demeter became enraged and planned

[4] Hesiod, *The Homeric Hymns*, trans. Hugh G. Evelyn-White (Cambridge, Harvard, 1914) p. 291, lines 17–41.

[5] Ibid., p. 295, lines 72–99.

[6] Graves, *Greek Myths*, 24.2.

"an awful deed, to destroy the weakly tribes of earth-born men by keeping seed hidden beneath the earth,"[7] She is the Earth Mother, "Unlaughing Demeter," and has the power to bring life and growth to a halt. In fact, with the northern hemisphere of Earth in darkness, nothing will grow on Earth until the sun returns from his half-year, half-day southern tour.

In the myth a bargain is struck with Hades so that Persephone will spend one-third of the year with him, two-thirds in the upper world. She thus becomes queen of the underworld.

It is a mark of the Silver Age, this threeness. The Greeks had a word for it, the division of the world into three seasonal parts. The Hecateia was representative of the triple Earth goddess: Hecate the crone, Demeter and her daughter Persephone. The cosmos was also divided by the Mesopotamians into the Ways of Enlil, Anu, and Ea, just as Zeus, Poseidon, and Hades were associated with Heaven, Earth, and Hell, respectively. The Earth Mother's division into three is mirrored by the division of the Graeai, Gorgons, Fates, Horai, etc. The three Muses were divided in three again, becoming nine.

Persephone's abduction to the Underworld is accepted as a myth of the seasons; there are not two opinions on the subject. What is not realized is that it is a myth of beginning, that it implies by its very existence a time when there were no seasons. For it is only the strange and unusual, the catastrophic, that excites myth-making. What we have here is the beginning of the Eleusinian Mysteries, which Demeter herself proclaimed, instructing the people to teach her seasonal rites. The same concerns are addressed by the Hopi priests turning the sun back at the Winter Solstice so it will return to make the crops grow.

Now we can unlock one of the strangest rites of Eleusis: that an initiate should bathe with a young pig in the ocean. It is the Hyades' animal again, new at the pole, and just below the ecliptic in the Watery Abyss. In performing the ritual, each initiate repeats the fall of the pole into the Chariot of Enmesharra with its circling marker of the Hyades, the beginning of a new set of world ages in the present solar system. At the commencement of the Silver Age, many new traditions began and new symbols arose. The wheat ear of the fruitful Earth was used at Eleusis, in the rites of the sacred king, and later in the Mithraic tradition.

Each ancient culture had a different fix on the end of the Golden Age and the beginning of that of Silver. The Eleusinian Mysteries were more Earth-centered, those of Mithras focused on the sky. A great concern was what had happened to the god who had formerly dominated the pole. Both the planet Saturn itself and the cometary Venus spinning off through the Saturnian band were involved. God and goddess were identified in different cultures as Osiris and Isis, Attis and Cybele, Adonis and Venus, Tammuz and Ishtar, Dumuzi and Innina. Confusion set in as some goddesses became identified with Earth, notably Isis, while Ishtar and Innina continued to represent the cometary Venus. But all these goddesses were searching for the "lost" god who had formerly been at the pole.

When "lordship had fallen" and the bright Saturn slipped down the solstitial colure to outer space, the long winter night fell and the stars came out in all their magnificence. The Egyptians encoded it in myth as Osiris-Saturn cut into pieces. Isis struggled to bring all the pieces together but never could find the phallus; the axis bond had disappeared. Plutarch reports in *Isis and Osiris*, ". . . at the time of the Winter Solstice they lead the cow seven times around the temple of the sun and this circumambulation is called the Seeking for Osiris."[8] This ritual is clearly modeled on

[7] Hesiod, *Homeric Hymns*, p. 315, lines 347–376.

[8] Plutarch, *Isis and Osiris*, LCL, Vol. V, p. 127, 372C.

the seven bands which surrounded Saturn. The sun of our present solar system is now Saturn's successor, and eventually a triple offering of incense would be made: "Resin at sunrise, of myrrh at midday, and of the so-called cyphi at sunset."[9]

The Mysteries of Isis have a seasonal cast to them similar to those of Eleusis. According to Plutarch, the disappearance of Osiris occurred in the month of Athyr (November), and at that time the priests "shroud the gilded image of a cow with a black linen vestment, and display her as a sign of mourning for the goddess, inasmuch as they regard both the cow and the Earth as the image of Isis."[10] Two of the things they mourn are "the days growing shorter than the night," and "the denudation of the earth together with the defoliation of the earth at this time," echoes of what occurred at the Fall.

Like the devotees of Eleusis, the Egyptian followers of Isis go down to the sea, but not to wash pigs. The priests bring forth a chest from the water and "a great shout arises from the company for joy that Osiris is found. Then they knead some fertile soil with the water and mix in spices and incense of a very costly sort, and fashion therefrom a crescent-shaped figure, which they clothe and adorn."[11] The crescent represents the illuminated crescent of the Saturnian band, but since the only crescent-shaped figure in the sky is now the Moon, Osiris becomes associated with the Moon.

This chest containing Osiris is mentioned in a number of myths, the famous chest that a tree grew up around. His association with the world tree was so powerful that the tree must follow him. Finding Osiris in the sea reflects his having fallen into the ecliptic, thus into the watery abyss.

The earthly, seasonal cast of the mysteries of Eleusis tended to keep the Greek civilization focused in that direction. Later, some Greeks would study in Egypt to fill in gaps in their knowledge. From the first, the Egyptian cognoscenti were well aware that Earth had fallen on its side and was lying in the ecliptic on the surface of the watery abyss. It was after the Fall that the hippopotamus goddess began to be depicted lying down, so that the crocodile could climb out of the water and on her back. Those familiar with Egyptian hieroglyphics are aware that many glyphs are sometimes given "lying down." This should add a new aspect to decipherment.

Returning to the ceiling of Senmut's tomb and the lineup of polar gods, we must examine the bird-headed god, fifth in line after hippopotamus Tau-ert. He is one of the four children of Horus, thus one of the four supports of Heaven or colures. This is Kebsenuf, widely represented with a long pole in his hands, often "spearing" the chain attached to Tau-ert. The pole is the solstitial colure which the bird-headed god represents. Tau-ert not only holds the chain, but can sometimes be seen holding the axis bond or menat. This menat is the "mooring post" that Earth is tied up to, the force that holds it in its precessional spin.

The Greeks knew the chain of Tau-ert as omphaloessa or umbilical cord, a name which eventually attached to the figure at the pole in the case of Ursa Major. The Egyptians apparently accepted the metaphor as well, accounting for the importance of "navel strings" in their culture.

The dissolution of the Saturn Configuration and subsequent spinning-off of all the planets into the ecliptic was, as we have seen, variously perceived. Heaven had disappeared, leaving only an empty place in the sky, but the location of Hell, spotted near the first-magnitude star, Canopus, was now easy to see. At the disappearance of Saturn, great lamentations began for his alter egos: Tammuz, Dumuzi,

[9] Ibid., p. 127, 372D.

[10] Ibid., pp. 95–97, 366E.

[11] Ibid., pp. 97, 366F.

and Osiris. The behavior is mentioned in the Bible as an abomination. "Then he brought me to the door of the gate of the Lord's house which was toward the north; and, behold, there sat women weeping for Tammuz."[12] This was serious cosmic business; the cometary Venus now identified with the mother goddess of Saturn's rings was now separated from the god, and apparently looking for him!

In "The Descent of Ishtar to the Underworld," the Babylonians recorded this journey. Unlike Persephone who was only abducted to the Underworld of Earth, Ishtar, the cometary Venus, went down to the real Hell or Eridu near Canopus. Exactly what made Venus a comet, following a path different from all the other bodies in the Saturn Configuration, can only be conjectured at present. But in this myth, "The Descent of Ishtar," she took up a path well outside the ecliptic, descending toward the southern pole of the precession.

The account of Ishtar's journey reads as if it were visible from Earth, and with our planet on its side in the Watery Abyss, the northern hemisphere in darkness, Hell would have been easy to spot near the bright star Canopus. The "descent" of Venus-Ishtar toward the south pole of the ecliptic would have been seen by anyone who cared to raise their eyes to the sky. To gain admittance to Hell, Ishtar made substantial threats.

> I will smash the door, I will shatter the bolt,
> I will smash the doorpost, I will move the doors,
> I will raise up the dead, eating the living,
> So that the dead will outnumber the living.[13]

The goddess is conversant with the metaphors of catastrophe, well knowing what it means to pass through the gates. The door post here is the tarkullu, the same as the menat of Egypt, both names for the force that flows between the poles of precession. Needless to say, Ishtar was admitted by Queen Erishkigal, and led through the seven gates of Hell.

At each gate, Ishtar loses an ornament or garment: crown, earrings, necklace, brooches, her girdle of birth stones, clasps around her wrists and ankles, and finally the cloth which covered her femininity.[14] Then, for a second time, she hangs or is hung on a wall until the great gods intervene and order Erishkigal to release Ishtar. During this time of her descent, we notice a parallel with the Greek myth of Persephone: "The bull springs not upon the cow, the ass impregnates not the jenny / In the street the man impregnates not the maiden." After the intervention of the gods, Ishtar is led back through the seven gates where her clothes and jewels are returned to her and she regains her former appearance.

The myth appears to record changes in the appearance of the comet-goddess as she moved on her new path as well as indicating the direction of her orbit. Ishtar does not find Tammuz, so that in the religion that incorporates this particular myth, the women will continue to weep and wail. The festival of the lost god is still celebrated by the Harranians;[15] it is el-Bugat, the weeping women. Ishtar-Venus went on to establish a cometary orbit intersecting with the ecliptic.

Isis, however, went on to find the parts of Osiris and put them together as recorded by Plutarch, and further illuminated in the *Pyramid Texts.* She puts them in a coffin, and

[12] Ezekiel 8:14.

[13] E. A. Speiser, *Ancient Near Eastern Texts,* "The Descent of Ishtar to the Nether World" (Princeton, 1950) pp. 106–109.

[14] One is put in mind of a striptease dance.

[15] Santillana, *Hamlet's Mill,* p. 282.

subsequently a tree grows up around it. It is an analog of the world tree, the terrestrial axis, but on its side in the ecliptic. The coffin is tantamount to the box found floating in the sea in the ritual described earlier. As might be expected, the Egyptians were extremely uncomfortable with their god lying on his side in the ecliptic, and kept exhorting Osiris to "arise" in the *Pyramid Texts.*

The Jews struggled mightily against astronomical apparitions in their religion, but even during the time of Isaiah and Jeremiah, Jewish women were still baking cakes in honor of Venus, the Queen of Heaven. There is an echo of the blighting of vegetation at the time of the Fall in *Legends of the Jews.* Eve complains that ". . . all the trees within my reach had cast off their leaves at the moment when I ate of the forbidden fruit."[16] According to Lactantius Placidus, "only upon the loss of Paradise, darkness and winter came over the earth."[17] But for the Jews, and later the Christians, the Fall would be held in conscious memory as a fall from grace.

Plutarch records an echo of the Descent of Lordship in an account of the religion of Zoroaster in *Isis and Osiris.* Areimanius (Ahriman) is "born from the darkness," while Oromazes is "born from the purest light." "Then Oromazes enlarged himself to thrice his former size, and removed himself as far distant from the Sun as the Sun is distant from the Earth, and adorned the heavens with stars."[18] Oromazes appears to be Jupiter, who would be a significant light in the heavens during the Silver Age, possibly a nova.

A happy aspect of the early Silver Age is that the Milky Way formed an ideal path from Earth's horizon to Heaven at the north celestial pole. Alas, the river Eridanus, the visible solstitial colure which had flowed through the Milky Way as it neared the ecliptic, had been "burnt up through Phaethon's fall."[19] Indeed, all the heavenly rivers, the four supports of Heaven, had disappeared, but it was still possible to travel from Earth to Heaven at the north celestial pole via the Milky Way.

A certain class of men who watched the stars soon deduced precessional motion, and understood that the ship of Earth was still tied up to the tarkullu or menat, the force that held the cosmos together. The problem was keeping connected to the Sky since the cosmic axis and heavenly rivers had disappeared. Fortunately the Milky Way provided a path to Heaven. But what of Hell? It was at this time that a new heavenly convention arose: the ferry to Hell, Tartarus, or Eridu recorded in classical culture. Hell became more important after the Fall since, as we shall see, all the measures were given from there.

Saturn was gone, thrown out of the chariot by Jupiter, who would now rule the Silver Age. In the following chapter, we will examine this first fallen age and see how the new ruler both initiated it and brought it to a close.

[16] Ginsberg, *Legends of the Jews,* p. 49.

[17] William F. Warren, *Paradise Found* (New York, Houghton, Mifflin, 1885) p. 281.

[18] Plutarch, *Isis and Osiris* (370), LCL, Vol. V, p. 115, 370A.

[19] Aratus, *Phaenomena* (360), referenced in Santillana, *Hamlet's Mill,* p. 254. ". . . the Galaxy was long known as Eridanus, the Stream of Ocean." Allen, *Star Names,* p. 474.

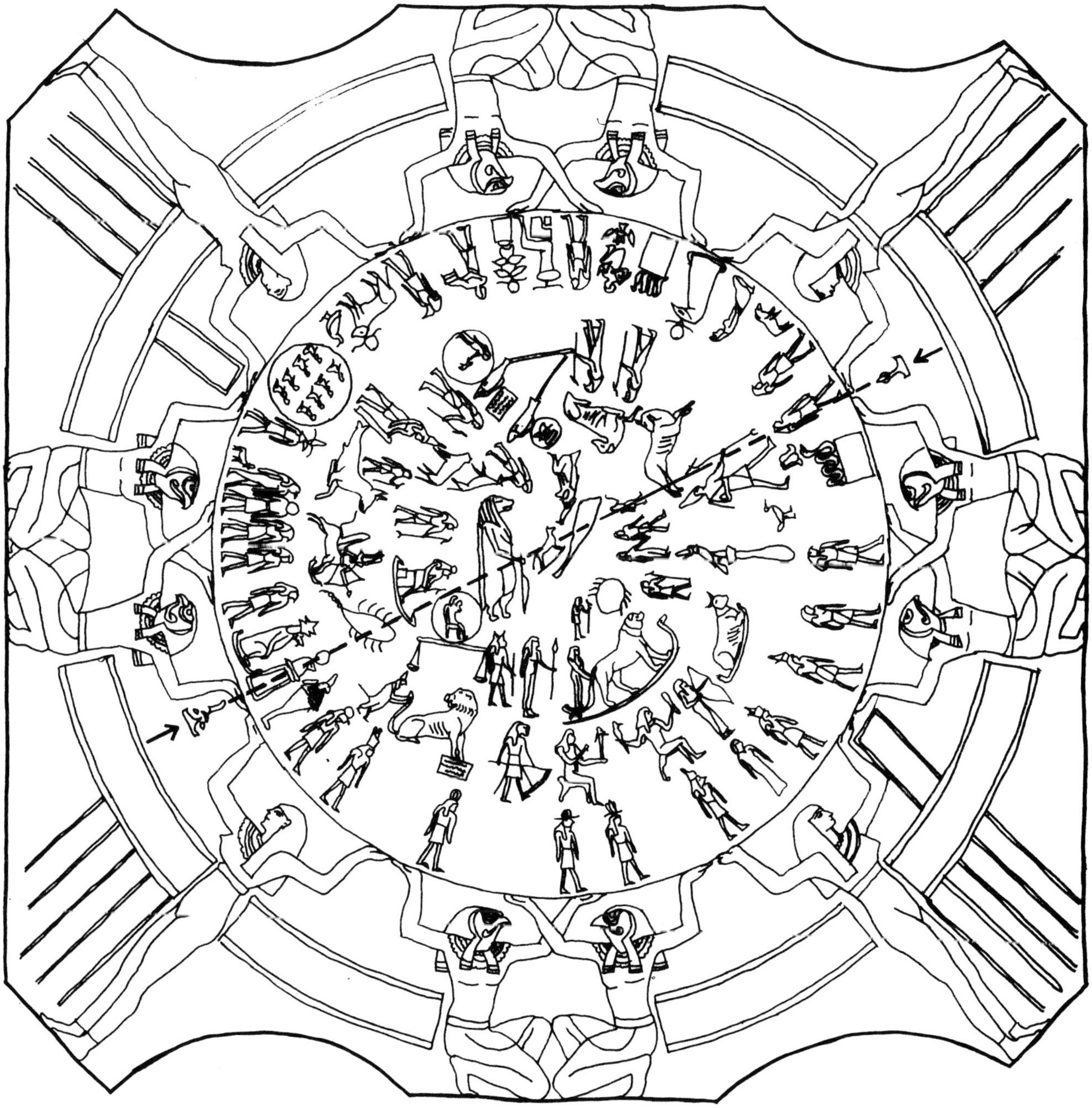

The Denderah Zodiac marks the solstitial colure at the Fall with hieroglyphs indicated by arrows. The heavy part of the line passes between Gemini and Taurus while Tau-ert (Draco) holds the mooring post at the exact center.

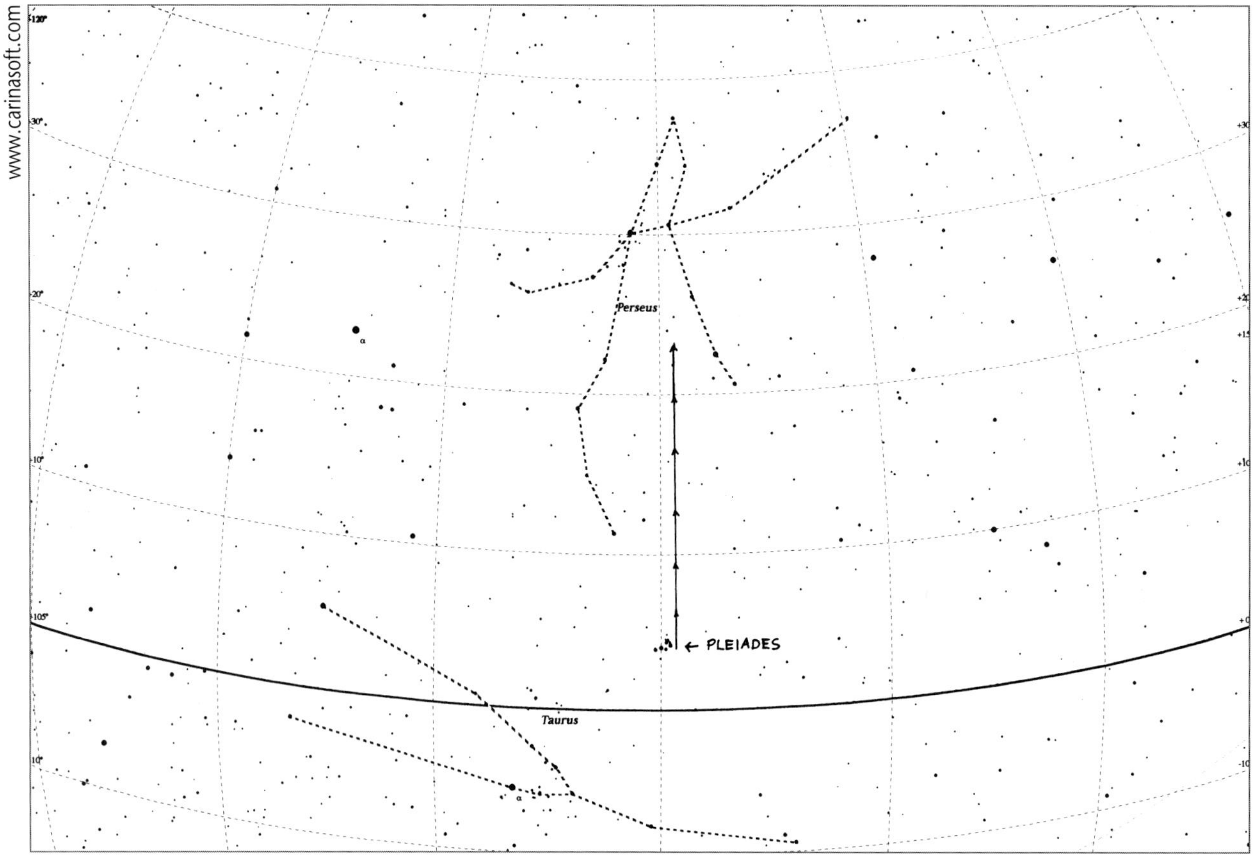

Earth tilts from the Pleiades to the constellation Perseus as the Silver Age goes down.

21. The Silver Age Goes Down

They [men of the Silver Age] were put away by Zeus son of Kronos, angry because they did not offer honour to the blessed gods who occupy Olympus.

– Hesiod, *Works and Days*

Zeus came upon Danae in a shower of gold and she bore him a son named Perseus.

– Robert Graves, *Greek Myths*

After Kronos was deposed, Zeus ruled the Silver Age with more flair than he has exhibited ever since, considerably brighter than we are accustomed to seeing him today. "When Jupiter was first born, he defeated the nakshatra Pushya by his brilliance."[1] When his orbit carried him through the pole, a horse (Hyades) pulled his chariot, a horse whose neck was "clothed with thunder." It was Pegasus pulling the thunder-car of Zeus. In Indic culture, the horse's head of the Rishi Dadhyank was dwelling in Lake Saryanavant (Watery Abyss), and "the bones of his skull were forged into a thunderbolt for Indra."[2]

Mithraic iconography represents Saturn handing over his sceptre and thunderbolts to Jupiter, suggesting that it was during the Silver Age that the ruler of Olympus won his reputation as wielder of the thunderbolt. Certainly it is Jupiter who is primarily associated with these bolts of lightning which now appear, during the Silver Age, to have been thrown when the planet god appeared in the sky close to the celestial pole of Earth.

With the dissolution of the Saturn Configuration, the planets spun off into the ecliptic around the present sun, approximating their situation in the present order of the solar system.[3] The fallen Earth lay with its axis nearly paralleling the ecliptic, North Pole pointing to the Chariot of Enmesharra between the horns of Taurus.[4] Thus from Earth, the planets could be seen to circle through its polar regions. Jupiter was very bright, and as the ruler of the new age he was carefully observed on his new path. As Marduk of the Babylonians, "He crossed the heavens and surveyed the regions."[5] Even toward the end of the Silver Age, Jupiter was more brilliant that we are accustomed to seeing him now, making the night bright for Abraham on his expedition to free his nephew Lot.[6]

In the fifth tablet of Enuma Elish, Jupiter as Marduk is busily arranging the new order of the cosmos, installing the stars, defining the year, setting up three stars for each of the 12 months (equivalent to the 36 decans of the Egyptians), and establishing the pole. Now

---

[1] Taittiriya Brahmana, 5.1.1, quoted in *Hamlet's Mill,* App. 39, p. 434, note 10.

[2] Santillana, *Hamlet's Mill,* App. 19, p. 393. With this weapon, Indra proceeded to slay the 99 vritras as Samson killed the Philistines with the jawbone of an ass.

[3] But for Mars which was an inner planet, and Venus which followed an irregular cometary path.

[4] The South Pole pointed to Scorpio, which in antiquity included part of Sagittarius. Hence we find Scorpion men on Babylonian boundary stones using a bow and arrow.

[5] Santillana, *Hamlet's Mill,* App. 39, p. 430.

[6] Ginsberg, *Legends of the Jews,* p. 106.

that we understand that the terrestrial pole of Earth follows the solstitial colure as it tilts on the equinoxes, lines 9 and 10 are clarified. "Gates he opened on both sides/Made fast the locks [equinoxes] to left and right."[7]

Dechend has assembled enough appropriate ancient texts to suggest that the problem was one of reorienting the Earth under a new Heaven. The Apsu, the depths of the Watery Abyss, the "chambers of the south,"[8] have suddenly become visible to ancient astronomers in the northern hemisphere since Earth has fallen on its side. Under the Saturn Configuration, the brightness of primeval Saturn must have interfered with star-watching. But in the Silver Age, a darker night began, revealing not only the brightness of the stars, but those parts of the heavens now in the southern hemisphere.

The poles of the precession were in apparent rotation around the Earth close to the celestial equator (the terrestrial equator projected against the vault of Heaven), and the ancient astronomers, those wise measurers of the sky, found them.[9] And given these two enduring points in the sky from which to measure, they chose the south pole of the precession. For naked-eye astronomers it was an obvious choice. At the north pole of the precession, the constellation Draco offers nothing substantial to get a fix on, the Earth being hung from "the empty place in the sky." Close to the south pole of the precession, however, lies the bright star Canopus.

Thus Canopus became a lodestar for the Mesopotamian civilizations, Nibiru, the ferryman, in charge of crossings or passages. And here begins the necessity, well recorded in classical civilizations, of taking a ferry to Hell, for the South Pole of Earth now points to Scorpius, almost 90 degrees away from the real Hell, the south pole of the precession near Canopus. The ferry ride will get shorter over the millennia, however, as catastrophe tilts the terrestrial axis into a more upright position.

Now Jupiter-Marduk set to work on his (actual) surveying of the heavens. He "squared Apsu's quarter," and "founded the station of Nibiru to determine their (heavenly) bands."[10] This is our clue as to why the sky was divided at latitudinal meridians into the "Ways" of Enlil, Anu, and Ea. Dechend paraphrases several German translations: "... the position of the Ways of Enlil, Anu, and Ea is a function of Nibiru."[11] Present expert opinion is that Nibiru is Jupiter. We must go back to Jeremias to learn that Nibiru is Canopus, but then Jeremias was aware that the Babylonians knew of the precession.[12]

Another clue is that Ea's Way is from the South Pole of Earth to 15 or 17 degrees south of the equator. Ea-Saturn was thought to have descended to Hell, and thus gets the bottom Way. He "gives the measures" at Eridu-Canopus.[13] What we must keep in mind is that the measuring of the Ways from Nibiru reflects Earth's position in the Saturn Configuration, upright between the poles of the precession. But this should not surprise us, given what we have learned about the necessity of bringing Heaven to Earth, and the accretion of primeval

7 Benno Landsberger, Wilson, Kinnier, "The Fifth Tablet of *Enuma Elish*," *JNES*, Vol. XX, Jan. 1961, p. 157.

8 Job, 9:9.

9 In the chapter on the precession, we will consider how this might have been done.

10 Santillana, *Hamlet's Mill*, App 39, pp. 430–431.

11 Ibid., p. 431.

12 This is a case where the baby was thrown out with the bath water. See the discussion on the reliability of early astronomy in *The Velikovsky Affair*, "Astronomical Theory and Historical Data" by Livio Stecchini.

13 Santillana, *Hamlet's Mill*, pp. 265–267.

media. Mesopotamian astronomers had their own style of saving the phenomena millennia before the birth of Plato.

"Nibiru shall be in control of the passages in heaven and on earth."[14] In Nibiru we have the first good shepherd, a title which would later attach to Mesopotamian kings and eventually to Jesus Christ. "May he shepherd all the gods [planets] like sheep . . ." The idea of finding and using the enduring southern point in the sky to order the heavens is reflected in Vedic text. In the Rig Veda, when Varuna "surveyed the first creation," he says "I fastened the sky to the seat of the Rita,"[15] which Santillana spots at Canopus.[16]

In such fashion did Jupiter survey and rule the Silver Age, a period that resulted in the acquisition of his names Pluvius and Pluvialis, implying the flooding that took place when the Hyades were circling the pole, and later the Noachian Deluge. But yet another catastrophe was in store for the men of the Silver Age, one that would tilt the Earth up for the first time in the present cycle of world ages. After the Deluge, within a few degrees on the precessional clock, when the Pleiades were still near the pole, a new cycle began within the solar system, marked by a 52-year Earth period and apparently involving both Jupiter and Venus.

When the cometary Venus spun off the Saturn Configuration, we temporarily lost sight of her. Fortunately, Dechend has spotted her, as Tiamat, the primeval dragon, in the region of Canopus. After taking three translators to task for editing out the information from the Epic of Gilgamesh, Dechend fills us in.[17] Paraphrasing Heidel would produce, "He (Nibiru) who crosses the midst of Tiamat without resting." This would be apparent motion, for the poles of the precession endure. Tiamat/Venus would have been moving. It is an unusual position for a member of the solar system to be 90 degrees out of the ecliptic, but it perfectly reflects Ishtar's descent to Hell. We must conclude that both records are describing the same event.

Although he never published the promised work on Saturn and Jupiter, Velikovsky remarked in *Worlds in Collision* that ". . . at an earlier time [before the Earth-Venus wars], Jupiter had already caused havoc on the planetary family, the Earth included, . . ."[18] In a short article, "The Destruction of Sodom and Gomorrah,"[19] I believe he has honed in on the end of the Silver Age. This is a good time to remember that the Earth tilted about 20 degrees at the end of the Silver Age, comparable to the tilt caused by Mars in –717, an event that slid below consciousness with hardly a ripple.

The 52-year Venus cycle is firmly embedded in Mesoamerican culture as the binding of the sheaves of years, a ritual reenactment of the fearful visits of the cometary Venus. The period defines a short world age for the Mexicans, located by Velikovsky's diligent scholarship in the 15th century B.C. The complete transformation of Venus from comet to planet at the close of that period suggests that all records of 52-year periods indicating actual visitation[20] or destruction (as distinct from ritual celebration) must apply to the previous age, that of Bronze. But, as we shall see, they actually begin at the end of the Silver Age in the time of the patriarch Abraham.

[14] Ibid., p. 433.

[15] Rig Veda, 8.41.10.

[16] Santillana, *Hamlet's Mill,* p. 265.

[17] Santillana, *Hamlet's Mill,* App. 39.

[18] Velikovsky, *Worlds in Collision,* p. 182.

[19] *Kronos* VI, 4, pp. 40-56.

[20] Ilse Fuhr, "On Comets, Comet-like Luminous Apparitions and Meteors," *Kronos* VIII, 1, pp. 38–39.

Velikovsky treated the 52-year period in *Worlds in Collision* as it applied to Venus, well realizing that comets do not return at exact intervals. "In the Codex Vaticanus the world ages are reckoned in multiples of 52 years with a changing number of years as an addition to these figures."[21] This would correspond with the work of Ixlilxochitl who also mentions "the multiple of 52-year periods played an important role in the recurrence of world catastrophes." Velikovsky argued persuasively that the 50-year Jubilee and Day of Atonement of the Jews reflected the same phenomenon. In fact, the Bronze Age was so short that it is probably accounted for by the four ages of 105 years mentioned by Censorinus.[22]

In *Mankind in Amnesia,* Velikovsky expressed concern that the 52-year cycle of destruction had become embedded in the psyche of mankind to the extent that we might compulsively repeat it in nuclear warfare. The phenomenon has been identified in the business cycle as the Kondratieff Wave[23] over the past 180 years. But it is also mentioned in *Legends of the Jews.* "For 52 years God had warned the godless; He had made mountains to quake and tremble." The warning issued was against the destruction of the cities of the plain, of which Sodom and Gomorrah are the best known. The calamity did not fail to occur, at dawn on the 16th day of Nisan.[24]

"The sun was risen on the earth when Lot entered into Zoar. Then the Lord rained upon Sodom and upon Gomorrah brimstone and fire from the Lord out of heaven / And he overthrew those cities, and all the plain, and all the inhabitants of the cities, and that which grew upon the ground."[25]

"And Abraham gat up early in the morning to the place where he stood before the Lord: And he looked toward Sodom and Gomorrah, and toward all the land of the plain, and beheld, and, lo, the smoke of the country went up as the smoke of furnace."[26]

Students of Bible history will recall how angels led Lot and his wife and two daughters out of the city (Sodom) before the holocaust and they escaped to Zoar (except for Lot's wife, who looked back and was turned into a pillar of salt). Zoar was spared, but apparently not without reason. "This city had been founded a year later than the other four; it was only 51 years old, and therefore the measure of its sins was not so full as the measure of the sins of the neighboring cities."[27]

The Biblical account has Lot and his daughters leave Zoar and retire to a cave in the mountains, where occurred the ugly incident in which Lot's daughters tricked him into impregnating them. The catastrophic event which they had witnessed was so horrendous that the young women thought they were the only three people left on Earth.

Before pursuing Velikovsky's account of the geography and history of the area, I would like to examine what appears to be a doublet or shadow of the 52-year period which can be traced in the Bible and *Legends of the Jews.* Abraham is a typically long-lived patriarch, 175 years, and the Bible is repetitive here, suggesting that some stories are given twice. One thinks of the two incidents in which

[21] Velikovsky, *Worlds in Collision,* p. 152, note 1.

[22] Liber de die natali, quoted in *Worlds in Collision,* p. 152, note 1.

[23] Shuman, James *The Kondratieff Wave* (N.Y. Dell, 1972). See discussion in Velikovsky, *Mankind in Amnesia,* pp. 143–145. The subject is now known as the Long Wave cycle.

[24] Ginsberg, *Legends of the Jews,* p. 118.

[25] Genesis, 19:23–25.

[26] Genesis, 19:27–28.

[27] Ginsberg, *Legends of the Jews,* p. 118.

Abraham and Sarah claimed to be brother and sister and Sarah was almost violated. Then there are the repeated interactions between Abraham and the Lord over Abraham's childlessness. More to the point, Lot is rescued twice, and Sodom and Gomorrah are not only overturned but suffer in a war between kings.

In *Legends of the Jews,* we discover that wise men read the fate of Nimrod's army being annihilated 52 years earlier at the birth of Abraham.[28] Now Nimrod is another name for Amraphael, king of Shinar, whose involvement in the war of the kings, "four against five," can be traced in the Bible. He is given a flaming character at some length in *Legends.*

"Since the Flood, there had been no such sinner as Nimrod." He was "... the first mortal to hold universal sway," and "the first among the leaders of corrupt men ..." Nimrod also has a cosmic cast to his character since he has been identified as the constellation Orion.[29] And if the bull Taurus dominated Heaven at the pole during the Silver Age, his appearance was insignificant as it is now; Orion was the powerfully visible figure. In many instances, the star clusters in Taurus are connected with the mighty hunter; both Hyades and Pleiades are considered to be his weapons or hunting nets in many cultures.[30] The Egyptian Orion held an ankh surmounted by a prominent star, Aldebaran in the Hyades; the Greeks saw the star as his weapon, a mace.[31] The Jews apparently saw the brilliant sky figure Orion as both sinful and threatening.

The 52-year doublet has the exaggerated character of a legend or myth. "When Abraham went into battle to free his nephew Lot, he was as tall as 70 men and marched in four-mile strides."[32] This was one rescue mission on behalf of Lot; another was, as we recall, performed by angels. Abraham defeated Nimrod on the 15th of Nisan, just as the seers had foretold 52 years earlier.

And now we must look upward to see what the defeat of Nimrod would imply in the heavens. In the catastrophic tilt in which the celestial pole of Earth moved from the Pleiades to the constellation Perseus, Nimrod-Orion would be moved 20 degrees away from the pole. To a people who had not yet broken with sky gods, who had a priest named Melchizedek, "Zedek (Jupiter) is my king," this sinking of Orion on the heavenly vault could be seen as his defeat.

Both the 52-year period ending in the overturning of the cities of the plain and the 52-year doublet ending in the defeat of Nimrod-Orion have another critical factor in common. Both periods come to a close on catastrophic dates, the 15th or 16th of the month of Nisan, a time when Earth is inclined to tilt on its equinoctial hinges, "the night appointed for miraculous deeds."[33] The Bible and *Legends of the Jews* appear to be telling the same tale twice.

Separating gods from men is still a problematical enterprise in the study of primeval media. Abraham, a prince or exalted one of the Elohim,[34] like many great leaders was a myth-attracting magnet, losing much of his humanity at the center of a catastrophic interaction when Heaven must be brought to Earth. But let us give him credit for being one of the first to hear a merciful god, and not sacrifice his first-born. The Phoenicians continued to practice child sacrifice to Tanit down to the time of their defeat by the Romans and afterward.

[28] Ginsberg, *Legends of the Jews,* p. 98.

[29] Santillana, *Hamlet's Mill,* p. 166.

[30] See discussion in *Hamlet's Mill,* p. 166.

[31] Bauval, *The Orion Mystery,* pp. 138–140.

[32] Ginsberg, *Legends of the Jews,* p. 106.

[33] Ibid., p. 105.

[34] Smith's *Dictionary of the Bible,* 1860, Genesis, XXIII:6.

Velikovsky's treatment of "The Destruction of Sodom and Gomorrah" is double-barreled, using both the mythical-historical approach of *Worlds in Collision* and the geological one of *Earth in Upheaval*. The latter work has never received the attention it merits.

"The rift in which the Lake of Galilee, the Jordan and the Dead Sea lies is the deepest depression on land. The surface of the Dead Sea is close to 390 meters below the level of the Mediterranean, and at its deepest is some 355 meters lower. The shore falls steeply from the Judean mountains on the west; on the eastern side of the rift rise the Moabite mountains. The walls of the chasm show sharp broken strata that remained horizontal, which proves that the breaking down was instantaneous.[35] The force which caused this slide movement must have been stupendous. The ground of the rift around the Dead Sea is covered with coagulated lava masses, taking the form of an immense herd of giant elephants with rough skin. These lava eruptions from fissures are ascribed to the second interglacial period.[36] To the south of the Dead Sea towers a big cliff of salt called Jebel Usdum (Mount of Sodom). 'It is absolutely impossible that salt sediment of a sea should precipitate in such a form.'[37] 'Only the rupture of the ground could create this site, singular in the entire world.' "[38]

In such a condition did the overturning of the cities of the plain leave the formerly well-watered plain of the Jordan[39] which had been fruitful, and supporting great and populous cities.[40] Classical historians have described the event, indicating phenomena such as lightning and thunderbolts.[41] Ascribing the lava flows to the second interglacial period reflects a geological dogma: NVBT, no vulcanism in biblical times.[42] Like the cities of the plain, it will be overturned. "Early in this century [20th] a Phoenician vase was found embedded in lava."[43]

The Dead Sea, a result of water flowing through salt deposits and into a deep depression, is first mentioned in the Bible at the time of the Exodus,[44] located herein when the first visit of Venus brought the Bronze Age to an end. Studies have been done of its accumulation of mineral salts that indicate a fairly recent age, perhaps 5,000 years.[45]

The Dead Sea, the Jordan, and the Lake of Galilee lie on one arm of a three-pronged crack in the Earth, one of the more enduring marks of catastrophe on the planet. One arm runs north through the Gulf of Aqaba through the overturned Plain of the Jordan to Lebanon; another northern arm forms the Gulf of Suez; the southern arm runs through the best part of a continent as the Great African Rift, totaling one-sixth of the circumference of the Earth.

"Africa was in tension and torn by north and south fractures [which] along with the sinking of a strip of the crust formed the

[35] Blankenhorn, "Entstehung und Geschichte des Todten Meeres," pp. 34–35, ref. in "The Destruction of Sodom and Gomorrah," *Kronos* VI, 4, p. 26.

[36] Ibid., pp. 41–42.

[37] Ibid., p. 34.

[38] Ibid., p. 35.

[39] "Lot desired to settle in the well-watered circle of the Jordan." Ginsberg, *Legends of the Jews,* p. 104.

[40] Tacitus, *Histories,* Vol. 7.

[41] Velikovsky, "Sodom and Gomorrah," p. 47.

[42] N. M. Sarna, *Understanding Genesis* (N.Y., 1976) p. 142.

[43] Velikovsky, *Earth in Upheaval,* pp. 197–198.

[44] Joshua 3:16, Numbers 34:12, Deuteronomy 3:17.

[45] Velikovsky, "Sodom and Gomorrah," pp. 43–44.

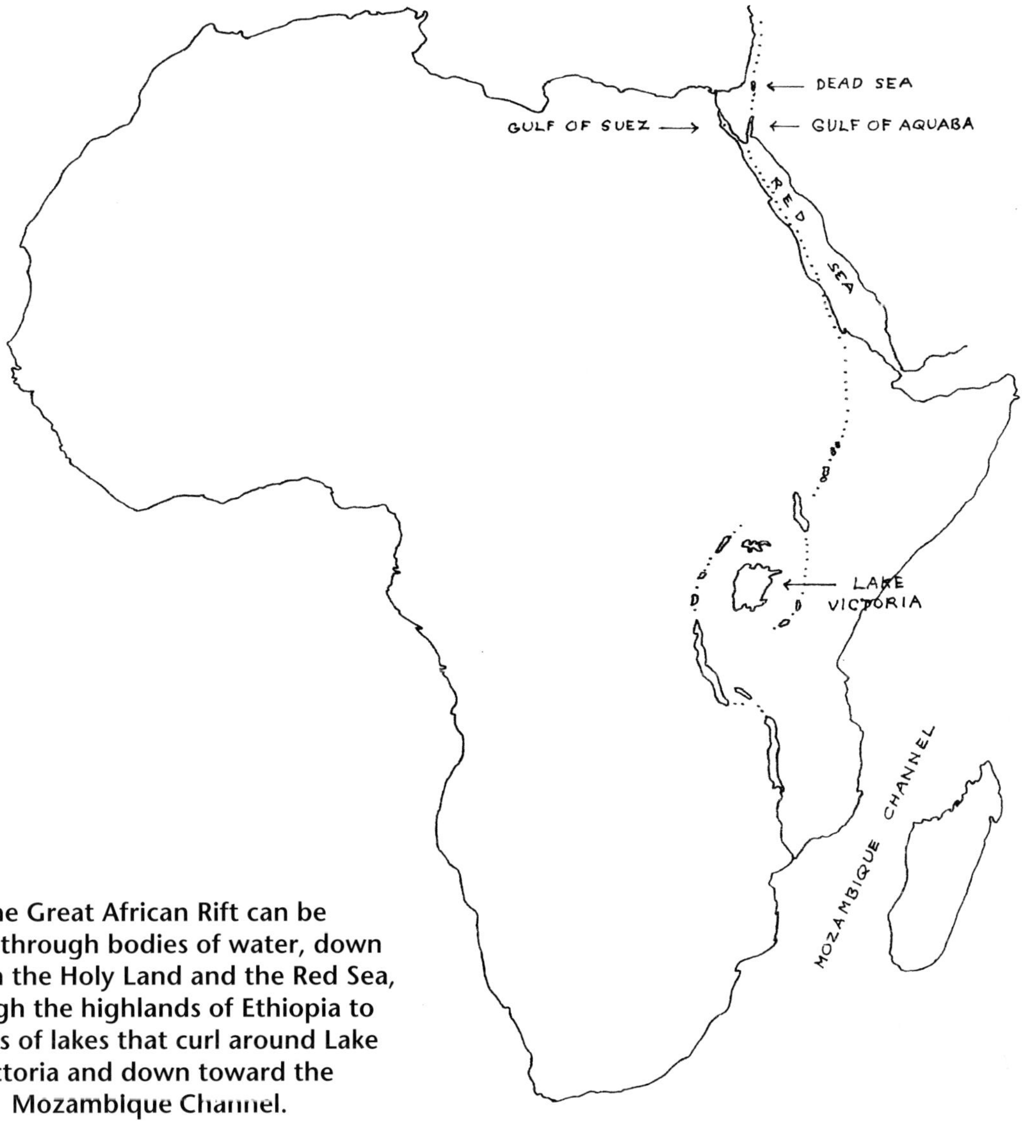

The Great African Rift can be traced through bodies of water, down through the Holy Land and the Red Sea, through the highlands of Ethiopia to strands of lakes that curl around Lake Victoria and down toward the Mozambique Channel.

longest meridional land valley on earth ... From Lebanon, then, almost to the Cape, there runs a deep and comparatively narrow valley, margined by almost vertical sides, and occupied by the sea, by salt steppes and old lake basins, and by a series of over 20 lakes, of which only one has an outlet to the sea. This is a condition of things absolutely unlike anything else on the surface of the earth."[46] "The opening of fissures of such magnitude can be explained only by ... vigorous volcanic action."[47]

Velikovsky finds it "probable ... that the Rift began in one great tension and increased in the next." Gregory found "Some of the fault-scapes are so bare and sharp that they must be of very recent date. This continuation of Earth-movements into the human period is one of the striking features of the district."

[46] J. W. Gregory, "Contributions to the Physical Geography of British East Africa," *Geographical Journal,* IV (1894), p. 290, from *Earth in Upheaval,* p. 89.

[47] Velikovsky, *Earth in Upheaval,* p. 91. Another possibility would be electrical scarring.

"All along the line the natives have traditions of great changes in the structures of the country." We must conclude that the most conspicuous sin of Sodom and Gomorrah was in being founded astride what would become one of the world's great fault lines, a focus for historical catastrophism.

We are left with a set of happenings which appear to go together: the overturning of the cities of the plain with an attendant rain of fire and brimstone, a doublet of the event as perceived in the sky as the fall of Orion-Nimrod, historical accounts of lightning and thunderbolts at that time, and the catastrophic myth of the birth of Perseus engendered by a "golden rain" from Jupiter. The ancients certainly believed in the thunderbolts of the ruler of Olympus. But can we believe that Jupiter cast his bolts across interplanetary distances? So it would appear.

Earth had fallen on its side. Uranus is still in that position. Perhaps all the planets experienced a similar "fall" and have been, like Earth, working their way toward a more vertical position. We may have to begin considering the solar system as electromagnetically all of a piece, view it as a whole. A more holistic view may eventually explain how Jupiter is involved in the 52-year cycle of Earth and Venus.

With a pattern of cyclical catastrophic Earth movements now in place, we can proceed with some of the fine points of the model. The history of the planet Mars in the cosmos is a colorful one, apparently at the root of mankind's current world view. In the following chapter we will discover how it formed a powerful archetype, that of the Magic Twin.

The Cacaxtla twins.

22. Mars, The Magic Twin

Nergal [Mars] . . . is frequently called the twin god, and as such his names are Lugalgirra and Meslamtaea.

– *Mythology of All Races,* Volume V

There is a mill which grinds by itself, swings of itself, and scatters the dust a hundred versts away. And there is a golden pole with a golden cage on top which is also the Nail of the North. And there is a very wise tomcat which climbs up and down this pole. When he climbs down, he sings songs; and when he climbs up, he tells tales.

– Tale of the Ostyaks of the Irtysh

Attentive readers may be able to translate the better part of the second quotation above, which demonstrates the ability of a fanciful fragment to suggest the appearance of the Saturn Configuration. The swinging mill of Earth's axis is an exacting touch.[1] To the prosaic minded, it will mean little, only the mention of the legendary mill. Here we are interested in the "wise tomcat," a metaphor for the planet Mars in its "twinning" travel up and down the axis bond between Earth and Saturn. As we will see, Mars had two decidedly different appearances: one when it was near the Earth, another when it was far away. Thus it came to be considered as twins, magically changing or transforming.

Ancient media covering the twins is vast, and the phenomenon continues into modern media as well, fields such as literature and film as well as religion and psychology. From the Dioscouri and the Magic Twins of the *Popol Vuh* to the yin-yang symbol, the concept is embedded in human culture, while the symbols and qualities associated with twins are so diverse that they defy sorting out. It is only by viewing the idea of twins through the glass of historical catastrophism that we can begin to understand the interweaving patterns of twinship.

Not surprisingly, much of the mystique of twins has devolved on dual births. Since the days of Florus and Laurus,[2] few parents have resisted the urge to give twins similar names, either rhyming or beginning with the same letter, often from the same root. The desire to dress them in similar fashion appears to be strong. They are the same, but different.

In other cultures twins tend to be surrounded by what we have chosen to call superstition, often with serious consequences. Twins can be lucky or unlucky. They can be made much of, or, alternatively, one or both can be put to death or exposed to the elements.[3] They are connected with a host of polar symbols down through the ages. The Shuswap Indians consider them to be under

[1] Dechend quotes Simrock's translation of the ferryman, Mundilfoeri, as "axis-swinger." Santillana, *Hamlet's Mill,* App. 14, pp. 376–377.

[2] Rendell Harris, *The Dioscuri in the Christian Legends* (London, 1903) pp. 1–19.

[3] Rendell Harris, *Boanerges* (London, Cambridge Univ., 1913) chapter V–XV passim. Also see *The Dioscuri in the Christian Legends.* The mother of twins is sometimes killed.

the protection of bear or deer.[4] Other cultures credit them with inventing the plow[5] or yoke[6] as well as the mast of ships.[7] The association with horses is a strong one.[8] Often the hair of twins is not cut and they are grouped with those born feet first.[9] Twins are differentiated as red and white, rough and smooth, strong and weak.[10]

Twins exist as a set of gods or heroes in virtually every culture: Horus and Set of the Egyptians, Castor and Pollux of the Greeks and Romans, Ahaiiuta and Matsailema of the Zuni Indians.[11] Commonly male, they can also be female like the Egyptian Isis and Nephthys, or male and female like Nut and Seb. A new and evocative set of male and female twins has recently been dug up in Cacaxtla, Mexico.[12] "Creirwy, the most beautiful girl in the world, and Afagddy, the ugliest boy . . ."[13] express opposite qualities. Many sets of twins are enmeshed in religion: Cautes and Cautopates of Mithraism, Florus and Laurus of the Syrian Church, James and John, the sons of Zebedee in the gospel of St. Mark. Jesus is sometimes thought to have had a twin, Thomas Didymus.

Although many sets of twins demonstrate attachment for each other, others do not get on well at all, and in some cases actually begin fighting in the womb: Set and Horus, Proetus and Acrisius (father to Danae),[14] and the biblical Jacob and Esau. Twins often express polar qualities: light and dark, good and evil.

The Dioscouri are the model twins of Western Civilization, seen in the sky as the constellation and zodiac sign of Gemini, its two brightest stars.[15] Their mythos shows them to be devoted brothers. When Castor fell mortally wounded in battle, Pollux (Polydeuces) could not bear to survive him. "Bid me also die, O king, with this my brother." Zeus answered, "If thou contendest for thy brother and art minded to have an equal share with him in all things, then mayest thou breathe for half thy time beneath the earth, and for half thy time in the golden homes of heaven."[16] The Dioscouri would thus spend alternate days above and below the Earth. They are generally held to represent the two halves of the Earth, one in sunlight, one in darkness.

The Mithraic twins, Cautes and Cautopates, also express light and dark, but at a seasonal level. Cautes with his torch pointing up represents the Spring Equinox and the attendant

4 Harris, *Boanerges,* pp. 145–148. The last two world ages have had bears at the pole while the deer would represent the Hyades animal of the Silver Age.

5 The Acvins (Indic twins) sowed grain with the plow in heaven. The plow has the V-shape of the Hyades.

6 The yoke is the omega-shaped Draco.

7 Harris, *Boanerges,* p. 226. ". . . *the first ship known to the Greeks was an oak tree* [cosmic axis] *with twins on board,* . . ." The reference is to the Argonautica of Apollonius Rhodius. As analyzed by Harris, the number of twins who shipped aboard the Argonautica is truly remarkable.

8 The horse-taming Castor does not break horses for riding, however, but for the chariot. Harris, *Dioscuri,* p. 12.

9 Harris, *Boanerges,* p. 56.

10 Harris, *Cult of the Heavenly Twins,* 1906. I have been unable to obtain this publication. See introduction to *Boanerges,* xxi.

11 Frank Cushing, *Zuni Creation Myth.*

12 See the following chapter.

13 Graves, *White Goddess,* p. 27.

14 Apollonius, *Bibl.* II. 2.

15 Harris, *Boanerges,* p. 229. The stars are also called Jason and Triptolemos, Apollo and Heracles, Zethos and Amphion, etc.

16 Pindar, *Nemean Odes,* 10.148ff.

approach of longer days; Cautopates with his torch pointing down represents the fall equinox which signals the approach of winter's shorter days. Thus Cautes brings light to the northern hemisphere, Cautopates to the southern. That the path of the sun is crossing the equator is underscored artistically by both twins having their legs crossed. Ulansey has compared them to the Dioscouri who sometimes have their legs crossed.[17]

The concept of celestial twins can be traced within the Saturn Configuration as the sunlit crescent of the Saturnian band revolving around the central orb. The crescent could be seen on one side or the other, above or below, in two different aspects. As mentioned earlier, the crescent was seen as the horns of bull or cow. This vision was brought to Earth in the Middle East as twin bulls, often on either side of temple gates.

The Bull of Heaven as twins. The bull is shown full face above.

In the Rig Veda, the twins are bull and cow who together comprise the ancient womb or Saturnian enclosure.[18] The bright crescent was also seen as opposed to the other, darker side of the configuration, the beginning of division of light and dark that would later devolve on the Dioscouri and other twins. Nephthys, mother of Set, would be the darker twin of Isis and Nephthys.

[17] Ulansey, *Origin of the Mithraic Mysteries,* pp. 112–116.

[18] Talbott, *The Saturn Myth,* pp. 377–378, note 51.

The notion of a central god with a twin on each side is repeated frequently in ancient art, myth and text: the Dioscouri on each side of a star, Cautes and Cautopates flanking the tauroctony.

Isis and Nephthys are winged kites (tcherti) who purify and protect the god between them.

The Dioscouri flanking a star.

The biblical twins, James and John, ask to sit on either side of Jesus in the gospel of St. Mark, "in thy glory."

That the revolving crescent could be either above or below the Saturnian orb is expressed by another set of Egyptian twins: Nut and Seb.

Isis and Nephthys as birds (detail from a lengthy frieze at Denderah where they are flanking an image of the cow goddess, Hathor).

Nut's body, resting on hands and feet, arches above while Seb reclines below. Saturn is most frequently represented with the crescent below because that was the "day" of Saturn when the failing sun appeared brightest, a time when the present sun was on the other side of the Earth. It is in this position, crescent below, that a hymn to Inanna compares the appearance of Saturn to the new Moon, a subject to which we will return presently. But needless to say, the Moon cannot duplicate other positions of the Saturnian crescent as depicted in ancient art when it is seen on top of the configuration.

The notion of a king having a twin is a very old one. In Egyptian hieroglyphics, the raised arms of the ka, which embrace the circle of the Aten from above or below, right or left, is an obvious analog of the crescent.[19] It is often translated "double" or "twin." "The ka of the king is his twin."[20]

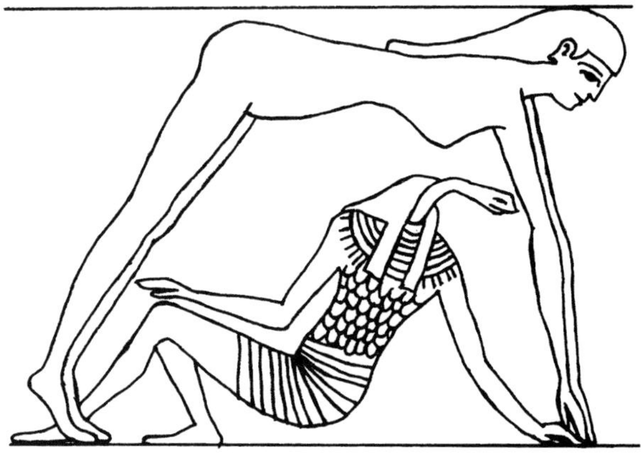

Nut above and Seb below.

Talbott and other catastrophist scholars locate the Saturn Configuration at the north terrestrial pole, and here we must revert to the pole of the Age of Heroes and the Bearson legend. A peculiarity of the legend comes into focus here. In several hundred versions of the Bearson legend analyzed by Panzer, the figure at the pole whom the Bearson and his companions must confront is exceedingly problematical, either a giant or a dwarf.

Now giants and dwarfs, magical twins, are standard roles in the repertory theater of myth and legend, as well as archetypes in the life of the mind. Hitherto we could only speculate as to what lay behind these fantastic figures, so that it is something of a surprise to learn that these denizens of fairy tales actually grow out of the reality of a previous order of the cosmos. In bringing one polar image into focus with another, the Saturn Configuration with the bear at the pole, the accretive nature of myth can be well illustrated.

[19] "Above," "below," "right," and "left" cause endless headaches for Egyptian translators since they cannot be understood in either solar or terrestrial terms.

[20] Frankfort, *Kingship and the Gods* (Chicago, 1948) p. 69.

In the chapter on the Bearson, we learned of the giant/dwarf who was eventually dispatched by our hero and his friends. The same pattern is apparent in the *Odyssey* with the monstrous one-eyed Cyclops in the role as giant. That the action takes place at the pole is apparent in the grisly scene where Odysseus and his four companions turn the sharpened stake (axis) in the eye socket (Saturnian bull's eye) of the monster, blinding him. Homer further clarifies the situation with a metaphor relating to a turning mill so we will know that the cosmic mill is under discussion.

Now when Homer counts, we must pay attention. Odysseus' four companions equate to the four colures, turning the cosmic mill while Odysseus, here reprising the role of Mars, pushes it into the eye of the monster. Though the Bearson legend and the *Odyssey* both relate primarily to the Earth-Mars Wars, they also include patterns drawn from previous catastrophes or orders of the cosmos. Notice that the episode of the Cyclops is part of the "flashback" books of the *Odyssey*, where the hero is relating past adventures. In such fashion does Homer distinguish between remote past and more recent catastrophes.

In contrast to the Homeric epics, the Bearson legends speak to us strained through the consciousness of the northern races over millennia. The possibility of confusion between giant and dwarf in a pattern that has otherwise been successfully transmitted is puzzling. In order to proceed, we must project a dynamic model in which the figure can be both giant and dwarf. Ancient media describing such a figure all points in the same direction, toward Mars and his analogs.

The archetypal strongman, Hercules, while not described as a giant, is distinguished by his powerful body, strong enough to hold up the cosmos for a short time to relieve Atlas. Another, less well-known aspect of Hercules is that of a dactyl-like dwarf. In fact, it is in the dwarf form that he founded the Olympic Games. And it is as a dwarf that Hercules is associated with the mother goddess Demeter, a tiny (one cubit) statue of Hercules next to a colossal one of Demeter.[21]

The dwarf Hercules is analogous to the Egyptian Bes, a wrinkled, bandy-legged dwarf who often sports a club and lion skin as well as the Hercules square knot. And in the iconography of Egypt we can more clearly see the dual forms of the god juxtaposed to one another. The image of the handsome young Horus, son of Isis and Osiris, is sometimes surmounted by that of the ugly dwarf Bes. Thus strongman Hercules is to dwarf Hercules as Horus is to Bes.

That the dwarf Bes is figured above the larger Horus gives the pattern "above is small, below is large." We are reminded of the many "bird nest" legends collected by Levi-Strauss

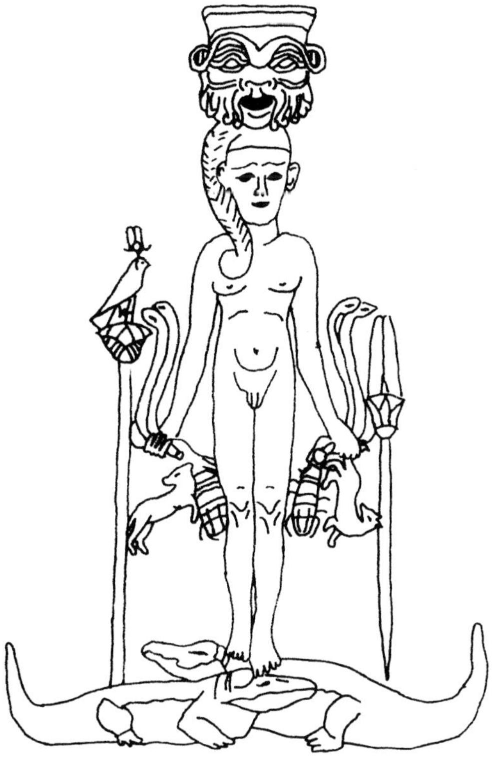

The dwarf Bes above Horus.

[21] Pausanias IX.19.5.

in which the shaman-hero climbs a tree and becomes small and withered.[22] Thorkill of the northern legends is another hero who becomes withered,[23] while Nergal becomes withered and unrecognizable upon climbing the heavenly stairway to the kingdom of the gods.[24] Ascension to Heaven is a theme common to the mythos of the warrior-hero.

On the other hand, certain heroes grow larger on occasion. In the Kalevala, Sampsa grows larger and fells the cosmic oak. Indra Pravrddha[25] swells to a giant form, and the Celtic Cuchulain does the same. Mars himself is surnamed Gradivus, from the Latin gradiviri, meaning to become big, to swell.[26]

Decided up and down motion is suggested by the dancing or leaping Salii, priests who served both Mars and Hercules in Rome. The "rise" and "fall" of Mars in the axis column is also suggested by the epithet Homer bestows on him in the *Iliad*,[27] "Ares alloprosallos" from allomai, meaning to leap, spring, or bound. In Africa, the Warundi dance and leap at the birth of twins.[28] Other atavars of Mars associated with the dance are Bes and Melqart. The dances appear to have been ritual and athletic with no suggestion of esthetics. In a description of the dances in honor of Tyrian Hercules, the celebrants "leaped high in the air, now crouched down to the ground, and now whirled their whole bodies about like men possessed." [29] Hercules, as cunctator is "the leaper," and the Indic Skanda, "the jumping one," has been identified as Mars.[30]

When we visualize the Saturn Configuration with the sun of Saturn stationary at the pole and Mars moving up and down the axis column, it is much easier to make sense out of ancient material, the ad hoc explanations for which have been strained. The Emperor Julian, in speaking of the solar (Saturnian) worship of Edessa, identifies the two "assessors," Monim and Aziz, with Hermes and Ares. He informs us that Ares (Mars) is the herald of the sun and marches in front of him,[31] an identification which makes no sense in the present order of our planetary system, but resonates easily with the Saturn Configuration. Looking to natural history, we find the Dioscuri referred to collectively as "castors" or beavers. And where do we find beavers but in a stream, swimming up and down.

Association of the Dioscuri with stars is a persistent artistic motif, often on coins.[32] They have been considered personifications of the morning and evening star,[33] difficult to understand since morning and evening stars are generally understood to be planets. But if we consider the twins as Mars traveling up and down the axis column, one would travel up as the morning star rises in the sky, while the other would travel down as the evening star sinks toward the horizon. The legend of their death when thrown into a well would reflect

[22] Levi-Strauss, *Naked Man,* trans. Weightman (N.Y. Harper & Row, 1981) pp. 31–36, 45–46, 232–235.

[23] Saxo Grammaticus, *The Danish History* (Wiesbaden, 1967) p. 355.

[24] O. Gurney, "The Sultantepe Tablets," *Anatolian Studies,* 10, 1960, p. 125.

[25] Pravrddha comes from a root meaning "to increase or swell."

[26] This epithet has been used to attach a characteristic of seasonal growth to Mars.

[27] Homer, *Iliad,* Book V, 831–889.

[28] Harris, *Boanerges,* p. 112.

[29] *Heliodorus,* trans. by Moses Hadas (Ann Arbor, Univ. of Michigan) chapter 4, pp. 102–103.

[30] Santillana, *Hamlet's Mill,* p. 157.

[31] Julian, *Oration* IV, pp. 195–200.

[32] Harris, *Dioscuri,* pp. 33–36.

[33] Ibid., pp. 13, 16.

the Fall when all the visible planets retreated to outer space. Subsequently, however, they were "discovered and shine out again like stars in heaven."[34] This would be as the two brightest stars in the constellation Gemini.

Witness the power of the warrior-hero archetype continuing into the present, as our modern hero Superman is "able to leap tall buildings at a single bound." Here we must begin to grasp the power of an archetype, not a moribund image of interest only to cultural anthropology, or a concept buried in the unconscious and studied by psychology, but a living pattern capable of erupting into present reality.[35]

Analogs of Mars are closely associated with the cosmic column. Nergal is "lugal-an-sa-gar," "lord of the pillar," as well as Meslamtaea, "the luxuriously sprouting Meshu tree." Etymology supports the identification of Sampsa with Sampo, the cosmic pillar, since the words are cognate, while the same parallel may be drawn between the Indic Skanda and the Skamba. The name of Gilgamesh contains the word for tree while meshu means "twin" in Akkadian. And who has not heard of the pillars of Hercules? Harris equates the two colossal pillars in the citadel of Edessa (Ourfa) to the twin-brethren.[36]

The axis column once visible between Earth and Saturn was by no means a foggy, miasmic apparition, but a lively expression of cosmic force powerful enough to hold a string of planets in line. If we are to credit the ancients, something very like bolts of lightning occasionally traveled through the cosmic column. A number of legends give us a picture of activity in the polar axis. Harris reports, "The thunder god of the Caribs, sometimes a star, sometimes a bird, blows the lightning through a great reed."[37] Among the Thompson Indians of British Columbia, "the Thunderbird shoots arrows using its wings like a bow." "Some Indians say that lightning is the twinkling of the thunderbird's eyes."[38] We read in the *Jesuit Relations* that "... the lightning is the flashing or winking of its eye, ..."[39] Here the "bird" is the Saturnian crescent, widely represented among Native Americans as the thunderbird, but also well known in other parts of the world.[40]

Other ancient media shows energy moving up the cosmic column. In the *Popol Vuh*, one of the Magic Twins shoots darts through a blowgun at Vucub Caquix (Saturn), giving a particular clear image. Certain myths of Heracles suggest the same sort of action. "Heracles distressed the Olympian gods with his arrows."[41] Elsewhere he "threatened Helios with his bow."[42] It is no wonder the axis is frequently associated with weapons while the twins have become universally known as sons of thunder.[43]

In the Gospel of St. Mark, we learn that Jesus ordained twelve apostles, two of whom were "... James the son of Zebedee, and John,

---

34 Ibid., p. 16.

35 See Rupert Sheldrake, *Presence of the Past,* passim.

36 J. Rendell Harris, *The Dioscuri in the Christian Legends*, p. 29.

37 Harris, Boanerges, p. 24.

38 Teit, *The Thompson Indians of British Columbia,* p. 338.

39 *Jesuit Relations* (X: 319, 320).

40 Harris, *Boanerges.* See chapter, "The Thunderbird," pp. 20–30.

41 Homer, *Iliad.*

42 In the change from Saturn Configuration to solar system, confusion between the sun of Saturn and Sol, our present sun, ensued.

43 J. Rendell Harris, *Boanerges,* passim.

the brother of James; and he surnamed them Boanerges, which is, the sons of thunder."[44] The twins were named immediately after Peter, signifying their importance.[45] Since they sit on either side of Jesus, they are comparable to the twin kites (birds), Isis and Nephthys, but minus the dark aspect of Nephthys. Yet the association with the crescent bow or thunderbird is sufficient for the thunder to devolve on them and other twins.

The concept of an exchange of electrical forces both along the axis column and within the Saturnian "eye" brings another set of twins into focus: Horus and Set. In the association of Horus and Bes we can see the twinship of a transforming identity; in that of Horus and Set, the twinship of polarity, good and evil. And these latter twins began fighting in the womb, in this case in the Saturnian "eye" within the circle band of the mother goddess. As Horus was the son of Isis and Osiris, Set was the son of dark Nephthys and Osiris. Set was a god originally considered to be good, who later became evil. As O'Neill notes, ". . . the figures of Set were destroyed or mutilated with a care *which makes them extremely rare in our museums.*"[46]

Set can be seen as one head of the two-headed representations of Horus, the dark animal head opposed to the hawk head more often associated with Horus. In passing, we note yet another bird associated with the Saturn Configuration. Set of the long ass ears becomes another representation of the Hyades animal at the pole. He is represented at the top of the rod (axis column) held by numerous Egyptian gods, reflecting his polar position. That a new, changed order of the heavens is in place is suggested by the crescent shape often at the bottom of the rod, very likely the crescent bird of his mother, the fallen Nephthys. Set also has ties to other "fallen" positions of the terrestrial pole. Ursa Major is the "thigh of Set," Ursa Minor the "jackal of Set."[47] Thus Set was considered to be the fallen, evil twin.

Horus and Set.

Now the Egyptians were familiar with the magnet, baa-n-pet, penipe or benipe in Coptic. They considered magnetized iron to be holy while nonmagnetic iron was believed to be evil.[48] Thus it must give us pause to discover that magnetic iron was believed to come from Horus while nonmagnetic iron was associated with Set.[49] This suggests a difference between

[44] Mark 3:17.

[45] Peter is from the Latin root meaning rock which equates with Mithras rock birth.

[46] O'Neill, *Night of the Gods*, p. 786.

[47] Lockyer, *Dawn of Astronomy*, p. 146.

[48] The natural magnet was called the Heraklean stone. O'Neill, *Night of the Gods*, pp. 146–150. Iron which fell from the sky was usually prized by the ancients.

[49] O'Neill, *Night of the Gods*, pp. 106–107.

the magnetic condition of the cosmos during the Golden Age and that after the Fall. More likely the difference was comparative, strong magnetic force followed by a weaker one. Conditions under the Saturn Configuration appear to reflect a more highly magnetized state.

It is impossible to understand the twin cult around the world unless we first grasp the terrible fear that the twins first engendered, such that at the birth of twins, children and mother were killed immediately to begin with. Later, purifying rituals would be developed to ameliorate this harsh situation, preserving first one, then all three of the parties. Eventually, the birth of twins would transform, in some societies, into a fortunate situation.[50]

Electromagnetic activity within the Saturn Configuration, both in the axis column and the central "eye" of the configuration, would have been the source of the fear of twins. The *Pyramid Texts* refer to a time "before that fear which arose on account of the Eye of Horus existed."[51]

In such fashion did mankind become fixated on twins early in the present cycle of world ages. After the Fall, when the pole appeared in the constellation of the bull Taurus, there was a natural concern to locate a set of twins nearby. As the precessional motion of the Earth was noted,[52] man looked back along the ecliptic to find his past and found two suitable stars just across the Milky Way from Taurus. They became the twins we know as Dioscouri, Castor and Pollux, the two brightest stars in the constellation Gemini. The ancient Peruvians called Gemini "creation time" while the starting point of the Olympic races was at the altar of Gemini.

No discussion of Mars can overlook the planet's color, red. Following Harris, red is often associated with twins. The Dioscouri wear red robes. On the ceiling of Senmut's tomb, illustrating the fall of Saturn, the small circle at the apex of the axis column represents the planet Mars, closest to Earth in the Saturn Configuration. After all the intervening millennia, it can still be seen to be red.

When it was closest to Saturn, Mars would have appeared from Earth as a circle within the circle of Venus, in turn within the orb of Saturn. This was the source of all the "eye" and "heart" imagery in ancient texts, the "pupil" of Mars within the "iris" of Venus, the ab heart within the hat heart. Thus we read in Egyptian texts that Horus is the ab en hat, "the ab heart of the hati-heart."[53] "I am Horus, that proceedeth from the eye of Horus."[54] "I am he who dwelleth in the middle of his own eye."[55] A more familiar simile, fitting for the small red sphere at the center of the eye image is "my origin is from the apple of his eye."[56]

The redness of the ab heart is revealed in its representations, always of a red material: red stones, red glass, red porcelain, red paste, and red wax.[57]

It is only necessary to affix the "bird wings" of the dark side of the Saturnian band to the central red orb to obtain a dark bird with a red head. This would inform the Roman wood-

[50] Harris, *Boanerges,* Introduction.

[51] Faulkner, *Pyramid Texts,* Utt. 486.

[52] With the pole so close to the ecliptic, it would have moved much faster that we are accustomed to seeing today, and been more easily detectable.

[53] Peter le Page Renouf, *The Egyptian Book of the Dead* (London, Society of Biblical Archaeology 1904) p. 67.

[54] Ibid., p. 128, chapter LXVI.

[55] Ibid., p. 164, chapter XCVI.

[56] Ibid., p. 143, chapter LXXVIII.

[57] E. A. Wallis Budge, *Amulets and Talismans* (New York, 1970) p. 138.

pecker myth in which Picus, king of Latium, was transformed into a woodpecker by Circe,[58] giving us yet another bird at the pole. In an ancient epoch, a Martian woodpecker led the Picentes to a new home (Picenum) whereupon they adopted its name and set up a pillar with the image of a woodpecker on top.[59] The bird of Mars spoke through an oracle at Tiora Matiene,[60] while all woodpeckers are now members of the genus Picus. When we turn to the chapter on heavenly gates, it would be well to remember that it was the woodpeckers or axe-birds who hewed out the gates of Cloud-cuckooland in *The Birds* of Aristophanes.[61] The poet also claims that the birds ruled before Zeus.[62]

Other thunderbirds identified by Harris are the robin, hummingbird, and eagle, but perhaps the cock with his bright red comb makes a particularly striking image of the thunderbird, sitting as he does today atop the weathervane with its directional pointers as analogs of the rivers of Heaven. The cock frequently attends the Mithraic twins, Cautes and Cautopates.

In the study of Mars and his atavars we can begin to grasp a developmental aspect of mythology. We see layers of information build up through successive catastrophes while occasional heroes attract enough energy to evolve on their own. In the mythos surrounding Mars, we can trace the warrior-hero through a number of world ages and catastrophic incarnations. In the Golden Age, he gives pattern to the cosmic twins; as the powerful Hercules, he holds up the heavens between the visits of Venus; as the five warriors of myth and legend, he attacks Earth in the 8th and 7th centuries B.C. while Hercules, his alter ego and now a protector of Earth, leaps to our aid.

The mythos of Mars is a classic accretive one, recorded worldwide in ancient media by ancient witnesses, elaborated by Homer and other mythographers, delineated in the Labors of Hercules as he carved out a new celestial equator for Earth. In these few pages, we can only give the general thrust of the warrior-hero's catastrophic story, one that has preoccupied mankind for millennia.

[58] O'Neill, *Night of the Gods,* pp. 40–41. By now, we can recognize Circe as yet another goddess of the Saturnian band. Her name gives her away. She was the one who transformed Odysseus' men into pigs (the boar of the Hyades, piglets following Graves) in the flashback portion of the *Odyssey.*

[59] Harris, Boanerges, p. 212. Also see Plutarch, "Quaestiones Romanae," on why the Romans venerate the woodpecker.

[60] Dionysus of Hallicarnassus, 1.14.5.

[61] Aristophanes, *The Birds,* trans. Gilbert Murray (London, George Allen & Unwin, 1950) p. 112, lines 1154–1170.

[62] Ibid., p. 40, lines 288–295. These would be the "birds" of the Saturn Configuration. Aristophanes notes that, following the tradition of bird rule, Zeus is represented by an eagle, Athena by an owl, and Apollo by a hawk; p. 61, lines 511–521.

The Mexican Star Sign from the Vienna Codex.
The five omega signs striking the base of the construction symbolize the five catastrophic interactions of the Earth-Mars Wars.

23. Order from Chaos: The Mexican Star Sign

. . . a mistake to try to reduce images to the concrete. They are multivalent as dreams are layered. The image is true only if we do not try to reduce it to just one of its meanings.

– Mircea Eliade, *Images and Symbols*

Maybe one should count for a change.

– Hertha von Dechend, *Hamlet's Mill*

The wind chimes that hang outside so many homes today are often constructed according to fixed mathematical ratios. When activated by a breath of air, they provide what appears to be a delightful interaction between nature and science. One of the great charms of number is that it can be connected both with music and the rational quality so highly prized by Western Civilization. Since the Greeks, the ability to grasp ratio is that of the rational.

"Things are numbers" according to the Pythagoreans, who made an esoteric study of the latter. "Ten thousand things" is the Chinese expression for the world. To the Maya, numbers were gods. Certainly things are intrinsically countable and the world yields to measurement. As Stecchini remarked, "The reason for which physics became the queen of the sciences is its consistent determination to introduce measurement into its discussions at the very earliest stage."[1]

An ancient symbol for the Earth is a stone in a netted bag, the net suggesting the meridians we draw on it in order to grasp it. In truth, we cast a net of numbers over all of reality, measuring the world in every possible way so that we can deal with it in our own characteristic style. Cosmic catastrophe pushed mankind along this path, for each one forced him to generate new sets of numbers to relate Earth and Sky. Days and months had to be counted anew, the angle of the ecliptic noted, solstice and equinox located. Thus in the *Popol Vuh* of the Quiche Maya, ". . . the measuring cord was brought, and it was stretched in the sky and over the earth, on the four angles, on the four corners, as was told by the Creator and Maker, the Mother and the Father of Life, . . ."[2]

Man has been hard at the business of counting and measuring time out of mind. Stecchini, the doyen of ancient measurement, tells us that "all units of length, volume and weight can be derived by a few simple arithmetical rules from a single lineal standard, the so-called Egyptian foot of 300 mm."[3] The order that number and measurement brought to culture was particularly remarkable in ancient Greece, for they tied their money to a system of measures developed in Mesopotamia which connected length, volume, and weight.[4] The Egyptians and Babylonians had already brought time into the equation.

Each civilization has its own style of ordering the world in a mathematical fashion. Roman numerals seem more of a stumbling block than an aid, but calculus is a Latin word, from a widespread habit of our forebears'

[1] Livio Stecchini, "Metron," *Kronos* VII, 1, p. 43.

[2] Recinos, *Popol Vuh*, p. 80.

[3] Stecchini, "Metron," *Kronos* VII, 1, p. 43.

[4] Ibid., p. 43.

drawing a grid on the ground and figuring with small stones or calculi.[5] The Arabs gave us the zero, while the decimal system appears to have come from Egypt, where the year of 360 days was divided into 36 decans of 10 days each. The Maya used the base 20. Their symbol for zero was sometimes a shell, suggesting that they had an organic perception of number. And while we like to run our calculations out to many decimal places, the Maya had a great concern with cycle.

Both God and man have a claim to be the measure of all things, for numbers first came from the sky but became Earth-commensurate, expressed in the homely metaphors of the human body. The concern for correct measurement generated almost a religious atmosphere, so that measures, including those of money, were often kept in temples or medieval churches.[6]

Although science, in its self-proclaimed role as arbiter of objective reality, preempts number, it belongs to art as well. We have already noticed that music is embedded in the natural order in a way that can be expressed by number. A 2:1 ratio produces an octave in lengths of metal or wood, tubes or stretched strings. A fifth, more subtle to the ear, is produced by a 3:2 ratio.[7] The beat of music relates to number just as the foot of poetry and dance. Proportion is intrinsic to the plastic arts, sensed as music is heard. Science and art still work together in architecture, number permeating structure to produce both strength and, hopefully, beauty. Thus number lends itself to both science and art.

The occasional impression we all have of living in an orderly universe is usually a fleeting one, obscured by the endless observable struggle of civilization to bring order out of chaos. In the past, the unitary character of knowledge allowed all of it to be brought to bear on historical catastrophe, the primary expression of chaos. We have seen the wedding of art and science in the bull-slaying of the Mithraic Mysteries, of myth and science in the Labors of Hercules. The Bearson folk tale brought order out of chaos by transmuting the Earth-Mars Wars into a generative new adventure.

Today we look to science to stand between us and chaos, largely because of the spectacular successes of physics in extending its models to change our world. This has had the unfortunate effect of intimidating other branches of learning, and we see them calling themselves sciences, aping science insofar as it is possible. The ancients remind us that art, myth, and legend can perform some part of science's task of making the world seem safe and orderly. One modern, Isak Dineson, has said that man can deal with anything if he can make up a story about it.

Art, like science, is a way of looking at the world, bringing order out of chaos and making life richer and more pleasant. A popular art that has endured for many centuries among women (and more recently, men) of Western Civilization is that of quilt-making. It is art, craft, meditation, and therapy, the distaff blues. Patterns for star and flower quilts abound. Often centered and quartered, they offer ample opportunity for expressing a mandala-like construction. One occasionally encounters a pattern for a quilt called Mexican Star, rather betwixt and between with its alternating starlike points and rounded petals: the former lending itself to piecework, the latter to applique. Like so many other patterns, Mexican Star is centered and quartered, but when we examine its prototype in ancient art, in codex and iconography, we find only half of this unusual star flower.

[5] Fell, Barry, *Saga America* (New York Times, 1980) pp. 314–330.

[6] Stecchini, "Metron," p. 44.

[7] John Backus, *The Acoustical Foundations of Music* (N.Y. Norton, 1997) 2nd edition, pp. 135–137.

The Mexican Star Sign (detail from the Vienna Codex).

Well documented in Mexico, less so among the Maya,[8] the Mexican Star Sign is variously drawn, but often with alternating rounded and pointed petals. The Venus omega glyph sometimes decorates an eye at the center of the flower, while other eyes can be sometimes be found within the rounded petals. Both lines and crossed bands can occasionally be seen on the rounded petals, while pointed petals can be distinguished by curved "teeth" or "claws." Like the crossed bands of the Olmec, the Mexican Star Sign was never seen in the sky, but derived from events which transpired there.

Tracing Mars in Mesoamerica is a highly problematical enterprise, for the trail is crossed and recrossed with the spoor of Venus. It is Venus who is considered dangerous, Venus who is associated with warfare, Venus whose very cycles determine when war should be waged.[9] That the straightforward cycles of 780 days described in the Dresden Codex are not admitted by orthodoxy to be those of Mars scarcely helps the issue. Though the Mexican Star Sign is associated with warfare and Venus, we will examine it for indications of the presence of Mars.

A few words are probably in order about the cycles of the planets as observed by the naked eye versus the net of numbers thrown up by Western Civilization. Venus, that paragon of regularity, has a historically well-observed favorable opposition to Earth: five synodic periods of Venus every eight Earth years. It yields a repeating set of numbers for the Venus periods: 580, 587, 583, 583, 587.[10] The highly irregular Mars has a mean synodic period of 779.9 days, but this can vary from 25 to 27 months! Mars' periods of favorable opposition occur every 15 years (seven synodic periods) but occasionally ratchet up to 17 years (eight synodic periods)! Thus it is immediately apparent that observers anxious about a close meeting of, say, Earth and Mars in space would be little aided by the Western habit of averaging out to lengthy decimal places. Cyclical rigor is sufficient to address the problem.

The ancient Mesoamericans were naked-eye astronomers with a habit of perception that differed from our own in being less linear, more holistic. Their great concession to the linear view was the Long Count, a record of actual elapsed days reaching millennia into the past. The shakeup in the sky during the first half of the 8th century B.C., that sent Mars off on a new destructive path, was a signal for them to start counting and recording new cycles. That they saw the Earth-Mars Wars as a string of five star flowers is as it happened, a curious record both linear and cyclical.[11]

[8] But see Virginia Miller, "Star Warriors in Chichen Itza," *Word and Image in Maya Culture,* ed. Wm. F. Hanks & Don Rice (Salt Lake City: Univ. of Utah Press, 1989).

[9] F. G. Lounsbury "Astronomical Knowledge and Its Uses at Bonampak, Mexico," *Archeoastronomy in the New World,* ed. A. Aveni (Cambridge U. Press, 1982) pp. 143–168.

[10] Robert W. Wilson, "Astronomical Notes on the Maya Codices," (Cambridge, Peabody Museum, 1924) Vol. VI, No. 3, p. 21.

[11] The Maya recorded the cycle of Mars on pages 43–45 of the Dresden Codex. Wilson treats the record in "Astronomical Notes" as well as anticipating tables for Jupiter and Saturn, and possibly Mercury. His work, naturally, is outside the paradigm.

Ehecatl raising up the sky with the five Mexican Star Signs.

In the Vienna Codex, Ehecatl holds up a skyband containing five Mexican Star Signs. Above are shells signifying beginnings. Ehecatl is generally thought to be an aspect of Quetzalcoatl holding up the sky. In fact, he is in the act of raising it. During the Jaguar Age, the sun would have wandered farther north and south, lending the illusion of the sky being lower. After the tilt of the Earth-Mars Wars, the sun stayed closer to the equator, giving the effect of a higher sky.[12]

The Dresden and Borgia codices took a different view of the five catastrophic events; in the former Quetzalcoatl throws five spears to Earth and there are five "piercings" in the latter. These seem to show Mars as a mere tool, Venus given credit for force and intent. The Olmec simply indicate five visits of a cometary body, five interactions of the jaguar at the pole with humanity, producing the five were-jaguar babies of the Altar of the Quintuplets. But the Mexican Star Signs, both singly and in groups of five, are the most highly patterned of all ancient media describing the Earth-Mars Wars. We will reduce some of the more carefully drawn signs to their component parts in an effort to penetrate their meaning.

At the center of a number of Mexican Star Signs is a circle or eye with the hair of the Venus omega sign hanging around it. Thus, although the sign will prove to be in part linear, it does have a center. In the 8th century B.C., Mesoamerica was still looking to the pole for its gods, so that the center would naturally be indicated. The new cycles of Venus and Mars would presently be sorted out,[13] but these two were already familiar to astronomers who had knowledge of the Saturn Configura-

[12] Thus Quetzalcoatl is analogous to Maui who raised the sky in Polynesian myth.

[13] The cycle of Venus was already familiar although it appears to have become more regular as a result of the Mars-Venus interactions. The object of study for the ancients would have been to record the new cycles of Mars and try to see a pattern in it.

tion.[14] The eye and heart imagery associated with a center is well attested in Mesoamerica. Round bands circle many an eye in codex and other art, repeating the bull's eye of the Saturn Configuration. Venus is well known as Nohoch Ich, "Great Eye." The focus on removal of the heart in sacrificial ritual, mirroring the dissolution of the Saturn Configuration, is too well known to require comment.

It is the sequence of the petals of the star sign that appears to express a linear pattern, a sequence in time. Since their roundness suggests the Venus omega sign and they often have an eye at the tip, the rounded petals can be taken for the eight-year cycle of Venus. Seeing Mars as the pointed petal requires more consideration.

Mars at conjunction is associated with the flint hatchet in the Dresden Codex,[15] suggesting that every near approach of Mars was viewed as potentially dangerous to begin with. But following Velikovsky and Olmec iconography, there was a period of grace of 15 years between catastrophic conjunctions. Thus, today's favorable oppositions mirror former catastrophic conjunctions,[16] suggesting that there has been relatively little change in the synodic period of Mars.[17]

Mars' mean synodic period is about 780 days, approximately three times the "mysterious" 260 day Mexican Tonalpoualli or "count of the planets or heavenly bodies which illuminate or give light,"[18] a little over two Earth years. Thus, three synodic periods of Mars would fit handily within an eight-year Venus cycle. And when we look at the pointed petal, recognize it as a flint, and notice the three teeth or claws, we see the pattern recorded. For the sake of artistic coding, one flint with three teeth equals three synodic visits of Mars. The codices are filled with Mexican Star Signs that display the pattern of alternating rounded and pointed petals, suggesting the importance of this particular pattern.

Again, we must keep in mind that patterns from the sky were serious business to be recorded and respected. En passant, the flint with three teeth or claws looks very much like the "knuckle-duster" of the literature, commonly thought to be held in the hand as an accouterment of the ball game, tlachtli.[19]

With the passage of time, the star sign described above would disappear as a phenomenon from the sky. The Venus cycle of 8 years, the Mars cycle of 15, do not mesh well; they would soon slide past each other, destroying the balance of the original image. But it would continue to be reproduced, probably idealized in the image of Ehecatl raising up the sky. The Mexican Star Sign can often be seen interspersed with the flint symbol in friezes and what appear to be skybands.[20]

Further sliding of the cycles of Venus and Mars against each other would eventually have

[14] The Maya recorded Saturn as Vucub Caquix in the *Popol Vuh*.

[15] Dresden Codex, 44, 45. Also see Mars glyph on 71 and 72 with stone hatchet.

[16] The 15-year catastrophic conjunctions of Earth and Mars at the Spring Equinox would have been more regular than the present favorable oppositions. On the other hand, a month was lost out of the calendar in 702 B.C., a week in 687 B.C.

[17] One change would have been the occasional cycle wherein Mars ratchets up one synodic period to give a 17-year cycle of favorable opposition.

[18] This is a period which properly belongs in the sky according to Motolinia, where its multiples closely approximate the synodic periods of the planets. See Zelia Nuttall, "The Periodical Adjustments of the Ancient Mexican Calendar," *American Anthropologist*, N.S., 6, 1904, pp. 497–499.

[19] This is typical Nahuatl poetic thought. Warriors are the "teeth" and "claws" of the state.

[20] See the Borgia Codex, pp. 5, 6, 9, 18 of the "ritual sequence."

resulted in the "favorable opposition" of both 8- and 15-year cycles drawing closer together in time. Thus in some representations of the star sign, we find the pointed petal within the rounded petal. Less frequently we can find instances where pointed and rounded petals are sorted out, producing star signs with all rounded or all pointed petals. All of these styles can be found at Chichen Itza where they have evolved into accouterments of war, notably a star skirt and star head ornament.[21]

The Mexican Star Sign continues to move and change, but most frequently as a half star. A fairly stabilized form is recorded at Chichen Itza on the Venus Platform. The half-star (point within rounded petal) is positioned next to a year bundle and eight eyes. These represent the 52-year and 8-year cycles of Venus. The center of the star sign is a study in the concentric circular symbolism, the "bull's eye" of the Saturn Configuration. It is all there:

Venus signs on the Platform of Venus in Chichen Itza.

the Saturnian band, Saturn itself, Venus within Saturn, and a central dot that can only be Mars within Venus. Also, each of the eight Venusian eyes carries the central Martian dot.[22]

The outer band of the Saturnian bull's eye is wrapped around an upright bar, and it has been suggested that this bar could be the Maya sign for five. Another possibility would be the axis column.

The Mexican Star Sign is also figured on a band above the Mercado Dais, connecting it more closely with the events of the 8th and 7th centuries B.C. Two bearded serpents confront each other on the narrow band, jaws outstretched, bodies curving away from the center in mirror image. A quincunx of five dots rests between them. Star signs are nestled in each curve of the serpents, five star signs awarded to each.

We suspect these two serpents to represent Venus and Mars, both of whom have a twin aspect that could account for this style of representation, and both of whom are enmeshed in the Mexican Star Sign. Venus is held to be Quetzalcoatl, which means "precious twin" as well as "bird serpent." She sometimes has a double eye glyph

The serpent band above the Mercado Dais at Chichen Itza.

[21] Virginia Miller, "Star Warriors." Here the rounded petal and pointed petal are called "lobe" and "ray."

[22] See Talbott for convergence of "eye" and "heart" imagery. "Mythical History of the Comet Venus" *Aeon*, Vol. II, No. 4, pp. 47–48.

(mirroring her double visit in the 15th century B.C.), and manifests as both morning and evening star. A separate chapter has already been devoted to Mars as twin.

An evocative mural which sorts out elements of the Mexican Star Sign was found at Cacaxtla in the Temple of Venus.[23] A giant figure wears a Venus skirt of five rounded petals, an omega glyph at the waist with eye and pupil. Five 5-pointed star signs are figured, one in the left hand of the image. This is the hand of Huitzilopochtli, "hummingbird from the left," but actually a jaguar's paw holding the pointed star sign of Mars with its five flints. The symbolism is transparent, the jaguar at the pole using Mars as a weapon.

The Cacaxtla twins.

With its head turned to one side, showing the single Saturnian eye which also represents Venus and Mars, this image is clearly that of a god for it is painted a light blue-green, a color which seems to have suffused the Saturn Configuration for long enough to be widely recorded.[24] It is a composite cosmic god, its scorpion's tail even reflecting the Summer Solstice between the two visits of Venus. More recently the female twin of this figure has been excavated; they are painted on adjacent pillars. Although much of the top of the female figure is gone, what remains is almost identical but for the lack of the scorpion's tail. Clearly both male and female, Mars and Venus are involved here.

The five star signs or flowers of the Earth-Mars Wars would continue to be celebrated in Mexico at Xochihuitl, a moveable feast honoring the divinity Macuilxochitl (five flowers), an event which lasted five days. Offerings were made of five corn tamales and S-shaped cakes were baked in honor of Xonocuilli (Ursa Minor).[25] In such fashion were the five cosmic interactions linked with a change in direction of the pole.

The Mexican Star Sign in its various appearances, singly and serially, is not the simple tale of the five visits of a war god related in other cultures. It is a holistic, patterned vision of the Earth-Mars Wars, knit from the very cycles of the planets, and including implications that reach back to the Saturn Configuration. Although foreign to our style of perception and representation, it is one place we can begin to study the ancients. For in order to perceive the alien skies of the past, we must be willing to look through their eyes.

[23] John B. Carlson, "America's Ancient Skywatchers," *National Geographic,* Mar. 1990, p. 98.

[24] See chapter on "Saturn Myth."

[25] Sahagun, *Historia,* Book I, Part II, cap. 14 (Santa Fe, Univ. Utah) pp. 31–32. In cap. 3 of I, VII Ursa Minor is described as S-shaped. Also see the translation by Fanny Bandelier (Nashville, Fisk 1932) p. 35.

An imposing gate at Karnak, ornamented with
the Sun of Saturn and its attendant snakes.

24. Heavenly Gates: The Solstices

The Gate of Heaven is at the North Pole in all ancient cosmologies.

– Max Muller, *Sacred Books of the East*

Gates he opened on both sides, made fast the locks to left and right.

– Marduk Installs The Stars, *Enuma Elish*, 5th tablet

Hera stepped upon the flaming car and self-bidden groaned upon their hinges the gates of heaven.

– Homer, *Iliad*, Book V

The vision of heavenly gates that open into Paradise enjoys a long, rich tradition with mankind, and gates in general a heavy symbolism. When passing through a gate or door, one should not speak according to Pythagoreans and Egyptian wise men. It is a matter of paying respect to the god who holds the beginning of all things. Even those who do not believe in Heaven and Hell must accept the reality of their gates, for they are still with us, the solstices in fact, as the phenomenon was understood by the ancients. For it came to be known that when the terrestrial pole tilted along the great circle of the solstitial colure, the Earth and all its inhabitants passed from one world age to another.[1]

Porphyry discusses gates in "De Antro Nympharum," referring to the cave of the nymphs on Ithaca in the *Odyssey*. The cave is analogous to the cosmos; its two mouths are gates to Heaven and Hell. We learn with some surprise that the upper gate is for men. Through it, they both descend into life and ascend out of it. The lower gate is exclusively for gods, which makes more sense when we recall the classic trip to Hell made by the gods at the time of the Fall, and repeated by gods, goddesses, and heroes ever since.

Following Porphyry, "The usage of referring to these two tropics in Cancer and Capricorn as 'gates' goes back to the theologians while Plato speaks of two mouths."[2] In fact, Plato called the 12 signs the Gates of Heaven,[3] and uniform precession would theoretically carry through the entire zodiac over 26,000 years. However, the Gates of Heaven, the ones corresponding to the Summer Solstice, have only been in five signs throughout the present series of world ages: Taurus, Scorpius, Leo, Cancer, and Gemini.

The Romans chose to celebrate the Gates of Hell, the Winter Solstice in Capricorn, honoring Janus, the god of gates, who gave his name to the first month of the year. Janus had two faces, an extra one on the back of his head so he could look both ways, symbolizing his power over gates. And here we should recall that the months of January and February only came into existence toward the end of the Earth-Mars Wars when years of ten months were replaced by years of twelve. The gates of Janus were ceremonially closed by Augustus to

1 Reckoning by tilts, ancient Greeks and Mexicans count five world ages. By adding precessional spins, their number may be increased.

2 Porphyry, "De Antro Nympharum."

3 O'Neill, *Night of the Gods*, p. 179.

proclaim peace in the Roman world, but the actual solstice festivities, the Saturnalia, continued to be celebrated. It comes as no surprise to learn that Janus, god of gates, loved Cardea, goddess of hinges, "of the socket in which the Universe turned on its Axis."[4]

It was entirely appropriate to hold a feast in honor of Saturn at the Winter Solstice, the Gates of Hell. Saturn, after all, was thrown out of the chariot after a wrestling match with Zeus and was last seen sliding down the river of Eridanus toward Hell. The upside-down rituals of the Saturnalia wherein masters and slaves changed places not only reflect the Fall, but also the good feeling and happy memories that attended the Golden Age of Saturn. Eventually this holiday air of brotherhood would devolve upon Christmas, which was not assigned a calendar date until early in the 4th century.

The Gates of Heaven are, of course, no longer in Cancer any more than the Spring Equinox is in Aries. These positions merely locate the period when mankind in general ceased to regard the precessional clock as a useful timepiece. The connection between Summer Solstice and the Gates of Heaven was ignored by Western science, and the largest part of astrology is content with signs rather than the reality of the moving heavens. As we will see, changes in the position of the gates were by nature too catastrophic for the developing Western Civilization to recall. However, they are so carefully marked in ancient media that, in order to set our past in order, we must learn something about them.

The background to gate symbolism can be seen in the Golden Age with the notion of twins on either side of the Saturn Configuration. Both concepts, that of gates and twins, are expressed in the Greek dokana, a construction of two upright parallel timbers joined by two transverse ones. The Spartans went into battle carrying these sacred crossbeams, the visible representation of the Twin Brethren, the Dioscouri.[5] The dokana is an expression of an opening to another world.[6] O'Neill sees it as a gate of Heaven.[7]

During the Golden Age, Earth and Saturn were connected by the visible axis column so that the way to Heaven was clearly perceptible. One traveled by way of the world tree. Gates imply gatekeepers, and it was at this time that Mars (and his analogs), moving up and down the road, bridge, or ladder to Heaven, became the guardian of the gates. Reseph, equated with Nergal who has been identified with Mars,[8] is described as "Gatekeeper of the sun."[9] As gatekeeper of Olympus, Hercules "stands ever before the gates."[10] In Mesoamerica, Cox'ol, the Mayan Mars who watches the mountain road, is tantamount to a gatekeeper. And Samson, whose extraordinary feat of moving the gates of Jerusalem has never been satisfactorily explained, can now be understood to be another gatekeeper.

Since Erichthonios established the Horae at the gates,[11] hours being a useful time connec-

4 E. R. Wharton, *Etyma Latina,* pp. 15, 23, referenced in *Night of the Gods,* p. 432.

5 Harris, *Boanerges,* p. 46.

6 O'Neill, *Night of the Gods,* p. 249.

7 Ibid., p. 245.

8 W. Fulco, *The Canaanite God Resep* (New Haven, 1976) pp. 34–38. Also see Ev Cochrane, *Martian Metamorphoses* (Ames, IA, *Aeon*) p. 5.

9 In the Golden Age, Saturn was considered to be the sun.

10 Callimachus, "Hymn to Artemis," trans. Lombardy & Rayor (Baltimore, Johns Hopkins, 1988) lines 175–185.

11 O'Neill, *Night of the Gods,* p. 418. Erichthonios is one of the names of the charioteer, the constellation Auriga, part of the horse and chariot configuration at the pole in the Silver Age.

tion, they have continued to be well attended. The Mikado of Japan is "divine door." And even though we have generally forgotten the location of the Gates of Heaven, there is still a guardian on duty, St. Peter the Rock.

Long before its attachment to an insurance company, the rock—with its aura of everlasting durability, holiness, and strength—was claimed on behalf of both Mithras and the Christian god. Mithras was held to have been born from a rock or egg, while Christian symbolism, begun in the Bible,[12] is memorialized in the hymn, "Rock of Ages." The metaphor continues in the first line, "cleft for me." Like so many other parts of the Saturn Configuration, the Rock would suffer "twinning," one part on either side of the gate. Not surprisingly, these rocks were understood to be dangerous and difficult to pass.

The gates of the Silver Age would develop doors, bolts, and hinges as they opened catastrophically, but in the Golden Age there was only an opening, well recorded in Egypt. Now the *Pyramid Texts* are thought to be the oldest corpus of Egyptian religious texts now extant. The historical period covered by these precious records includes the Golden Age, the Fall, and the Silver Age. They are a record of sky phenomena which were eventually brought to Earth as funeral ritual. When an earthly ruler died, the event was related to the death of Osiris-Saturn at the end of the Golden Age, alien skies and catastrophic events incorporated into ritual.

The ritual of the "opening of the mouth" is well known to students of the ancient culture. The mouth is opened by the "iron fingers" of the four children of Horus. Now the "Eye of Horus," the planet Mars, hung from the hands of his children. Combining this with the knowledge of Mars traveling up and down the axis column puts the action of "opening the mouth" at the heavenly spring at the center of the Saturn Configuration, source of the four outflowing springs which were also the supports of Heaven, the visible colures or four children of Horus.

Open mouth and protruding tongue can be seen in the face of Medusa, an image already shown to be congruent with the Saturn Configuration. *Pyramid Texts* describe the "Eye of Horus" as sitting on the "tongue" of Osiris. Here we will leave behind all eyes, hearts, mouths, etc. which relate to the human body. What we have here is the "eye" of the concentric circles of the Saturn Configuration when Mars is at its highest point in the cosmic column. It is the "eye" of Saturn, "iris" of Venus, "pupil" of Mars.[13]

Now the "Eye of Horus" is not always in position at the center (mouth) of the Saturn Configuration because Mars, "pupil" of the "eye" moves up and down the cosmic column. Hence we read:

"O Osiris the King, make the Eye of Horus come back to you . . ."[14]

"O Osiris the King, Horus has given you his Eye; provide your face with it."[15]

". . . take the pupil which is in the Eye of Horus, for your mouth is split open by it . . ."[16]

". . . take the Eye of Horus which hung from the hands of his children . . ."[17]

---

[12] Deuteronomy 32:18.

[13] The "eye" is, of course, considered to belong to all three: the bull's eye of Saturn, the Great Eye, Nohoch Ich or Venus, of the Mayas, and the Egyptian "Eye of Horus."

[14] R. O. Faulkner trans., *Egyptian Pyramid Texts* (Oxford Press, 1969) Utterance 66.

[15] Ibid., Utt. 21.

[16] Ibid., Utt 155.

[17] Ibid., Utt. 710.

"... split open your mouth by means of the Eye of Horus ..."[18]

"... he will be pleased with Horus who is on his tongue."[19]

The *Pyramid Texts* should be read in their entirety for the full effect. These are not occasional references, but a theme.

Exactly what is happening as the "Eye of Horus" moves closer to the "mouth" of Osiris is not clear. Perhaps it is an electromagnetic effect. Some sense of the attraction involved is suggested in Utterance 35: "... your mouth is the mouth of a sucking calf on the day it was born," a proper metaphor for the "Bull of the sky,"[20] "the Bull with radiance in the midst of his eye,"[21] "... the lord of the affairs (?) of the place of the four pillars."[22]

But now comes the catastrophe: "... the Eye of Horus is gouged out ..."[23] "... the Eye of Horus which Seth has trampled ..."[24]

"There is tumult in the sky; 'We see something new,' say the primeval gods."[25]

"This King is Osiris in a dust devil."[26]

"... this King has dropped his cords [pillars of the sky, the four children of Horus] to the earth ..."

"... this King's meal is in the Abyss ..."[27]

"The sky is overcast / The stars are darkened / The celestial expanses quiver / The bones of the earth-gods tremble, ..."[28]

In Utterance 303, the king is described as flying to the sky like a bird, flapping his wings. The bird wings would be those of the illuminated crescent of light, still visible as Saturn leaves its polar position.

We have come to the time "... when the sky was separated from the earth, when the gods ascended to the sky,"[29] and it is at this point that "the sky is opened, the earth is opened, the apertures of the celestial windows are opened, the movements of the Abyss are revealed."[30] As the Gates of Heaven, the solstitial colure, are opened for travel, the gods slide down the "river" of Eridanus with the Milky Way in the background as it approaches Earth. The terrestrial pole also tilts down, stopping just short of the ecliptic in the new apparition of the bull of the sky, the constellation Taurus.

The head of Taurus, the Hyades, can now be understood to be the Apis bull. "I affixed the head of Apis on the day when the ox [Saturn] was lassoed."[31] There is insufficient space here to go into it deeply, but we are now in the myth of Isis and Osiris, where Osiris has been killed by the evil Seth (Horus' twin) and Isis and her sister Nephthys are hunting about for the parts of his body. "O Osiris the King, you have gone but you will return, you have slept, [but you will awake], you have died, but you will live."[32] Amazingly, the parts are found and Osiris is put together, but he will not look the same.

[18] Ibid., Utt. 93.

[19] Ibid., Utt. 504.

[20] Ibid., Utt. 273–274.

[21] Ibid., Utt. 319.

[22] Ibid., Utt. 217.

[23] Ibid., Utt. 540.

[24] Ibid., Utt. 111.

[25] Ibid., Utt 257.

[26] Ibid., Utt 258, also 259.

[27] Ibid., Utt. 259.

[28] Ibid., Utt. 273–274.

[29] Ibid., Utt. 516.

[30] Ibid., Utt. 503. The Watery Abyss, the celestial vault below the zodiac, is now visible from the northern hemisphere of Earth.

[31] Ibid., Utt. 254.

[32] Ibid., Utt. 670.

"O King, your head is knit to your bones for you, and your bones are knit to your head for you. The doors of the sky are opened to you, the great bolts are drawn back for you, . . . your face is that of a jackal."[33] The Hyades have long been associated with Seth, who has an animal head, and his head and that of the jackal-headed Anubis are often seen in hieroglyph at the top of a long rod which would indicate the terrestrial pole.[34] We suggest that the evil Seth is credited with moving the terrestrial pole to a new association with the Hyades. This location is also suggested by Utterance 699: ". . . be alive! Be young beside your father, beside Orion in the sky." Orion holds a star in his hand in many representations; it is the bright Aldebaran, the prominent bull's eye of the Hyades. The *Pyramid Texts* are unequivocal on the new jackal-face of Osiris.

"O King, whose form is as mysterious as Anubis, receive your jackal-face." ". . . Anubis who presides over the God's Booth, . . ."[35]

"O King, you are clad as a god, your face is that of a jackal as Osiris."[36]

"Re summons you into the zenith of the sky as the Jackal, . . ."

"O King, your shape is hidden like that of Anubis on his belly; receive your jackal-face and raise yourself . . ."[37]

Another theme of the *Pyramid Texts* can be seen in the endless pleas addressed to the king to raise himself, for "This Great One has fallen on his side."[38] Although the new figure at the pole is high in the sky as seen from Earth, the Egyptian astronomers knew which end was up; they were well aware that Anubis was flat on his belly in the ecliptic. They knew that Osiris-Saturn had formerly been at the pole of the ecliptic in the constellation Draco, and that the cosmic axis stretches between the two poles of the ecliptic. Thus another theme, "The Great Mooring Post calls to you . . ."[39]

But at least Osiris has his eye back. "O Osiris the King, the gods have knit together your face for you and Horus has given you his Eye, . . ."[40] "O, Osiris, take the Eye of Horus which Seth has hidden."[41] But there are now two eyes in the head of the knit-together face of the Hyades, the brilliant Aldebaran and another smaller star which balances it. Thus we find ". . . great and mighty sisters [Isis and Nephthys] who gathered your flesh together, who raised up your members, and who caused your eyes to appear in your head, . . ."

The Egyptians, like many of the ancients, naturally concentrated on what was occurring in front of the Saturn Configuration as it came apart. As the powerful magnetic force that held the Saturn Configuration in alignment changed to a lesser one, Giant Jupiter appeared for the first time from behind Saturn. At first it would have been seen as a crescent, another broad-winged bird, the eagle that has always been identified with Jupiter. And it was Jupiter as Marduk whom the Babylonians credited with opening the Gates of Heaven. The "locks to left and right" which he made fast would be the equinoxes, lending great stability to the new

[33] Ibid., Utt. 355.

[34] I have not had an opportunity to examine the original hieroglyphic text, but this is probably the point at which the various rod-like axis symbols begin to appear on their side, flat down on the zodiac as it were.

[35] *Pyramid Texts,* Utt. 468.

[36] Ibid., Utt. 610.

[37] Ibid., Utt. 677.

[38] Ibid., Utt. 442.

[39] Ibid., Utt. 483.

[40] Ibid., Utt. 638.

[41] Ibid., Utt. 752.

solar system. For while the equinoxes (the two points where the equinoctial colure crosses the terrestrial equator) act as hinges in a tilt, and spin along with solstices in a sudden precessional movement, they are like the zodiac and its poles; they endure.

Thus, at the end of the Golden Age, the Gates of Heaven opened for regular and catastrophic motion. The opening would no longer be at the axis of the universe, the Great Mooring Post, but at the solstitial colure, one of the four pillars of Heaven, but no longer visible. Now the regular precessional motion of the Silver Age could be seen as the new sun Sol made its slow, apparent way backward through the zodiac. It is at this point that constructing imposing gates became a popular activity on Earth. Both the terrestrial pole and the point where the solstitial colure crosses the zodiac were in Taurus, the latter determining who guards the gates. So we find bulls or bull-men guarding the gates during the Silver Age.

And since the solstitial colure was also a pillar of Heaven, bulls can also be found on either side of pillars.

Bull-men at the gates.

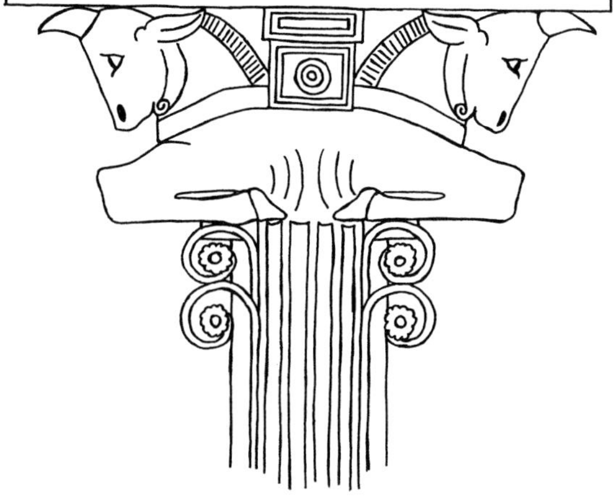

Bulls at pillars.

Regular precessional motion, then, carries the Gates of Heaven through the constellation of Taurus to a point just past the Pleiades. Then the gates open again for an upward tilt which moves the celestial pole of Earth to the constellation Perseus. However, the point where the solstitial colure crosses the zodiac remains in Taurus, so the gate guardians remain the same.[42] The bull, then, guards the gates throughout both the Silver and Bronze Ages.

The next time the Gates of Heaven open is at the 115-degree tilt from Perseus to the constellation Hercules at the horrendous first meeting of Earth and Venus. This event ends the Bronze Age and opens the brief interim period when Hercules was at the pole. Earth's pole tilts, following the solstitial colure through the pole of the ecliptic so that the solstices are exchanged, summer for winter. Now the Summer Solstice is in Scorpius[43] and Scorpion-men guard the gates in the Epic of Gilgamesh, as well as being found on boundary stones.[44]

Heavenly gates open again for the Age of Heroes at the second visit of Venus, but this

[42] Taurus was much larger in antiquity, but the Jews and others considered the Pleiades to be the rump of Aries. Thus we sometimes find rams on either side of a pillar or world tree.

[43] The great circle of the solstitial colure runs through both Taurus and Scorpio, but now the pole of Earth is closest to Scorpio, which determines the Summer Solstice.

[44] The period when Hercules was at the pole and Scorpius at the Gates of Heaven lasted only 52 years and is minimally recorded as far as the Summer Solstice is concerned.

time the catastrophic motion is different, now more like sliding doors. Earth takes a sudden precessional spin while it slides along its revolutionary path. The armature of the colures moves with it in relation to the celestial vault, and the heavenly gates of the Summer Solstice slide three zodiac signs from Scorpius to Leo. Now the familiar lions are at the gates to guard them throughout the Age of Heroes. Lion gates are well recorded from Mycenae on Crete, through Egypt and the Middle East, to Puri in India. At Mycenae, they actually flank a column above the gate, mirroring the solstitial colure as a support of Heaven. Plutarch remarks in *Isis and Osiris*, "They [Egyptians] also hold the Lion in honour, and they adorn the doorways of their shrines with gaping lion's heads."[45] "In Hittite art lions occur chiefly as guardians of portals or supporting statue bases.[46]

After 700 years of a remarkably close-knit Indo-European civilization,[47] largely due to peace in the heavens, the heavenly gates open during the Earth-Mars Wars as Earth tilts catastrophically to inaugurate the Iron Age. The pole tilts into the constellation of the Dragon, to Alpha Draconis, but the gates of the Summer Solstice remain in Leo.

The last two interactions of the Earth-Mars Wars produced two more movements of the Gates of Heaven, sliding ones or sudden precessional spins. We have already discussed the movement of the Spring Equinox from Taurus to Aries. At the same time, because the armature of the colures maintains its integrity as it moves along with Earth in precessional motion, the heavenly gates moved from Leo to Cancer. Fifteen years later, the last precessional spin moved Earth and its armature of colures only about nine degrees on the precessional clock and did not change the sign of the Summer Solstice gate.

Crab gates never caught on, though students of art history should be alert to them.[48] Western Civilization was rising, with its terrible urge to control; we preferred to forget the reality of catastrophic gates for they are patently uncontrollable. The slowly turning circle of the solstitial colure transmuted to tropics, following the Greek word meaning to turn, and came to indicate positions of latitude. Cancer and Capricorn would remain the tropics throughout the Age of Iron, despite the real Heavenly Gates being presently far along in Gemini.

Now we can grasp why gates appear so problematical in myth and legend, their opening and closing dangerous in the extreme. They represent the narrow, difficult place we have to get through, when we are between a rock and a hard place. The cleft rock has become clashing rocks.[49] The entrance of the heavens-river (solstitial colure) is guarded by the κυανεαι πετραι, dark-blue rocks.[50] The good ship Argo came through this entrance, as did Ulysses on his "flashback" journey.[51] These rocks are tantamount to Scylla and Charybdis as well as the wandering rocks in the Argonautica. Dangerous rocks also figure in the death of Ajax in the *Odyssey*. Dual rocks thus represent the catastrophic celestial gates, but a single rock can stop up the cave of the

45 Plutarch, *Isis and Osiris*, 366:38.

46 Ekrem Akurgal, *Art of the Hittites* (Harry F. Abrams, Inc., New York) p. 113.

47 John Dayton, *Minerals, Metals, Glazing and Man.*

48 Lions continued to be used at gates and entrances, and also moved to tombs. They stand at either side of the entrance to the New York Public Library, which still contains books that we cannot read, secrets that we have not penetrated.

49 O'Neill, *Night of the Gods*, p. 38.

50 Ibid., p. 382.

51 Ibid., p. 381.

cosmos at the pole as in the Bearson legend. One needs special knowledge to pass through the entrance, a talisman or password such as "open sesame" in the tale of Ali Baba and the Forty Thieves.

Despite the difficulty which attends the passing of gates for the sturdiest of heroes such as Jason and Odysseus, the gods of Olympus sail through with ease. Twice in the *Iliad* Hera and Athena determine to interfere in the Trojan War on the side of the Argives. In Book V, Ares has been carrying the day on the side of Hector and the Trojans. Hera herself yokes the horses to the chariot while the goddess of war prepares herself for battle.

Athena slips off her soft linen robe, puts on the tunic of Zeus, and sets the goatskin aegis with the awful Gorgon's head upon her shoulders. On her head is a helmet with two horns and she carries a spear, "... so heavy, huge, and strong, with which when she is angry she vanquishes battalions of fighting men." Athena and Hera get in the chariot, usually described as fiery or flaming, and Hera takes the reins as charioteer. "The celestial gates opened of themselves, groaning upon their hinges, ..." Homer has done his best to describe the start of a catastrophe. When the celestial gates groan upon their hinges, we had better be paying attention.

Joining the battle on the plain of Troy, Athena plays charioteer for Diomedes and guides his spear into Ares' belly, "where the kilt was girded." Although belt-wrestling never became popular in Greece as it did farther east, belts yet carried significance. In Homer's epics, the metal belt worn by fighting men is "mitra."[52] We recall that Mithras is given credit for holding the cosmos together, and in the present configuration of the heavens, that force is expressed by the belt of animals, the zodiac. In such fashion does Homer make clear that the action on the field of Troy is, at least in part, in the heavens.[53]

This first foray of Athena-Venus against Mars is carried out with the approval of Zeus. Homer always defers to the ruler of the Olympians. But in Book VIII Zeus changes his mind and Homer lets him speak, describing how he could hang all the gods by a golden chain from a peak of Olympus.[54] He is in the middle of a tirade against the gods interfering in the Trojan War, but this does not discourage Hera and Athena, who plan yet another sortie on behalf of their favorites, the Argives.

The two goddesses prepare as before and mount their chariot. Again the heavenly gates open of themselves to let them through. But this time they are caught out by Zeus, who sends Iris to them with a frightening message of what he will do to them if they proceed. Thinking better of their plans, the goddesses turn back and the Seasons unharness their horses and lean their chariot against the wall. Thus we see heavenly gates open twice during the Trojan War.

We cannot overlook a quarrel between Athena and Ares which took place on Olympus, for it must have been a celestial apparition. Hearing that one of his sons had been killed, Ares called for his horses, Deimos and Phobos, and began putting on his armor. In this instance Athena "left her seat," dashed after Ares, disarmed him, and "pulled Ares back to his seat still furious."

It is not until Book XXII of the *Iliad* that Athena and Ares meet on the battlefield, one on one. Ares lunged at the goddess with his spear against the "tassled aegis cape," while

[52] J. Puhvel, "Mitra as an Indo-European Divinity," Acta Iranica, p. 341.

[53] See *Aeon* Vol. VI: No. 2 for Felice Vinci's persuasive argument that the scene of the *Iliad* and the *Odyssey* was in the Baltic region.

[54] This metaphor describes the Saturn Configuration.

Athena stepped back and "picked up a great stone that stood in the ground, black and rough, an ancient landmark, and threw it, and hit Ares on the neck." This is the action that indicates a change in Mars' boundary line or orbit, tossing him out of the inner circle of planets and into an orbit intersecting with that of Earth, thus setting the stage for the Earth-Mars Wars.

The incident in which Mars received a blow from the "ancient landmark" thrown by Venus raises an interesting issue, for Homer goes on to describe Aphrodite the Moon trying to lead Ares-Mars from the battlefield. Seeing this, Athena pushed Aphrodite on the breast and knocked her down. In order to involve our Moon, we might assume that this incident took place very close to Earth. Exactly what happens in planetary interaction is not clear, but it can happen over great distances as in the case of Jupiter bringing down the Silver Age. On the other hand, it seems that planetary bodies can also approach close to each other and interact selectively or not at all. Mars made approaches to Earth which did not result in moving the Heavenly Gates. One could imagine that the degree of electric charge of the bodies would be critical in any event.

The above phenomena recorded by Homer appear to have taken place in the first part of the 8th century B.C., and not to have affected Earth. But the descriptions of moving gates alert us to the possibilities of catastrophic interactions between Venus and Mars. The spectacle of planets moving suddenly along their revolutionary path is just the sort of event that would have caused the Greeks to reestablish the races of the Olympic Games. More of them presently, but first we will examine the catastrophic art and monsters generated by the appearance of the heavens during cyclical catastrophe.

Constructing imposing gates has been a popular activity throughout the current "fallen" series of world ages.

The griffin leading lions to the gate of the Summer Solstice.

25. Catastrophic Symbols and Monsters

In the no-man's-land between Aegean and Euphrates we find in the second millennium B.C. a monster unknown in earlier time. Without antecedents it succeeded in penetrating the homeland of monsters, Mesopotamia, where it became so thoroughly acclimatised as to out number within a few centuries all but the most vigorous native breeds. It also played an important part in the religious imagery of Minoan Crete and in the decorative art of Mycene, whilst occasionally penetrating Egypt and Anatolia . . . To complete our hypothesis we might add that we should like to regard the griffin as an "Angel of Death."

– Henri Frankfort, "Notes on the Cretan Griffin"

. . . the Egyptians . . . hold the Lion in honour, and they adorn the doorways of their shrines with gaping lions' heads . . .

– Plutarch, *Isis and Osiris*

In our examination of the tauroctony of the Mithraic Mysteries, we learned how hard information can be contained in a so-called work of art. Our linear habit of thinking and dependence on the printed word makes interpretation of art and symbol seem difficult, but this is because we are trying to translate them. The development of trademark, logo, and signage in our own culture demonstrates that the message can move without words. Some of the most powerful mythological symbols introduce our films, which are then brought to life by "stars," while insurance companies seem to prefer those symbols formerly associated with the Almighty. The study of art and symbol, their translation into linear language, is an exercise in linking up the rational and the intuitive. We will find them to be not as estranged from each other as we have chosen to believe.

The most powerful raw symbols, those which least appear to mean anything, come to us from the Saturn Configuration: the sphere, the circle or wheel, the crescent of the Saturnian band, the cross of the heavenly rivers.[1] Talbott remarks that the pictograph of the enclosed sun, a small circle within a larger one, is ". . . a straightforward portrait of Saturn and his legendary home."[2] More Earth-connected is the mountain, pillar, or tree of the axis bond. Fuhr has made a study of the omega symbol in the Old World and we have found its prototype in the heavens and extended it to the New World. These primeval symbols have provided the beginning of written language in many cultures.

Powerful symbols come from the sky as a rule, but they can attach to familiar Earth objects which can then be put back in the sky. In order to understand them, we must allow them to move, turn, transform, attach themselves in new context. Symbols are dynamic, and efforts to grasp them firmly will result in their slipping away. The illuminated crescent

[1] Modified into the Christian cross, the symbol of the heavenly rivers both carries forward and gains additional meaning. The crescent of the Moon saves much Venus and Saturn phenomena.

[2] Talbott, *The Saturn Myth*, p. 62.

of the Saturnian band continued as the illuminated crescent of the Moon and Venus while earlier information was handed down from crescent to crescent.

Myth and symbol can both be seen to be accretive. The Phaethon myth of the sun chariot begins with the sun of Saturn and keeps going, attaching catastrophic information through history each time the present sun changes its path. The Venus omega symbol can be widely traced by itself, yet it can double into a Hercules knot, quadruple into the Venus Lamat glyph with a central point making a quincunx which can also represent the five visits of Mars. These inquiries could lead far, but we can already recognize a general effort to save the phenomena, making what we can perceive as a historical record. Our own compulsion to write and rewrite history springs from the same impulse. Like science, art and developing symbol are an attempt to set the world in order.

In the style of ancient media, the tauroctony of the Mithaic Mysteries can be seen as possibly the last conscious link with historical catastrophism. But between the raw symbols of the Saturn Configuration and the art of the Mithraic Mysteries are millennia of symbol and art to explore in a catastrophic context. Rather than attempt an overview of this vast body of material, we will concentrate on the history of Venus in the cosmos. As recorded in the art of the ancients, the goddess attached a number of symbols while making her monstrous appearances in the skies of Earth.

A catastrophic monster with whom we are already familiar is the Gorgon Medusa with her snakelike hair. This monster has always been connected with Athena-Venus since the goddess carries the head of the Gorgon on her aegis. Now that we have seen that the Medusa image is comparable to the Mithraic lion-headed god, both drawn from the same Saturnian phenomena, we can appreciate what is probably the earliest connection between Venus and the lion.[3]

The association of goddess and lion is extensively recorded in art over a long period. To begin with, the goddess was the lion, "lioness of the Igigi gods," "raging lion," etc.[4] Labbatu, "lioness," was one of Inanna-Ishtar's most common epithets.[5] This vision began as the spiraling Venus, forming what turned into the snakelike locks of Medusa or curly mane of the lion. Later, as hanging goddess, she formed the uraeus (snake) draped over the orb of so much Egyptian art, the primeval Saturn. But after the dissolution of the Saturn Configuration, these images disappeared, replaced by a multitude of stars. It was the effort to find the lost images in the star patterns of the night sky that created some of the constellations we are familiar with today.

Draco must have been the easiest pattern to follow, curled around the pole of the ecliptic exactly where the cosmic serpent had been draped around Saturn. Although the image of Draco is not perfectly symmetrical, it was taken as a symbol for Venus as mother goddess and stylized into the omega sign, each side curling in mirror image.

Venus symbol.

[3] The inspiration for the well-recorded lion-bull combat can be attributed to the dissolution of the Saturn Configuration, the lion-Medusa visage of Saturn obscuring the original appearance of the Bull of Heaven. See Chapter 19.

[4] B. Foster, *Before the Muses: An Anthology of Akkadian Literature* (Bethesda, Md., 1993), p. 59.

[5] W. Lambert, "Labbatu," *Realexikon der Assyriologie* Vol. VI, edit. Ebeling & Meissner (Berlin, 1980–1983) p. 411.

Finding the lion was more problematical, for that image was 90 degrees away on the ecliptic. But Leo, like Scorpius, Cygnus, and a few other constellations, genuinely resembles its namesake. The backward question mark of the constellation suggests the full mane and forehand of the lion, while the following triangle can be the hind legs and tail.

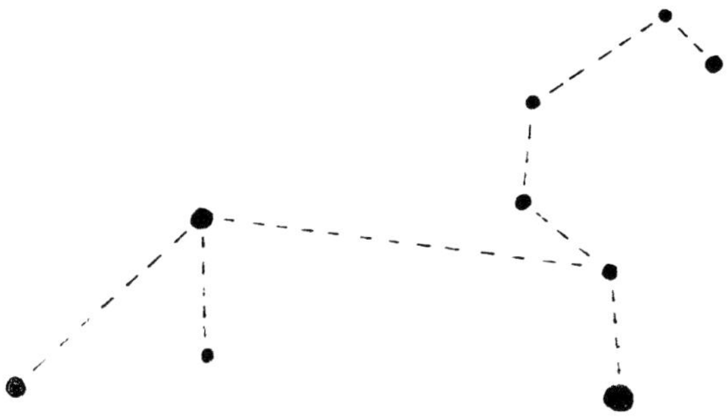

The stars of Leo.

When the lion was noticed on the band of the zodiac, he was naturally assigned the same symbol as Venus. But given the lesser stature of lion versus the mother goddess, the astrological symbol for Leo was never stylized. It remains a perfect copy of the constellation Draco, the side at our right drooping down.

Astrological sign for Leo.

Thus, at the dissolution of the Saturn Configuration, Medusa and a particularly frightening lion were the first catastrophic monsters to be associated with the goddess. By that time she had also acquired two symbols: the spiral[6] and the omega sign. But the cometary Venus would go on to make other appearances in the skies of Earth and incite destructive motions of the planet. And in the course of these later visitations, the goddess would show other monstrous faces.

In the Bronze Age when the constellation Perseus was at the pole, the cometary Venus was recorded by the Sumerians as the reed bundle of Innana. This equates with the sickle-shaped appearance of the comet Typhon which Velikovsky associated with the first visit of Venus. And it is at that time that other monsters begin to appear in the skies of Earth.

Now Typhon, as recorded by Hesiod, was something of a family man, and as the mythographer recounts his progeny, we enter a world of ancient monsters. "Typhaon the terrible, outrageous and lawless, was joined in love to . . ." ". . . the goddess fierce Echidna who is half a nymph with glancing eyes and fair cheeks, and half again a huge snake, great and awful, with speckled skin, eating raw flesh beneath the secret parts of the holy Earth." This unattractive couple produced a brood of horrible monsters, two of which have already been identified in the sky, the Hydra of Lerna and the brazen-voiced hound of Hades. Somewhat later Echidna "was subject in love to Orthus," one of her monster sons, and brought forth the sphinx.[7]

It would be easy to dismiss these monsters as having escaped from someone's nightmare, but the celestial identity of some of them is all too plain. They are, in all likelihood, apparitions formed during the planetary catastrophes caused by Venus, and brought to Earth in myth and art by the ancients. Sorting them out may take a little time, but we will make a

6 Although the spiral and associated maze are figured all over the world, perhaps the most charming representations are the wrapped headdresses of many cultures, notably the xqunq'a kumatz, "rainbow serpent head wrap" of the Tzutuhil Maya.

7 Hesiod, *Theogony*, trans. Evelyn-White, LCL (Cambridge Harvard, 1914) pp. 101–103, lines 295–330.

beginning with the sphinx whose lion-like attributes and connection with gates lead in a direction in which light is available.

The portal sphinx from Hilani II at Zincirli has a human head with curly hair, one lock curling forward in front of the ear, as well as wings, cloven feet, and an unusual motif for either a natural or mythical animal: three circles set within another larger circle on each hock.[8] Another sphinx on a relief from Sakcegozii is male and sports curly hair and beard as well as wings. The style owes much to earlier winged bull-men, but it is clearly lion-like, again with triple motif on both hocks. Portal lions from Zincirli and Sakcegozi also have the triple motif on their hocks.[9]

The Sphinx Gate at Hatussas is ornamented by sphinxes not only female but truly feminine, winged, each with long spiral hair resting on the breast. Six spirals also rise above the head, each containing an eight-petalled Venus rosette. A pair of sphinxes form a pedestal for a statue from Sam'al (Zincirli). They are elaborately feathered as well as winged.[10] The sphinx is thus a lion-like monster who has attached marks of Venus.

Portal lion with triple hock motif from the inner citadel gate at Zincirli.

A pedestal of Sphinxes from Sam'al.

We have already discussed paired lions as guarding the gates of the solstice during the Age of Heroes while the bull Taurus was at the Spring Equinox. Simply noting their positions at either side of portals or gates, supporting columns or statues in pairs or singly, establishes the equivalence of sphinx and lion. It would appear that some of the ancients saw a sphinx-like monster on the ecliptic, with essentially the same lion-like profile.

As the lion Leo moved to the solstice at the second visit of Venus, so the sphinx is a second-generation monster just as Hesiod claims. Typhon-Venus and Echidna produced their first offspring at the first visit of Venus, while Echidna and her son Orthus produced the sphinx at the second.

In the horrific battles that took place in the sky during cosmic catastrophe, it was generally perceived that two celestial monsters were interacting. Historical catastrophes are always

8 Akurgal, *Art of Greece,* p. 55.

9 The mark of triplicity is certainly that of the goddess, while the hock is a joint capable of producing powerful motion.

10 Innina on her lion throne wears a feathered skirt which reminds us of Quetzalcoatl, the feathered serpent of Mesoamerica. Andre Parrot, *Sumer, the Dawn of Art* (New York, Golden Press, 1961) p. 169. Also see pp. 118, 122, 130.

interpreted by their survivors who tend to see the event as a genesis or rebirth. Following Hesiod, Echidna "was subject in love to Orthus," apparently in the sky. It thus appears that Echidna gave birth to a new figure at the solstice at the second visit of Venus, the lion-like sphinx.

But Venus also has a wide pictorial representation as the griffin. We have already seen how Mars and his analog Hercules fought in the Earth-Mars Wars, Hercules defending the Earth against the red planet. In the Earth-Venus Wars we see the same dualistic pattern. The griffin, quintessential monster of the coming age, plays a dual role, has a dual character and representation: a destructive monster and a humanized counterpart who fulfills a beneficial function. Frankfort stresses "the common occurrence of these two related figures in Assyrian art from the 14th century onwards." [11] "Perhaps the consciousness of a great cosmic force might find double pictorial expression, namely in its purely terrifying aspect as the griffin, in an aspect more accessible to man's prayers as the griffin-demon." [12]

Both griffin and griffin-demon carry the marks of Venus. The human figure of the griffin-demon wears the omega sign as a cap of hair. The griffin is four-legged with the same distinguishing mark which seems to carry through all its artistic representations, no matter how stylized. Both have the head of a bird of prey, yet another bird descended from the band of the mother goddess. Depiction of the griffin-demon underscores one aspect of the relationship between Venus and lion. In at least two monumental stone sculptures, the griffin-demon is shown between two lions, holding and apparently leading them.[13] Thus Venus as griffin is credited with bringing the lions to the gates of the solstice,[14] as depicted on page 232.

Innana-Venus commonly sits on a lion throne. The sculpture in the Louvre is supported by six lions, two sitting, two standing, two running. We are indebted to Willy Hartner for noticing that they are all adorned with a sunlike symbol[15] found on the shoulder of uncounted lions in Middle Eastern and Egyptian art. This "shoulder ornament" became an issue for a time in the *Journal of Near Eastern Studies,* which served to bring up a large number of representations.[16]

Lion with shoulder ornament.

There are several distinct expressions of the sunlike symbol and lion. In one of them the symbol is above the lion or engraved on the lion's back above the shoulder. This would relate to the goddess sitting on the lion or lion throne, fighting "the battle having climbed on

[11] Henri Frankfort, "Notes on the Cretan Griffin," British School at Athens, annual, No. 37, 1936–37, p. 108.

[12] Ibid., p. 120.

[13] One of these, showing the griffin-demon full face, is in the archaeological museum at Istanbul, the other is in the archaeological museum at Ankara.

[14] We will examine the mechanics of this motion in the chapter on precessional force.

[15] This particular form has been tentatively identified as a "winged sun disk" by Anne Vollgraff, "The Lion with Body Markings in Oriental Art," *JNES* 1953, Vol. 12, pp. 40–49.

[16] Helene Kantor, "The Shoulder Ornament of Near Eastern Lions," *JNES* 1953, Vol. 12, pp. 250–274.

the shoulders of Leo." In another, the symbol is shown on the shoulder or (occasionally) hindquarters of the lion, and can range from the Saturnian bull's eye through starlike images to rosettes of curved lines.[17] In the instance of the curved lines, the symbol acquires a sense of motion which then attaches to the lion, who as Leo has indeed moved, from the Spring Equinox to the Summer Solstice at the second visitation of Venus.[18]

The rosette of curved lines resolves into a more elaborate form in Scandinavia where it is as common as the shoulder ornament of the lion in the Middle East. Davidson refers to it as the "usual whirling disc motif."[19] It is often depicted in four or eight segments, recalling the quartered image at the pole. In discussing a series of Bronze Age (orthodox dating) stone carvings, she has this to say: "One very prominent object is a disk-like shape, which may be shown as a plain or decorated circle, or as a wheel with four or eight spokes. Men raise their hands to this disc in worship, or hold it up, and it is sometimes set on a support or stand, or carried on a ship."

Following orthodox reasoning based on the present order of the universe, Davidson does not doubt that this whirling motif is the sun, but any sun that sits on a stand or in a boat is the sun of Saturn. When this same disc exhibits motion, however, such as "the famous disc from Trundholm in northwestern Zealand, plated with gold and elaborately decorated with circles and spirals," we are seeing the forces of the precession at work. They held the Saturn Configuration together during the Golden Age and now, in a weaker form, govern both uniform and catastrophic motion in the present solar system.

Whirling disc motif.

The Egyptian monster, Amemit the Devourer.

[17] Other symbols attaching to these lions are well worth study from a catastrophic viewpoint.

[18] This occurred when the catastrophic precessional motion carried the celestial pole of Earth from the constellation of Hercules to that of Ursa Major.

[19] H. R. Ellis Davidson, *Pagan Scandinavia* (Thames & Hudson, London, 1967) p. 199, notes on No. 46.

Mithraic cap with spiral and omega sign.

The whirling rosette can frequently be found on the flank or shoulder of an animal, notably that of the monster Amemit the Devourer in the *Egyptian Book of the Dead.* It is allied to the triskeles and the swastika, all of which express the same turning motion. Nuttall made an extensive study of these symbols that imply motion, always assuming that they were an expression of polar rotation. However, it is the unusual, the catastrophic, that impresses itself on the mind of man and is most likely to be recorded. Her collection of "whirling" symbols is well worth attention.[20]

But there are two forms of motion that precessional force imparts to the Earth in the present order of the solar system: an axial tilt and a sudden movement along its precessional path. Can symbol differentiate between the two? Certainly the whirling type symbols would indicate the spin of a sudden movement of Earth along its precessional path. But what of an axial tilt? What of the particular axial tilt of the first visit of Venus that carried the celestial axis of Earth 115 degrees out of its place, making it seem that the Earth had turned upside down?

The spiral S is another symbol associated with Venus, a link clearly seen in the art of Mithraism. In the 7th grade of Mithraism, the double spiral is connected to the Venus omega sign figured on the Phrygian cap. This particular symbol has been recorded in mosaic, so there is no question about the intent of the depiction. At first glance, it could be easily lumped in with the other whirling symbols, but a particular style in which it is sometimes depicted suggests otherwise.

A bronze caldron at the Hermitage Museum in St. Petersburg records a number of world ages in its symbolism. The enclosed sun of Saturn alternates with bulls' heads at the top of the container while a row of mirror-image Venus spirals, first in one direction, then in the other, fills the space immediately below. This style of putting the spirals together can be widely traced. One does not have to be a seer or an art critic to see that following the spiral turns the pattern upside down. This mirrors Venus reversing 115 degrees of the sky at its first interaction with Earth, to a casual observer turning the world upside down.

Bronze caldron with Saturn and Venus signs.

[20] Nuttall, *Fundamental Principles*, see Index. The question of direction of rotation deserves further investigation. The ancients carefully distinguished between clockwise (sunwise) or counterclockwise, deisul or widdershins. Homer is deliberately redundant in Book I of the *Iliad* describing Hephaistos carrying nectar around to the gods "moving round rightways," "as the sun goes, as the clock goes."

Catastrophic symbols can often be seen to cluster. An exquisite Celtic coin gives an obvious representation of the catastrophic Venus.[21] The goddess herself is on the front of the coin, framed in her own spiraled hair, while an omega sign is poised on top of a mountain on the reverse, a secondary spiral (recalling the symbolism of the 7th grade of Mithraism) spinning off to the right. The surprising 45-degree X on her cheek, mirroring lion sculptures with X's on the shoulder,[22] describes the angle of the sun in the Age of Heroes closely. And, of course, the goddess brought this new shape of time into existence at her first visit just as she led the lions to the solstitial Gate of Heaven at the second.

A Celtic coin illustrates the catastrophic Venus.

An elaborate gold headband of the Scythian culture was found in Barrow three at Kelermes. The head of a griffin with extra spirals is the central ornament while eight-pointed Venus rosettes cover the remainder of the band. A crown would also relate to the goddess as Saturnian band.

Because catastrophic imagery comes out of the sky, ceilings were often richly furnished with Venus symbols. Running spirals are very common as well as the eight-rayed or petaled Venus rosette, and the horns of the goddess as Saturnian band. The ceiling of the palace of Amenhotep III at Thebes has a pattern of running spirals, a bull or cow's head and crescent horns below a sphere, while running spirals and the eight-petalled Venus rosette adorn another ceiling there.

The Medusa with a goose in each hand is a common image of a goddess carrying dead birds by the neck. Following Graves, sphinx means "throttler,"[23] suggesting a dearth of oxygen at the second visit of Venus. This particular image haunted me for months, but it was not until I decided to use it that goddess and bird began to give up their secrets.

Medusa's head, we recall, ended up on Athena's aegis, so that we can consider her equivalent to the goddess here. The dead geese were puzzling, even more so when I noticed catastrophic gate animals, lions, sphinxes, etc. with the head of a goose or water bird on the end of their tail! This sort of representation does not just happen, however. The Pleiades were seen as a flock of birds by the Greeks, and we see them sitting just above the ecliptic (surface of the Watery Abyss) like water birds, solidly connected to the pole at the end of the Silver Age.

[21] See *The Celts*, Sabatino Moscati (New York, Rizzoli, 1991).

[22] Some lions carry a 45-degree X on their shoulder, reflecting the short period between the interactions with Venus as well as the subsequent Age of Heroes.

[23] Graves, *White Goddess*, p. 417.

Medusa with dead birds the goddess "killing" the Pleiades.

Gate animals with waterbird tails.

It was some time before I recalled that the Pleiades were also connected with the Summer Solstice. Even the catastrophic tilt of the terrestrial pole from Taurus to Perseus did not move the Pleiades from their solstice position, and the Bronze Age was so short that they were still relatively close at the first approach of Venus, certainly well within a 30-degree "gate" position. The double wings of Medusa apparently carried the goddess through both Venus events, for the short-lived Scorpion (only 52 years) at the Summer Solstice is not taken into account here. Thus the lion-like gate animals have a goose at the end of their tails to recall the Pleiades at the Summer Solstice gate previous to the Venus interactions! The goddess herself has killed the geese of the Pleiades. An examination of the figure of Athena facing Chapter 14 reveals the head of a water bird looking out from under her shield!

It is significant that the most destructive, warlike behavior of Venus did not affect her identification with the original mother goddess, for many of her analogs can also be found carrying dead geese by their necks: Artemis, Demeter, etc.

Association of lion (or sphinx) and goddess continues through the Age of Heroes; they are often depicted on either side of the goddess. In the archaic Artemis temple on Corfu, Medusa is flanked by lion-like figures.

The second interaction between Earth and Venus has transformed the freewheeling comet into a predictable member of the solar system. As depicted in a seal published by Frankfort, the former monster is now securely bound to the chariot of the sun.[24]

[24] H. Frankfort, "Gods and Myths on Sargonid Seals," Iraq, Vol. I, part I, Plate III, f, text on p. 21.

Medusa with lions.

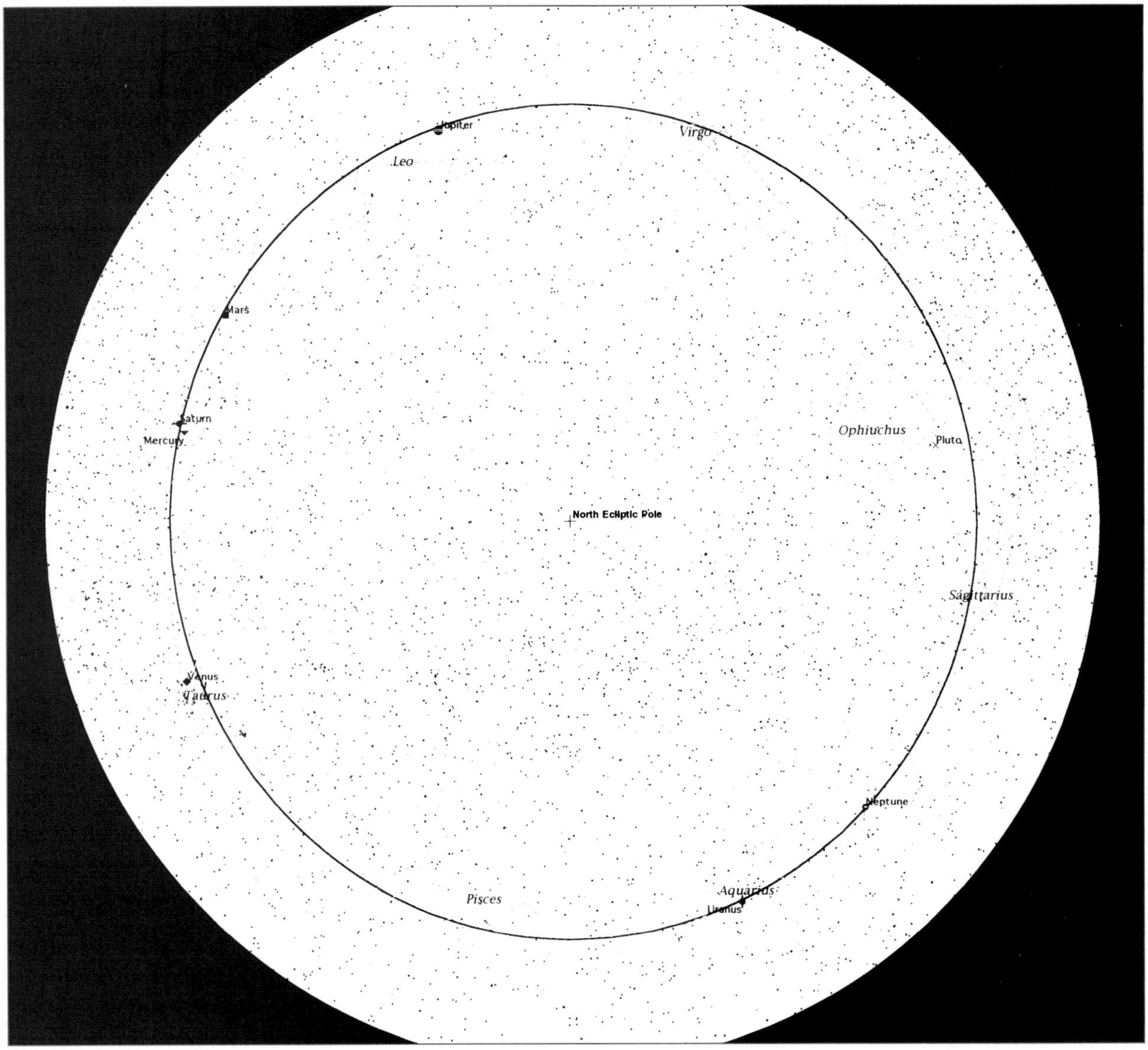

The ecliptic-zodiac is the racetrack of the planets.

26. Olympic Games: Chariot Races In the Sky

Many a wondrous sight may be seen, and not a few tales of wonder may be heard in Greece, but there is nothing on which the blessing of God rests in so full a measure as the rites of Eleusis and the Olympic Games.

– Pausanias, *Description of Greece*

As the Eleusinian Mysteries are rooted in alien skies and cyclical catastrophe, so are the same celestial marks written large on the Olympic Games, their stadium and environs. Olympia is located in Elis, land of the Eleans, in the Greek Peloponnese. This important historical area is named after Pelops, a warrior-hero at the center of a number of myths.

Since the institution of the modern games, more people are aware of the ancient series of Olympic Games which began in 776 B.C. and continued through the flowering of Classical Greece and Rome.[1] What many are unaware of is that the ancients measured time in these four-year periods. Our own conception of centuries, years B.C. and A.D., is a system imposed on the past. Time passing in Classical Greece and Rome was reckoned in four-year Olympiads. Thus the initiation of the classical games in 776 B.C. was an important beginning, but compared to the overblown modern games, a modest one.

Pausanias makes it clear that the Olympic Games which began in 776 B.C. were a re-institution of games which had been carried on in previous eras. "When Iphitus renewed the games, as I have said before, people had forgotten the ancient customs, and they only gradually remembered them, and as they remembered them, piece by piece they added them to the games."[2] One might draw a parallel to *Mankind in Amnesia*, in which we gradually recall previous world ages and their attendant information. In fact, it would appear that the games of earlier times had not been entirely forgotten since women had been celebrating a similar event, the four-year Heraea, with foot races at Olympia for some time previous to 776 B.C.[3]

That the men were moved to reinitiate the games was not happenstance. As suggested earlier, their memory was jogged by the sight of Venus and Mars fighting it out in the sky in their race around the sun[4] at some time in or slightly before 776. That the classic games were imitative of the Heraea can be seen by the choice of date (either two weeks before or two weeks after the women's feast) and that the games consisted of only foot races for many Olympiads. Following Pausanias, the double foot race was instituted in the 14th Olympiad while the pentathlum and wrestling

[1] The classical Olympic Games are thought to have been abolished by Theodosius in 394 A.D.

[2] Pausanias, *Description of Greece,* trans. J. G. Frazer (London, N.Y., Macmillan, 1898) Book V.8.5, p. 247.

[3] Ibid., Book V.16.2, p. 260.

[4] We must assume that Venus and Mars would both experience precessional spins as Earth did, resulting in sudden movements along their regular revolutionary path.

were added in the 18th, and boxing in the 23rd Olympiad.[5]

Any discussion of the Olympic Games must deal with their having been a moveable feast since the end of the Earth-Mars Wars, the start of the present Iron Age. Easter is another: the first Sunday after the first full Moon after the Spring Equinox. Like other Christian feasts, Easter was imposed on earlier pagan festivities with the intent of eradicating them, so that the Church had little interest in shedding light on the roots of the phenomenon. The dating of the Olympic Games has a similar ring: the eighth Moon after the first full Moon after the Winter Solstice.[6] This puts it approaching the fall equinox.

The definitions for a moveable feast carry a force that we can still sense, the weight of a time when cycles of Earth and Sky bore on mankind with more immediacy. The effort to bring equinox and solstice into focus with the phases of the Moon can be seen as an attempt to honor an old astronomical cycle and bring it forward into the present one. Cornford puts his finger on the crux of the problem when he refers to "the earlier system of time-reckoning by the moon."[7] It is the system Velikovsky elucidated in his chapter, "Year of 360 Days," in which was reckoned 12 moons of 30 days, new Moon on the 1st of each month, full Moon on the 15th.

The classic Olympic Games were founded in the Age of Heroes when sun and Moon were synchronized in their movements, "married" as their relationship is frequently described. As related in the chapter on the goddess, the games were reinstituted as a masculine celebration of the planets in their courses, racing around the sun. Now that we grasp the concept of sudden precessional motion, it is easy to see how sudden movements of Venus or Mars along their regular path,[8] particularly after an interaction between the two, would remind skywatchers of the race in the sky and the necessity of bringing it to Earth by reinstituting the games.

Since the older women's games were firmly in place at Olympia, the men had to schedule around them, either two weeks before or two weeks after,[9] putting their celebration at the full Moon. The additional light provided would have agreed well with the dawning of masculine logic and reason.[10] Thus the classic Olympic Games were never a settled calendar event at any time.

The Heraea, in contrast, was a regular calendar feast during the Age of Heroes, celebrated every four years at the new Moon on the first of the month of Parthenios, and close to the Autumnal Equinox.[11] The concatenation of astronomical cycles bears looking at. The same forces that moved the Jewish New Year from the first to the seventh month (at the 115-degree tilt of the Earth) can account for the Heraea being located near the fall equinox. It was a matter of choice for the ancients, and

---

[5] Ibid., Book V.8.7, p. 247.

[6] F. M. Cornford, "Origin of the Olympic Games," *Epilegomena and Themis* (University Books, New Hyde Park, N.Y., 1962) p. 225.

[7] Ibid., p. 229.

[8] Sudden precessional motion naturally includes orbital motion as well as precessional spin, as will be discussed in the chapter on precession. Sudden orbital slide of Venus could account for the irregularity of the planet's cycle as described in the controversial Venus tablets of Ammizaduga.

[9] Cornford, "Olympic Games," p. 230.

[10] Similarly, the dating of Easter ensures a full moon during Holy Week, useful for night ritual. Interestingly, the Uayeb, the five days of nothing, are celebrated during Holy Week by the Tzutuhil Maya.

[11] Cornford, "Olympic Games," pp. 230–231.

they chose variously to retain or move their feasts. The four-year interval of the Heraea corresponds with a half Venus cycle, but it is clearly the new Moon which dominates calendar placement. With what we have learned of the Saturn Configuration, let us reexamine the phenomenon of the new Moon in greater detail.

The new Moon makes only the briefest of appearances in the sky. After following the sun closely and inconspicuously throughout the day, it "comes out," begins to shine against the darkening sky when the sun slips below the horizon. For less than an hour we see the new Moon just above the western horizon before it follows the sun below it. It is a slim crescent of light where the sun illuminates the bottom or lower sides of the orb. Sometimes the dark portion of the Moon is slightly illuminated by "earthlight," sunlight reflected from the Earth. It is at this time that the Moon bears resemblance to Saturn as it once appeared in our skies.

The new Moon with its crescent at the bottom recalls the "day" of Saturn, the Moon coming out as Saturn once "came forth by day." That the dark portion of the Moon can be slightly illuminated by earthlight creates a more striking resonance,[12] and we begin to grasp the importance of the new Moon to the ancients. During the Age of Heroes the appearance of Venus as evening star added a complementary element.

It was during the 700-odd years of the Age of Heroes that Venus retained enough of her cometary substance to exhibit "hair" and "horns": hair when the entire planet was in full sun before and after superior conjunction, horns when in crescent phase before and after inferior conjunction. Thus, as evening star nearing inferior conjunction, Venus would exhibit "horns" extending from a crescent at the bottom of its orb,[13] matching those of the new Moon in the western sky after sunset.

Returning to the "Hymn to Ishtar," certain portions of the text begin to clear. Innini stands "in the heavens like the New Moon and the Sun."[14] The crescent of Venus bears obvious likeness to that of the new Moon as noted above, but it can in no way be compared to the present sun. And we suddenly realize that the sun referred to is the ancient sun of Saturn with illuminated crescent on the bottom, elsewhere in the hymn "the king who is a god and dwells therein with her."[15] In fact, the reference is to Idin-Dagan who is Tammuz and Saturn. The Egyptians also recorded the resonant images of Saturn and Moon in their *Pyramid Texts*. "Osiris . . . may you be pure at the monthly festival, may you be manifest at the New Moon."[16] Given three orbs with illuminated crescents below, of course Innini stands "in the heavens like the New Moon and the Sun."[17]

Venus as Innini was closely tied to the cycles of the Moon, "decrees fates for her land / Monthly at the new moon."[18] And again,

[12] The phenomenon is known at present as "the old moon in the new moon's arms."

[13] Venus was brighter during the Age of Heroes, more highly visible. Even at present it is brighter at inferior conjunction when the illuminated crescent can be seen with a good pair of binoculars.

[14] S. Langdon, "A Hymn to Ishtar as the Planet Venus and to Idin-Dagan as Tammuz," *Journal of the Royal Asiatic Society*, 1926, p. 19.

[15] Ibid., p. 35.

[16] Faulkner, *Pyramid Texts*, Utterance 483.

[17] Because the path of the Moon does not follow the ecliptic exactly, but is at an angle to it, the crescent of the new Moon is often askew, raising the possibility that the Moon may have traveled in the ecliptic during the Age of Heroes. This would have put the crescent exactly at the bottom of the new Moon.

[18] Langdon, "Hymn to Ishtar," p. 20.

"That she fulfill the decrees at the dark of the moon / At the festival of the New Year."[19] The association of Venus and new Moon was also tied to the equinox, when the New Year began for many ancient civilizations, either in spring or fall. Given the timeframe of both Heraea and Olympic Games, that they both honored the eight-to-five period of Venus to Earth in the period of their celebration, the catastrophic, alien-sky credentials of both feasts appear to be impeccable.

The Heraea, then, was held every four years, every 48 lunations during the Age of Heroes. This is echoed in the 9th Pythian Ode where Pindar tells of Danaus of Argos who sought to marry off 48 maidens by means of a foot race. Mythology must be adjusted later to reflect the present order of the heavens, and we presently discover other myths with 50 daughters of a god or hero.

The starting point of foot races in the Olympic stadium was at the tomb of the fallen Endymion who married Selene, the Moon. Cornford remarks, "The most cautious scholars accept Boeckh's view that the 50 daughters of this marriage are the 50 moon months of the Olympiad."[20] One view was that the full Moon marry the midsummer sun, but the moveable feast of the Olympic Games is far too late to fit that scheme. But 50 moon months makes an imperfect fit. There must be 50 alternating with 49 to fit the present order of the heavens.

Greek calendar makers of the 8th and 7th centuries B.C. and modern scholars both struggled with the same concept of marriage of sun and Moon. It begins to appear that this "marriage" began when the sun of Saturn and the "moon" of the illuminated crescent of the Saturnian band were integral, the primeval god/goddess of the Golden Age. The "marriage" was still in good order in the Age of Heroes when the cycles of sun and Moon were in sync. But after the Earth-Mars Wars, the efforts to retain a marriage between sun and Moon were doomed; a year of 365¼ days and a lunation of 29½ days looks more like divorce.

When the heavens settled into their present order after the last visit of Mars, the work began: setting out myth to reflect the new order, creating moveable feasts to deal with the present calendar while continuing to honor previous orders of the heavens,[21] counting and measuring, putting Earth and Sky in right relation. In the rise of Western Civilization, much old science was lost. The developing media of written language never interfaced with ancient media. An excellent example would be the confusion surrounding the Canopus decree of Egypt, written in three languages. We are still arguing about what it said.

When we learn that Pindar dates the Olympic Games from the victory of Pelops over Oinomaus in the chariot race, we must conclude that he describes an earlier series, for the chariot race was not "remembered" in the classical series until the 25th Olympiad.[22] The Elean informants of Pausanias said that "Pelops celebrated the games in honor of Olympian Zeus in a grander way than all who had gone before him." Here we stumble upon a central problem in interpreting myth: sorting out real men from heroes and gods. Besides being a polar figure, Pelops sounds

[19] Ibid., p. 36.

[20] Cornford, "Olympic Games," p. 228.

[21] The same struggle can be observed in the New World. Nuttall notes that ". . . the New Cycle, which began in 1507 with the year Acatl, had commenced on March 14th three days after the vernal equinox and that this delay had obviously been intentional, in order to wait for the new moon, which fell on March 13th at 11.40 A.M., and the planet Venus, . . ." *Fundamental Principles,* p. 179.

[22] Pausanias, *Description of Greece,* Book 5, VIII.2

too much like the other "eye" figures, Cyclops and Kekrops. We are not inclined to grant him mortality.

Gods are generally credited with giving the alphabet and writing to mankind, so it is sometimes useful to examine their names in writing. Ops in Greek has the look of elements of the Saturn Configuration, Ωψ. The omega sign has already been dealt with. The trident-like appearance of the last letter is reflected by the presence of Poseidon, the trident-bearer, in the myth of Pelops.

Oenomaus the Arkadian, who ruled over Pisa and Elis, had devised an unusual way to keep his daughter, the fair Hippodamia, from ever leaving his house.[23] He challenged each of her suitors to a chariot race, and since his horses were a gift from his father Ares, and swifter than the North Wind, Oenomaus had never failed to win the race and transfix the suitor with a spear, another gift from Ares.

Enamored of the fair Hippodamia (horse rider), Pelops "came near to the hoary sea, alone in the darkness, and cried aloud to the Lord of the Trident in the low-thundering waves."[24] Poseidon promptly appeared and Pelops went on, ". . . speed me on swiftest chariot to Elis, and bring victory to my embrace." Pelops would definitely need help, for the crafty Oenomaus had no intention of marrying off his daughter, having already dispatched 13 of her suitors.

It was the habit of Oenomaus to give the suitor and Hippodamia a head start in their chariot, on a course leading from Pisa, across the river from Olympia, to the altar of Poseidon at the Corinthian isthmus.[25] After the couple set off in their chariot, Oenomaus sacrificed at a column, then leaped into his own chariot driven by his charioteer, Myrtilis. The plan was to overtake the fleeing couple and stab the suitor in the back.

Pelops, however, had a powerful ally in Poseidon, who "glorified him with the gift of a golden car and horses with wings unwearied."[26] Hippodamia, in addition, had persuaded Myrtilis to remove the linchpins of her father's chariot, so it was a foregone conclusion when a wheel fell off the chariot and Oenomaus was "accidentally" killed by the spear given to him by Ares-Mars. The myth thus contains two important elements: young king succeeding the old and sacred marriage.[27]

The myth appears to cover the dissolution of the Saturn Configuration. The event took place in the sky: sacrifice at the pillar of the cosmic column, the wheel of Saturn flying off the chariot, the old god Saturn impaled on the weapon of Mars. Removing the linchpins represents the cessation of the force that held the Saturn Configuration together, after which the celestial pole of Earth promptly fell into another chariot, that of the Babylonian Enmesharra within the constellation of the bull, Taurus. Although the charioteer in the sky, the constellation Auriga, is often held to be Erichthonios from another myth covering the fall of Saturn, he has also been identified as Myrtilis.[28] The death of Oenomaus thus

[23] Graves, *Greek Myths,* 109 passim.

[24] Pindar, Olympian Odes L.69, quoted in Cornford, "Olympic Games."

[25] This is a course that links the Olympic with Nemean and Isthmean Games.

[26] Pindar, Olympian Odes L.86.

[27] Another important myth of Pelops in which he was boiled and eaten, then resuscitated, can be traced in the mythos of Mars in the cosmic column. The burning or boiling of a number of gods and their rebirth with youthful vigor is examined by Cochrane. "The Death of Heracles" *Aeon*, Vol. II, No. 5, pp. 61–63.

[28] Graves, *Greek Myths,* 109.L.

mirrors the Phaethon myth where Saturn is tumbled out of the chariot.[29]

Efforts to perceive a chariot at the pole do not end at the Silver Age, however. Ursa Major was the wagon of Odin or Woden[30] as well as a wagon, mar.gid.da, to Babylonians. The south-pointing compass (Ursa Major) of the Chinese was usually mounted on a wagon or cart. To the Greeks, both sky bears bore the name αμαξα, chariot. The most obvious apparent motion around the pole is rotational, making a figure of a wheeled conveyance appropriate.

The marriage of Pelops and Hippodamia reflects the link between Golden and Silver Ages: Pelops is a polar figure of the Golden Age, Hippodamia tantamount to the horse-headed Demeter of the Hyades who appeared at the pole in the beginning of the Silver Age. Following Pausanias, it was out of gratitude to Hera for her marriage to Pelops that Hippodamia assembled 16 women and initiated the Heraea. Thus the women's games not only enjoy the distinction of being celebrated before the reinstitution of the Olympic Games in 776 B.C., but were actually founded by the same set of sky events that inspired the original Olympic Games.

Another series of games were apparently held in Olympia 50 years after the flood of Deucalion, which Velikovsky placed at the first visit of Venus. Pausanias gives the record of Olympic victories at that time, but it is a very mythical-looking cast of characters: Pelops, Hercules and his charioteer Iolaus, as well as the twins Castor and Pollux. Subsequently, after the reign of Oxylus, the games were discontinued down to the establishment of the classical series in 776 B.C.[31]

We have already discussed the connection between races run on a circular track and the path of the planets around the sun, and can now better understand the arrangements at the stadium of Olympia. The oldest of the classical races, the foot race, began at the tomb of Endymion, the fallen sun who married the Moon. These foot races were run in the Age of Heroes when the sun and Moon were truly "married" in astronomical cycle, 12 lunations to a year.

The chariot race began in the sky: first Saturn in his "wheel," second the charioteer Auriga driving the Babylonian Chariot of Enmesharra, originally pulled by the Hyades, then as the pole moved, by the pair Hyades and Pleiades. The powerful image of the horse was assimilated to the goddess, horse-headed Demeter, but it was too good to be passed up by other gods who also had polar associations during the Golden Age. Pausanias notes altars "... in the open air, just about the middle of the starting-place; of Horse Poseidon and Horse Hera." "At the entrance of the so-called Wedge there is an altar of Horse Ares on the one hand, and an altar of Horse Athena on the other."[32]

West of the stadium and hippodrome, a small group of buildings and altars are generally considered to be the oldest structures at Olympia: the Heraion, the Metroon (site of an ancient cult of the mother goddess), and an altar of the Kouretes. Their situation close to the sacred hill of Kronos suggests their antiquity. Farther out is the Pelopion and the temple of Zeus, the latter dwelt on at length by Pausanias.[33] A golden shield featuring the Gorgon Medusa was hung at the middle of the

[29] A parallel myth describing the same event would be the wrestling match between Zeus and Kronos (Saturn) which was thought by some to date the initiation of the Olympic Games.

[30] Allen, *Star Names,* p. 428.

[31] Pausanias, Book V.8.1–6, pp. 246–247.

[32] Pausanias, *Description of Greece,* Book V.15.5,6, p. 259.

[33] Ibid., Book V.10, pp. 249–251.

rear gable, while the front gable of the temple was decorated with an elaborate sculpture representing the chariot race between Pelops and Oinomaus.

Here, at Olympia, if anywhere, we should expect to find plenty of evidence of the Earth-Mars Wars which were waged throughout the regular celebration of the games in the 8th and 7th centuries B.C. But Pausanias and his Elean informants tell us nothing of the catastrophic visits of Mars. In a real way, history as we know it had not yet been invented, and Olympia speaks only in ancient media, in this instance, artistically.

It was in 717 B.C. that the celestial equator changed drastically as a result of the terrestrial pole tilting from Ursa Major into Alpha Draconis. Like their brothers, the Olmecs across the sea, the Greeks knew what to do: bring Heaven to Earth, make a record. In Greek mythos, Hercules was their champion, protecting them from Mars. Thus Hercules was seen as carving out the new celestial equator in his Labors, identifying for posterity the constellations that it passed through. Pausanias says that most of the Labors of Hercules are represented at Olympia,[34] an excellent record of catastrophe in ancient media.

In 702 B.C. the Hercules equator would disappear in the catastrophic precessional spin that moved the Spring Equinox from Taurus to Aries. It is a more subtle movement than an axial tilt, but the Greeks managed to record it in the the ritual of the Laconian Karneia. This is a festival somewhat akin to the Olympic Games in which a rite race was held. One of the young men of the Karneatai was singled out to lead the pack; the rest were in pursuit. Now Karnos means ram, so that what we have here is actually a ram-racer or ram-runner, echoed by Olympian Apollo, who is both Karneios and Dromaios. And it was the ram Aries who replaced Taurus, racing into a new position at the Spring Equinox.[35]

In the early history of the Olympic Games can be traced the same adversarial relations between the sexes that were initiated by the celestial confrontations between Venus and Mars. Although Heraea and Olympic Games were held in the same location, they remained sexually exclusive. Indeed, when men commanded the environs, women were forbidden to enter the precincts of Olympia. Pausanias, in speaking of a nearby mountain with high cliffs, says, "It is a law of Elis to cast down from this mountain [Typaeum] any women who shall be found to have come to the Olympic Games, or even to have crossed the Alphaeus on the forbidden days."[36]

The ancient media of Greece tells the story. The Olympic Games are yet another instance of the ancients bringing Heaven to Earth, drawing patterns from alien and catastrophic skies. In the following chapter we will map the path of the celestial pole of Earth against that most enduring timepiece, the precessional clock.

[34] Ibid., Book V.8,9, p. 251.

[35] Cornford, "Origin of the Olympic Games," *Epilegomena and Themis,* p. 233–235.

[36] Pausanias, *Description of Greece,* Book V.6.7, p. 244.

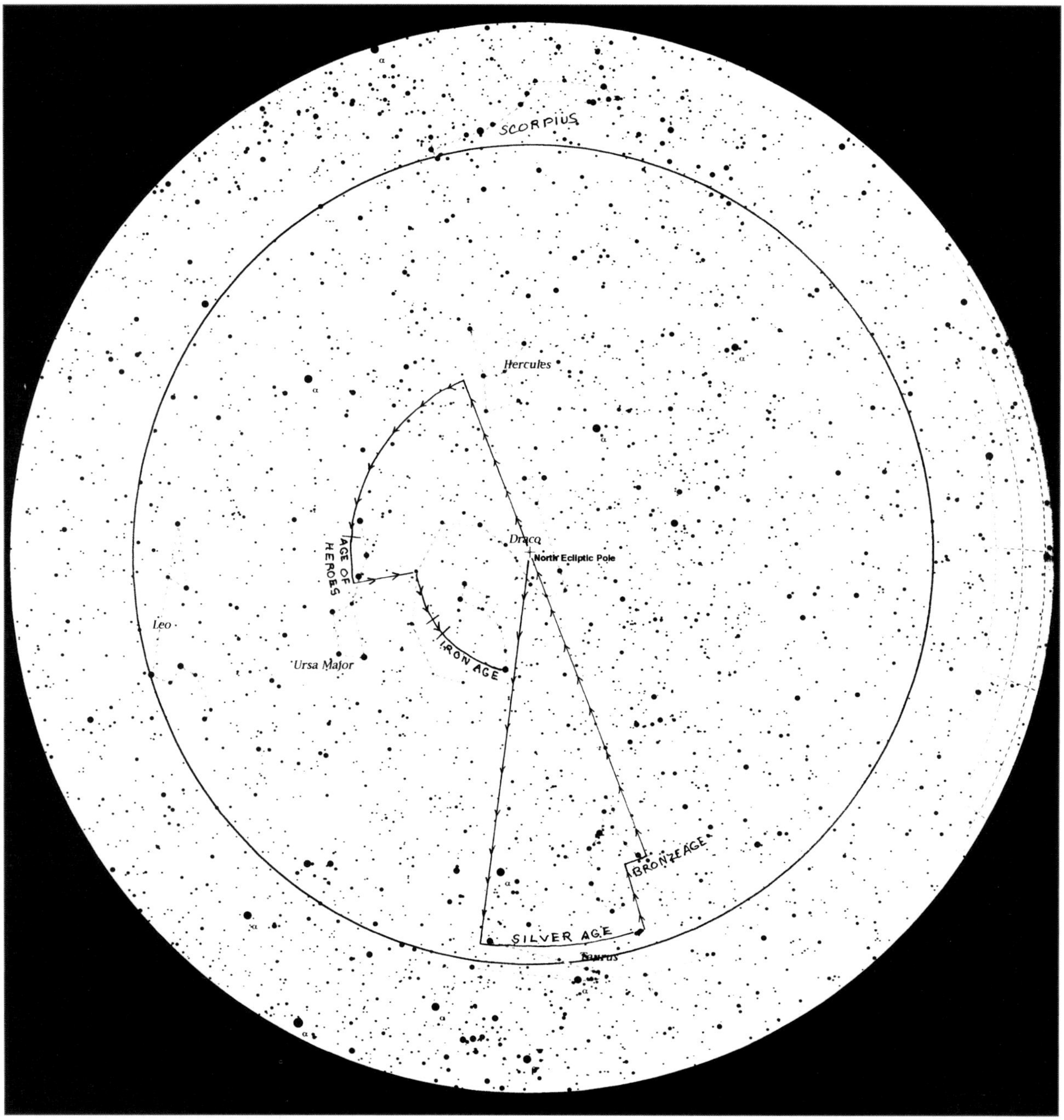

Arrowed lines indicate the catastrophic tilts and spins of Earth traced out in the sky by its celestial pole. Begin at center.

27. Mapping the Pole In the Frame of Time

Le ciel meme, malgre l'ordre de ses mouvements n'est pas inalterable.
(The same sky, despite the order of its movements is not inalterable.)

– Laplace, Oeuvres Complete

Bring out number, weight and measure in a year of dearth.

– William Blake

It is not the particular style of the times that matters, but rather effective contact with reality.

– George Robert Talbott

If the ancients disagree with us in their vision of the past, at least they agree with each other, even across oceans. A vast worldwide body of myth, legend, art, ritual, and architecture supports an Earth history of world ages separated by catastrophe. The Greek mythographer Hesiod named them for us: gold, silver, bronze, Age of Heroes, and iron. And ancient media allows us to plot the movements of Earth against the starry vault of the sky. The mythological model begun by Velikovsky and continued by his followers has, hopefully, been made more cohesive within these pages. We will now map the territory, chart the movements of the celestial pole of Earth against the clock of the precession.

The vault of stars which surrounds Earth should be no stranger to us. It has shown above us since the beginning, an imperishable link with the past that will show us where we have been, marking past ages as clearly as the hands of an analog clock. The sky is our closest link with the past. Stars have been born and others have burned out; some have grown brighter or dimmer; a few have even moved or changed color during historical times. But the pattern of the constellations remains. It is the Earth that has moved: tilting on its equinoctial hinges, spinning-sliding on them in a kind of time travel.

The ancients made their record of catastrophism in their own style, and now we will make ours: not as artistic as the Mithraic bull-slaying, lacking the force of myth, and without the flow of narrative. Yet it is our style, and thus more comprehensible to us.

The precessional clock is an excellent background for our map, satisfying our linear sense of time and enabling us to set the history of Earth in order, both uniform and catastrophic. We will trace the path of Earth's north celestial pole and give a brief synopsis of events. For once, we begin at the beginning, or at least the one we can now perceive. It is in the Golden Age, the First Time of the Egyptians, the Krita Yuga of India, the Garden of Eden. The period is described in all cultures in the most idyllic, even spiritual terms. It was when Saturn, the primeval king of many names, loomed over the North Pole of Earth.

Saturn and Earth were connected by a visible axis bond, described variously in many cultures, for it was a time before Heaven and Earth were separated. We remember Saturn as Kronos or Father Time with his curved sickle and association with death. What we have forgotten is the reality behind that curved blade, the illuminated crescent of the Saturnian band, constantly circling around the central orb. For Saturn and our present sun once revolved around each other in a double star system, so that part of Saturn and its band were always lit by the sun.

The revolving crescent of light was brought to Earth in many other symbolic forms: the horns of the Bull of Heaven, the god boat, the wings of the Native American thunderbird. It is Saturn's weapon, the harpe with which he castrated his father, Uranus, and was in turn beheaded as Medusa by Perseus. It is the "flaming sword which turned every way" at the gates of Paradise.

We begin at the center of the map with the terrestrial axis pointing through Saturn to that enduring point within the constellation Draco, the pole of precession or of the ecliptic. How long it was there is a question we will not address within these pages. But it was not so long ago, for Egyptians and Mesopotamians recorded the era in hieroglyphics and cuneiform writing. Because the texts describe an alien sky, they are impenetrable to orthodox scholarship. What is one to make of a statement like "I descend in the ship of the morning."[1] Following *The Saturn Myth,* the crescent god boat descends from the left to the "coming forth by day" of Saturn, the time when the sun of Saturn was brightest.[2]

The Golden Age came to an end at the dissolution of the Saturn Configuration when the pole of Earth suddenly tilted almost 90 degrees toward the ecliptic. This catastrophic event is marked on our map by a straight line from the pole of precession into the constellation of the bull, Taurus. Having grown used to the Bull of Heaven, Saturn, at our North Pole, we very naturally found another bull there after the tilt. The event was well marked in primeval media, occasionally in early text. For those who watched Saturn slide down the solstitial colure, the river of Eridanus, it was the "Descent of Lordship." For us it was the Fall, not from grace, but a very real fall into the ecliptic,[3] a fall into time as we know it. There would be a new succession of day and night, a Moon with phases, changing seasons.

Both the Bull of Heaven, Saturn, and the sky bull Taurus are well recorded over millennia in the art and iconography of cultures around the Mediterranean, sometimes as a sacred cow as in India and Egypt. The Bull of Heaven would naturally have a circular sun indicated between his horns, for Saturn was a sun, if a dying one. The wide representation of the bull in ancient art is thought to be a celebration of Taurus at the Spring Equinox, retrocalculated to the 3rd and 4th millennium B.C., but this is simply an error of uniformitarian thinking.

As we see on our map, when the north terrestrial pole fell into Taurus, it began the slow, regular precessional movement with which we are familiar, remaining in the constellation for over 18 degrees on the precessional scale. This would represent over 1,200 years at the present rate of precession, a period covering both the flood which gave the Hyades their name and the Biblical Deluge which occurred when the Pleiades were at the pole.[4] This latter star cluster gained its notoriety while in or near this position, becoming associated with a day of the dead celebrated all over the world early in November. This would have occurred in part because of the Deluge, but also because the Pleiades were associated with a rain of fire during the time they were at the pole. This does not appear to be the one caused by Jupiter, however.

[1] Coffin Text, 118.

[2] Talbott, "Saturn's Day," *The Saturn Myth,* pp. 320–327.

[3] The position of Earth during the Silver Age, vis-à-vis the ecliptic, was similar to that of the planet Uranus today.

[4] But was the rate of precession the same? Successive catastrophic spins might have increased the rate of regular precession.

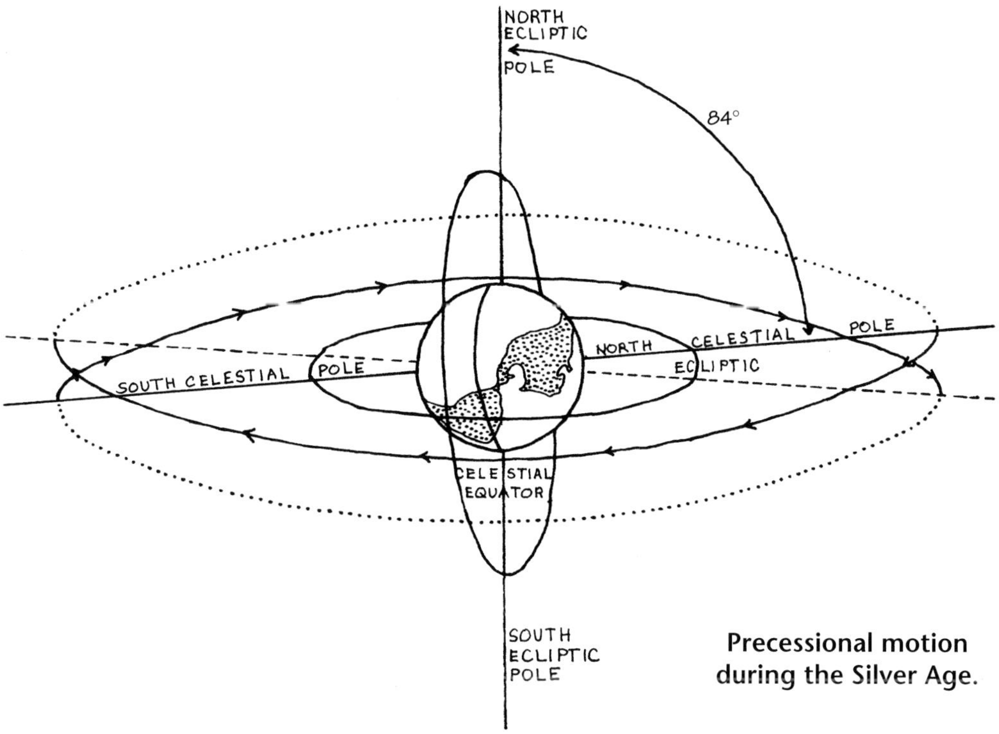

Precessional motion during the Silver Age.

This was the Silver Age when the planet Jupiter ruled the skies as Zeus of the Greeks or Marduk of the Babylonians, and won his reputation as wielder of the thunderbolt. It was toward the end of this age that a 52-year cycle began which is associated with Venus. But it was Jupiter who overturned the cities of the plain, destroying them in a rain of fire and brimstone. The site of these cities, of which the best known are Sodom and Gomorrah, lies on a major geological fault extending south as the Great Africa Rift through one-sixth of the Earth's circumference.

The Silver Age came to an end in this catastrophe, as the pole of Earth tilted upward from the Pleiades into the constellation Perseus. It is here that the pattern begins of tilt toward the pole of the precession, indicating force present in that direction. And as one age comes to an end, another begins as Zeus (Jupiter) descends upon Danae, engendering Perseus in a "golden rain."

Now the Bronze Age of Hesiod begins, another period of uniform precession as the terrestrial pole moves through the constellation Perseus. It is considerably shorter than the Silver Age, probably accounted for by the four ages of Censorinus mentioned by Velikovsky.[5] These would correspond to four double Venus periods, echoing those kept by the Meso-americans, something well over 400 years. Perseus can be linked with Mithras here: his "killing" Taurus catastrophically can be equated to Mithras stabbing the bull in the "shoulder" of the Pleiades in the tauroctony of the Mithraic Mysteries.

Perseus is the first great hero to dominate the pole in the history of the cosmos and his story is embedded in European folklore. Many of us will remember the classic tale in which a hero slays the dragon and saves the princess.[6] This follows Greek myth in which Perseus rescues the princess Andromeda from a sea-monster that has been threatening her country.

[5] Velikovsky, *Worlds in Collision,* p. 152, note 1.

[6] Talbott has treated the theme as primary to historical catastrophism.

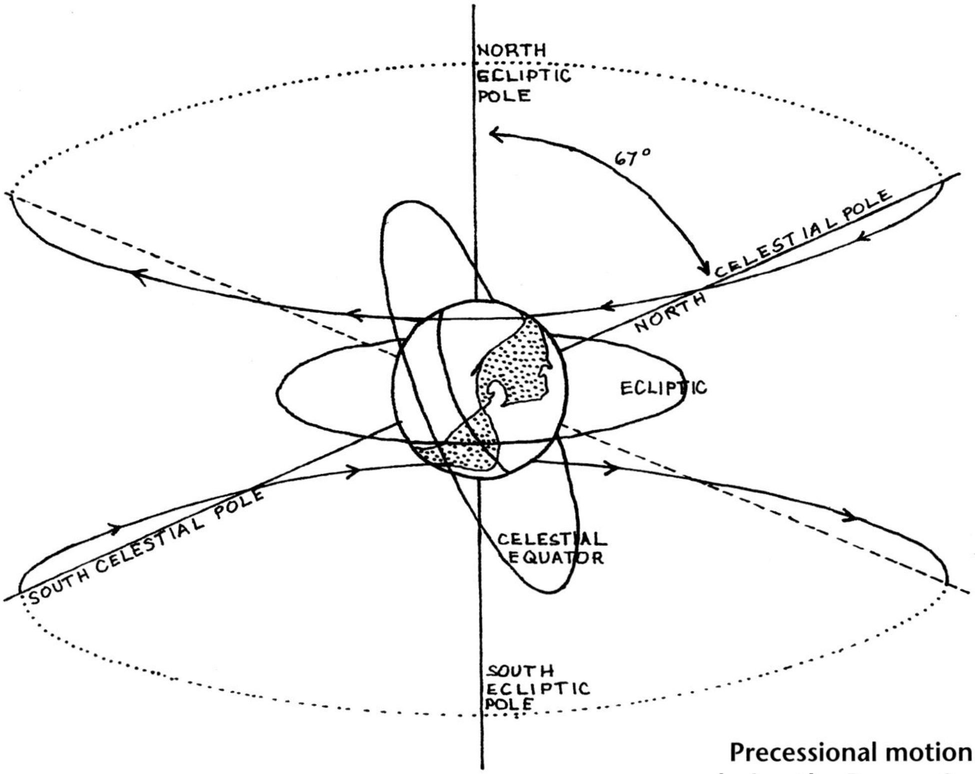

Precessional motion during the Bronze Age.

The 52-year periods continue during the Bronze Age,[7] recorded in Mesoamerican legend where they are associated with the then cometary Venus.[8] These visitations culminate in the Earth-Venus wars, two disastrous interactions which brought civilization to its knees as described in *Worlds in Collision.* The first visit of Venus was associated by Velikovsky with the Ten Plagues of Egypt at the time of the Jewish Exodus.

Looking at our map, we see that the tilt of Earth's pole at this time is enormous, the longest tilt to occur during the present series of world ages. The forces exchanged at the first catastrophic encounter with Venus carried the pole from Perseus up to the pole of precession, through it, and down to the constellation Hercules. Note that Earth's pole comes to rest in a position closer to the pole of precession than it began. Thus it can be seen as working its way closer to the pole of precession.

This horrendous tilt exchanges the solstices, Scorpius for Taurus, winter for summer in the northern hemisphere. Since the order of the seasons is naturally preserved, winter followed by spring, summer by fall, the equinox positions are exchanged as well. Because of the tilt we have records of the seasons being disarranged, "the winter is come as summer." Since the Exodus occurred at the Spring Equinox, the tilt plunges the northern hemisphere back into winter again.

The best part of the sky visible from Earth, 115 degrees of it, is now disoriented from the North Pole, giving mankind the impression that the world has turned upside down.
The Andromeda episode of the Perseus myth preserves the memory of the tilt as far as the

[7] Fuhr, "On Comets, Comet-like Luminous Apparitions and Meteors," *Kronos* VIII, 1, pp. 38–39.

[8] Velikovsky, *Worlds in Collision*, p. 152, note 1.

pole of precession, for the pole of Earth moves through the constellations named after two of the principals in the story, her parents, Cassiopeia and Cepheus. Perseus dies by the axis-sword of Medus, whose name expresses the true middle, the pole of precession. The threads of myth are now drawn together, for Medus is the son of Medea, the Venus analog or Gorgon Medusa whom Perseus slew at this same point.

Now the Bronze Age is over and Earth's pole points to the constellation Hercules, grandson of Perseus. This situation is reflected in the myth in which the strong man briefly took over from Atlas, holding up the cosmos. A 52-year period begins in the darkness of the Shadow of Death, the period when the Jews wandered in the desert, for the Earth was so devastated by the encounter with Venus that it was unable to support life. No crops could grow and mature.

The Jews were sustained by the problematical manna as those in other lands were fed by by madhu, honey-lash, morning dew, or ambrosia. Earth's atmosphere had become mixed with that of Venus, recalled in myth as Hera's ambrosial robe. The unusual atmospheric mix precipitated as an edible carbohydrate, falling in such enormous quantities that it gave rise to legends of lands flowing with milk and honey. This phenomenon, a headache to Western scholarship, is supported by another name of the constellation at the pole, for Hercules was also Melkarth and Melicertes, and Melicertes is a reference to honey ("sweet power" following Graves). The manna eventually ceased to fall, but its qualities devolved on honey with a wide spiritual reference among the ancients, frequently found in caves, an image of the cosmos.

The brief period when Hercules is at the pole is too short to register on our precessional map, but it was one of uniform precessional motion. Sandwiched as it is between the two catastrophic encounters with the cometary Venus, it reflects the close association of Hercules and Athena (Venus) throughout the extensive mythology of the Greek superman. She is always at his side with guidance and support.

The second visit of Venus produced a precessional spin-slide, a sudden movement of Earth along its regular precessional path. The Spring Equinox is carried through three signs of the zodiac while the pole moves from Hercules to a new configuration of stars, thought to be a bear because of that animal's powerful association with honey. Ursa Major would have climbed the world tree to get the honey which was still presumed to be there.

Now, finally, in 1500 B.C., the Spring Equinox is in Taurus, but too late to do orthodoxy any good. By retrocalculation, it should be in Aries. It is Hesiod's Age of Heroes, the Jaguar Age in Mesoamerica, a remarkable flowering of civilization throughout the world, largely due to the 700-odd years of regular precessional motion indicated on our map. Trade is carried on from Britain to India and beyond, but the back of this cultural matrix will presently be broken by the most recent series of historical catastrophes, the Earth-Mars Wars of the 8th and 7th centuries B.C. These were first described in modern times by Velikovsky in *Worlds in Collision*.

The final transformation of Venus from comet to planet occurred at its second interaction with Earth, when it also lost the greater part of its tail. Its orbit changed from one of roughly 52 years to one very close to its present cycle. Unfortunately, the cycle of Mars during the Age of Heroes must have intersected with that of Venus in order for the interactions described by Velikovsky to have taken place. Although the battles between Venus and Mars do not appear on our precessional map, they are clearly involved in a working model of the solar system, setting Mars off on a destructive new path intersecting with that of Earth.

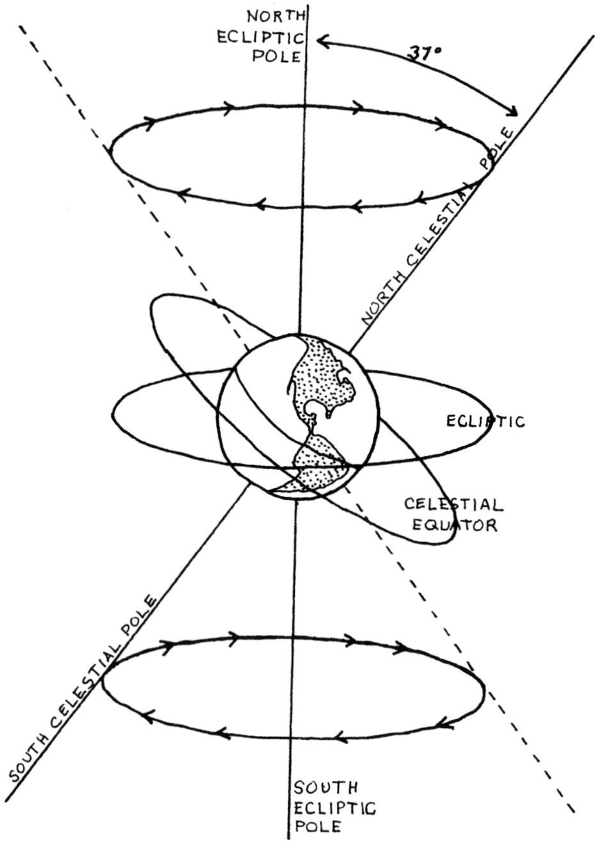

Precessional motion during the Age of Heroes and the period when Hercules was at the pole.

Velikovsky described the Earth-Mars Wars in *Worlds in Collision,* where he noted that they appeared to occur at 15-year intervals, the present period of Mars' favorable opposition. Within these pages I have added support for the 15-year period of grace, plus a sizable body of evidence for there having been a total of five interactions between the two planets. These fit handily in the 60 years between the best-attested dates, 747 and 687 B.C. Thus I have added the date of 732 B.C. to complete the pattern, discarding the date of the founding of the Olympic Games in –776 as having to do only with the Venus-Mars Wars. Catastrophic movements which register on the precessional map occur only on the last three meetings of Earth and Mars, however.

The first visit of Mars, "the descent of the fire god," occurred at the commotion (raash) in the days of Uzziah. The Moon is disturbed at that time, necessitating the institution of a new calendar in the "Days of Nabonassar." A serious perturbation of the Moon's cycle must have occurred, since only ten months are recorded in the days of Romulus. The first month is named after Mars, the second after Venus, indicating a new order of perceived threat from the heavens. Mars is now the war god, the resuscitated Huitzilopochtli in the New World.

I have posited a visit of Mars in 732 B.C., 15 years after the first one,[9] since ancient media supports five interactions. Isaiah, before the third meeting of Earth and Mars at the burial of Ahaz, refers to two previous occasions on which the land was "afflicted." And Olmec iconography supports two "jaguar attacks" before the Earth was tilted.

The third bout in the Earth-Mars Wars occurs in 717 B.C. at the commotion at the burial of Ahaz. In a serious catastrophic interaction, the north polar axis of Earth tilts from Ursa Major to Alpha Draconis, again upward toward the pole of precession. Fenrir "swallows the sun" and Fimbul Winter begins in the north (and extreme south) with year-round snowfall, for the sun no longer moves as far north and south. The tilt of Earth now agrees with its present angle, but more catastrophe is to come.

The most important result of the tilt for our purposes is that the new celestial equator is recorded twice in ancient media: in the tauroctony of the Mithraic Mysteries, and in the Labors of Hercules. This was fortunate, for the Hercules Equator will be in place for only 15 years. Mithraism, which once rivaled Christianity as the religion of the Roman Empire, can now be seen as an elucidation

9 Here I part company with Velikovsky.

of historical catastrophism, an interesting example of "the road not taken."

The first Sargon of Akkad, who also begins to look like the second Sargon of Akkad,[10] was much struck with the new star at the pole, Alpha Draconis,[11] at the celestial pole of the Hercules Equator. And it appears that it was also noticed in Egypt, for a shaft in the Khufu pyramid was pointed to it.

A substantial precessional spin of Earth took place in 702 B.C. when Mars approached for the fourth time. The Spring Equinox moved over 30 degrees from the middle of Taurus to the middle of Aries, and the Hercules Equator ceases to exist. The Babylonians recorded the new rising times of the constellations in Mul.apin, an astronomical text, and Hezekiah doubled the month of Nisan to keep the seasons in order. The Egyptians recorded the new celestial pole with another shaft in the Khufu pyramid. Since there is no prominent star at the pole, they angled the shaft toward Kochab in Ursa Minor, on the solstitial colure.

What remarkable faith the Egyptian astronomers had in those who would come after to grasp what was obvious to them! Having marked the solstitial colure, one need only deduce where it crossed the path of precession to see where a precessional spin ended! And how long it has taken us moderns to understand it!

In 687 B.C., at the fifth and last round of the Earth-Mars Wars, the third precessional spin-slide occurred, moving the Spring Equinox to agree with the present order of the heavens. A disturbance of the sun's movement was noted when people began to sing the Hallel prayer of the Passover service. Following the regular movement of precession, it would have moved backward. That night the "blast" which destroyed Sennacherib's army was recorded. An equinoctial movement of nine degrees was noticed in two places by Velikovsky. The Greeks marked the sky at the Spring Equinox, for at least seven temples are oriented toward the star Hamal, Alpha Arietis, at or close to the equinoctial colure after its final catastrophic motion.

Records of this nature are regularly swept under the rug by orthodoxy, assuming that the ancients can neither count nor describe what they see. But, like Velikovsky, we will take the ancients at their word. Thus, based on ancient media within these pages, and assuming a regular rate of precessional motion since the Fall, we place that event at approximately 3360 B.C. And given the catastrophes which have occurred since then, periods of darkness, changes in year length, etc., we begin to grasp why the Maya instituted the Long Count, a record of actual elapsed days. More careful astronomical measurement and historical research may bring the above date into closer alignment with Mayan day zero in 3114 B.C.

The last segment of curved line on our map represents the regular precessional movement of the terrestrial pole through Ursa Minor since 687 B.C. This is all we know of stability, of uniform motion in the solar system. Yet we have extended it into the remote past, far out of sight, and denied our catastrophic history. In *Mankind in Amnesia,* published after his death, Velikovsky, a trained psychiatrist, airs his views of this phenomenon. We have included some of his ideas in the chapter of the same name.

In the following chapter, we review the two styles of our cosmos, Saturn Configuration and solar system, and suggest the outline of a theory that will account for both of them.

[10] The "two" Sargons can now be seen to be separated not by time but by the terrestrial tilt of 717 B.C. complicated by subsequent precessional spins.

[11] Allen, *Star Names,* Alpha Draconis.

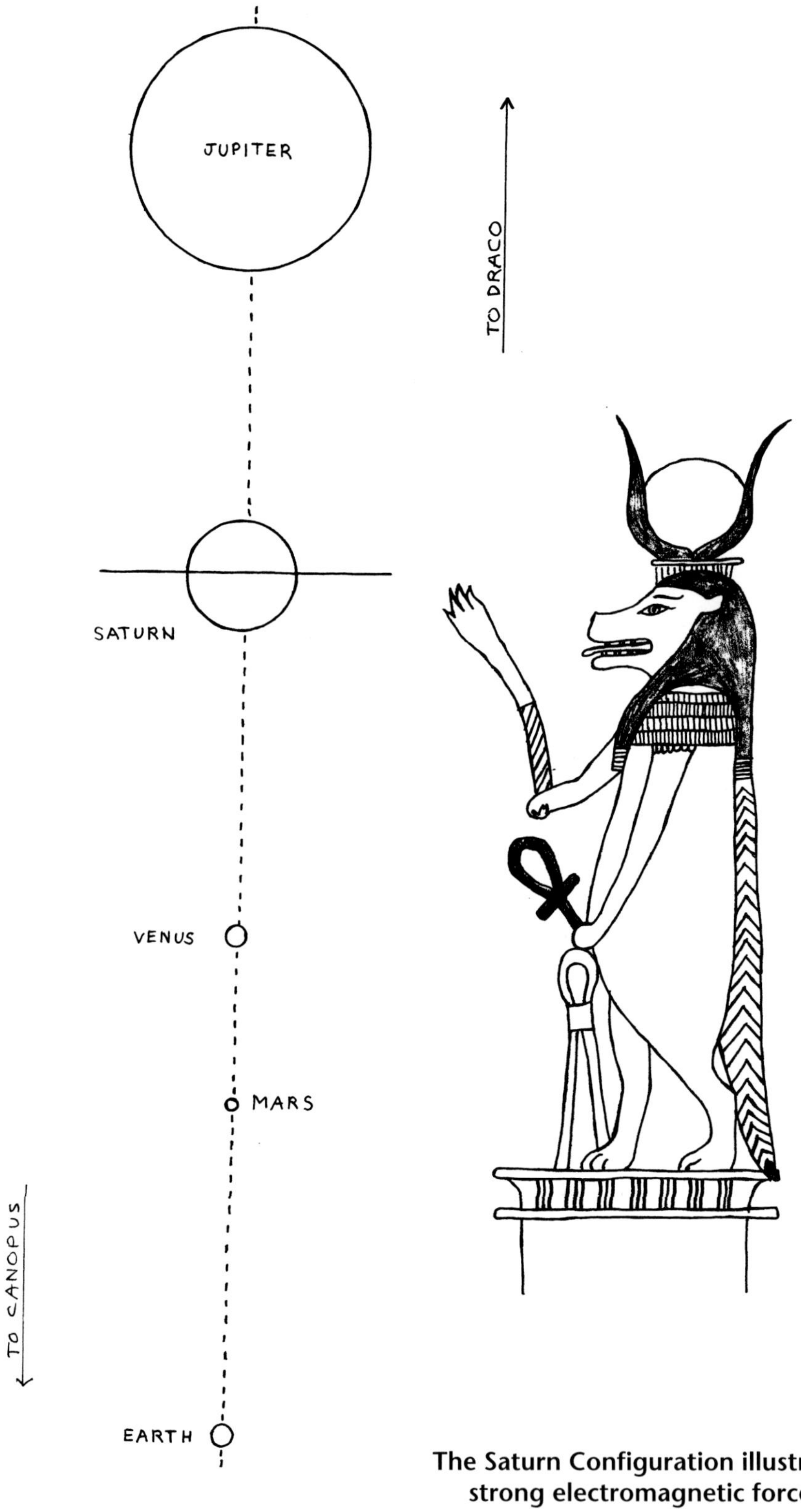

The Saturn Configuration illustrates strong electromagnetic force.

28. The Force Behind Precession

... the doctrine of the eternal stability of the solar system since its creation eons ago, is a theological dogma for which there has never been presented scientific evidence.

– Livio Stecchini, "Astronomical Theory"

I have granted thee the knowledge of thy father Mithra. Do thou keep his commandments and thus secure for thyself a cable and a secure anchorage throughout thy life.

– Emperor Julian, "Caesars"

In his essay, "The Inconstant Heavens," Stecchini deals with a puzzle of Western Civilization, that "... the organization of the heavenly bodies came to be considered the foundation of ethics." After a discussion ranging over the history of astronomical theory well salted with that of catastrophism, he comes to the crux of the matter, that the issue of whether or not the solar system has a history stands or falls on the historical evidence. Repeating a litany of scientific dogma will not do. However, now that we begin to grasp man's habit of bringing Heaven to Earth, it is obvious that it is the present regularity of the heavens that is brought to Earth as the foundation of ethics. Historical catastrophe understood as "accidents of celestial traffic" would not serve the purpose.

The regularity of the movements of the heavens can be seen in the order of the planets around the sun, the rotation of the Earth on its axis, the dependable waxing and waning of the Moon. The longest cycle of all, the background against which all the others take place, is the Precession of the Equinoxes, the slow, top-like spin of the Earth that completes its cycle over a period of 26,000 years, a process going on as you read these words.

Upon opening a textbook on astronomy early in the 21st century, we read that the precession is caused by the gravitational attraction of sun and Moon on the Earth. According to the Cambridge Dictionary of Astronomy, "The Earth is not quite spherical but is slightly oblate. Because of this, the gravitational forces between the Earth and the Sun produce a couple which causes the axis of rotation of the Earth to precess [go backward]." This is where the *Principia* of Newton and the mathematics of Laplace have left us. Having read the preceding chapters, one would begin to suspect otherwise, for the ancients have left us an historical record in primeval media which tells a different story.

We have watched imaginatively while the Saturn Configuration dissociated, the planets falling into an orbit around the sun, the Earth's axis nearly parallel with the ecliptic. Subsequent catastrophes tilted the axis upward again and again, each time closer to the pole of the ecliptic in Draco. We are about 23.4 degrees away from it now, and anyone who has followed the argument thus far could predict the probable outcome of another tilting catastrophe. There is something up there exerting force on the Earth, and something below us as well.

The force, very likely electromagnetic, that held the Saturn Configuration upright between the poles of the precession, is still at work, though in lesser degree, combining with the circular motion of the solar system to hold the Earth and other planets in precessional spin. Considering the electromagnetic force flowing between the poles of the precession, it is almost as if the cosmos has an "on" and an "off" position, or perhaps simply two different alternating forms.

The Egyptians tried to communicate some of the above information to us. Their style is not ours but it still can be seen to demonstrate "effective contact with reality." We return to the astronomical representations on the ceiling of the tomb of Senmut. Earlier we discovered that the "objectionable orientation" of the southern panel was due to the 115-degree tilt of the terrestrial axis which took place during the first visit of Venus. This tilt moved the Orion-Sirius group from the northern to the southern hemisphere. Simply turning around from facing north to facing south to view the group in the sky put Sirius and Orion on different sides of each other.

Velikovsky assumed, following orthodox astronomy, that the northern panel of Senmut's ceiling described the sky shortly after the time of Senmut.[1] In fact, it goes back much earlier to the end of the Saturn Configuration. The scene on the northern panel of the ceiling is virtually bisected by a tall, narrow obelisk.[2] It is the axis bond of the cosmos which formerly held the Saturn Configuration together,[3] the Menat, Great Stake, or Great Mooring Post. The Bull of Heaven has just toppled over into the ecliptic. He is at the end of a long rod (solstitial colure of the river Eridanus) held by Kebsenuf, summer solstitial child of Horus, incorrectly identified as "An, the falcon-headed deity who spears Mes on many ceilings." Kebsenuf also appears in the lineup of the four children of Horus behind Hippo-Draco and Isis. Alone of all the "northern gods" in similar representations, he can sometimes be found without the sun symbol on his head!

Most important, the Menat is still in place, dominating the scene even though the bull has fallen. Hippo-Draco stands next to the Menat, also holding a symbol of the Menat in her hand. The crocodile can crawl up on her back at this point because Earth has fallen into the Watery Abyss.

Saturn, the Bull of Heaven, falls from the axis bond, slides down the solstitial colure of Kebsenuf.

[1] Velikovsky's few errors can sometimes be seen as associating particular phenomena with the wrong catastrophe, something which he realized was likely to occur in the first reconstruction of historical catastrophism.

[2] Neugebauer and Parker, *Egyptian Astronomical Texts,* Plate 1.

[3] Eric Burrows, S.J., "Some Cosmological Patterns in Babylonian Religion," S. H. Hooke, *The Labyrinth* (New York, Macmillan), p. 48, note 2. Ninurta's (Saturn's) title in Assyrian was "Holder of the Bond (*markasu*) of Heaven and Earth."

Much the same scene appears on the Harendotes coffin lid where the bull falls from the symbol of the Menat held in Hippo-Draco's hand. Now that the Bull of Heaven has fallen and "lordship has descended" as the Mesopotamians put it, we will occasionally find Hippo-Draco (Tau-ert or Isis-hippopotamus) holding a chain or cord attached to a polar constellation. "It is the office of Isis-hippopotamus to guard this chain."[4] She often holds a chain/cord attached to the leg of a bull, the Egyptian image of Ursa Major (again a polar figure reflecting the Bull of Heaven). Here she represents the pole of the precession, "holding" the axis of Earth.

The Egyptian texts described here are repeated again and again throughout the volume of plates in *Egyptian Astronomical Texts.* Thus was preserved the most significant sky event ever to have been witnessed by their culture.

As for "An, the falcon-headed deity who spears Mes [Ursa Major] on many ceilings,"[5] orthodoxy has confused two different polar images of the bull: that of primeval Saturn with Ursa Major, the Bull of Heaven with the leg of the bull. The history of Egypt can begin to be sorted out when these two figures along with Hippo-Draco and the other so-called northern constellations are recognized as polar.

The Menat, Great Stake, or Great Mooring Post of the Egyptians is the same as the Akkadian "tarkullu," the Sumerian "dimgal" or "great binding post." This is the once-visible bond that runs between the poles of the precession, between Draco, the "yoke star of the sky" and Canopus, the "yoke star of the sea." Thus did the ancients describe the two enduring points in the sky and the visible force that flowed between them. Later cultures seemed only aware of the point in the northern sky:

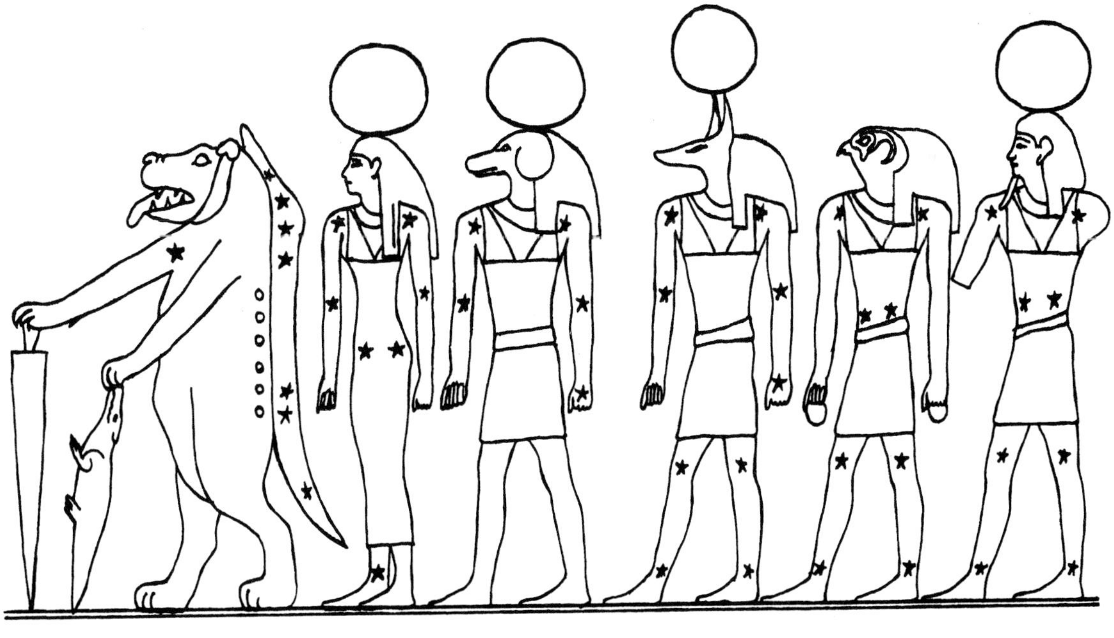

Draco, Isis, and the four children of Horus at Pedanope.

[4] Brugsch, *Thesaurus Inscriptionun Aegyptiacorum* (reproduced 1968) p. 122.

[5] In fact, it is Kebsenuf, the summer solstitial child of Horus, skewering Saturn, who will then slide down that particular colure.

the "nail of the north," the "peg in the middle of the sky," the "empty place in the sky" from which God hung the Earth. Orthodoxy confuses these with the celestial pole of Earth, but the ancients understood that Earth was "yoked," that the pole of the Earth was tied to the pole of the precession. Later, of course, this was forgotten.

The yoke is a remarkably mutable word and symbol. We have already discussed the Babylonian constellation of the Chariot of Enmesharra at the celestial pole of the Silver Age, complete with the horses of the Hyades and Pleiades, the charioteer of Auriga. The chariot was brought to Earth along with the wheel in the sky (Saturn's ring), and "Trojan chariots had upon the yoke a boss called omphalos,"[6] again repeating a sky pattern. "Omphaloessa," umbilical cord, was a Greek name for Ursa Major, at the pole during the Trojan War in the Age of Heroes. The Romans called both bears Septemtriones or "seven threshing oxen." Here the concept of a mill (the visible axis bond of the Saturn Configuration) is joined to that of a yoke, for oxen must be yoked to engage their strength, while an ox is a castrated bull, reflecting the Bull of Heaven, Saturn.

The goddess is also connected with the yoke. Hera was worshipped at Argos as "goddess of the yoke."[7] Although it has slipped below consciousness in the modern machine age, the yoke against which a draft animal pulls has historically had the shape of the Venus omega sign. As Burrows reminds us, "The omega-like thing represented on the Kudurrus [Babylonian boundary stones] is identified after Zimmern with the markasu rabu (great bond) of the 'holy' house[8] mentioned on the Nazimaruttas kudurru in the Louvre . . . Since the object naturally suggests a yoke, and was listed as such by Hinke, it may reasonably be taken for a stylized representation of the constellation (Draco) which was actually called the Yoke, and must in fact have been seen as a somewhat omega-shaped yoke."[9]

Burrows has distinguished himself by looking at the sky, but it is well to recall here that human perception is at issue. The symbol is most useful if it is not nailed down, if it is allowed its natural mutability. Later the omega symbol doubles into the Hercules knot, again suggesting "holding" or "yoking" power.

The Menat, Great Mooring Post, Tarkullu or Dimgal is the axis bond of our solar system, a force governing precessional motion of which Western science is totally ignorant. It held the Saturn Configuration in line and it governs both uniform and catastrophic motion in the present order of the cosmos. In a catastrophic tilt, Draco "pulls" on the north terrestrial pole while Canopus "pulls" on the south terrestrial pole.[10] Because Earth is held firmly in its spin, it does not move but tilts on its equinoctial hinges. In a sudden precessional spin, Earth is held steady in its precessional attitude while it moves suddenly along its preordained path around the sun. Tilting in place is more easily grasped than the precessional spin in which the Earth's revolution around the sun must also be taken into account.

6 G. Radet, Revue des Etudes Ancienne, XIX, 1917, pp. 98–100, referenced in *American Journal of Archaeology,* Vol. XXI, 1917, p. 454.

7 Mircea Eliade, *History of Religious Ideas,* p. 278.

8 Following Talbott, this indicates the Saturnian band.

9 *Mythology of All Races,* Vol. 5, Burrows' review in *Journal of Royal Asiatic Society,* 1933, 699, 63.

10 Of course, we do not assume that these particular stars exert the force. They simply show the direction that it is coming from.

In a precessional spin-slide, the Earth moves in two different frames of reference; it spins in the top-like precessional movement while it slides along its orbital path. This is absolutely necessary in the model described in these pages. In order for the sun to move suddenly backward through the signs of the zodiac as the ancients describe and is interpreted on the precessional map, it is the Earth that must move, suddenly and with unimaginable speed over enormous distances. Whether it moves 90 degrees on the precessional clock as in the second Venus event, or either 30 degrees or 9 degrees as happened during the Earth-Mars Wars, makes little difference. The movements happen within hours as described by the ancients, and all are equally impossible under present theory.

In his historical reconstruction of the Earth-Mars Wars, Velikovsky was much struck with the movement of the Spring Equinox recorded by the ancients. Whether he thought the problem through and, like Plato, was unwilling to commit his best thoughts to writing, or simply had an intuitive flash, is unclear. He contented himself with the following statement in the preface to *Worlds in Collision*: "... modern physics, of atoms and of the quantum theory, describes dramatic changes in the microcosm—the atom—the prototype of the solar system; a theory, then, that envisions not dissimilar events in the macrocosm—the solar system—brings the modern concepts of physics to the celestial sphere."

In order to equate the phenomena of historical catastrophism to modern science, we must move from the macrocosm with its associated gravitational theory to quantum theory and the electromagnetism of the microcosm. We have been assured by orthodox science that the same laws do not operate in both spheres, but developing catastrophic theory suggests otherwise. Primeval media and the records of the ancients tell a story that implies the regular phenomena indicated on our precessional map. Velikovsky may yet have another prediction confirmed. "If the activity in an atom constitutes a rule for the macrocosm, then the events described in this book [*Worlds in Collision*] were not merely accidents of celestial traffic, but normal phenomena like birth and death."[11]

The dialog style of Plato which enables him to be quoted on both sides of the issue of historical catastrophism can be seen to grow out of two visions of the cosmos: the static, uniformitarian one of the Saturn Configuration, "... an era in which the god himself guides the universe and helps it on its way," and the lively, catastrophic one of the present order of the heavens, "... an era in which he releases control."[12] Plato's model of the Saturn Configuration gives rise to speculation on how the ancients gained their remarkable knowledge of the cosmos.

The axis bond of the Menat was visible, described again and again by the ancients. The same is true of the "rivers" of Heaven, although the confluence of the rivers, Eridu or Hell, is less well attested. Plato's vision also indicates visible colures. And the cosmic mill is so well described in the sources of Santillana and Dechend in *Hamlet's Mill* that the whirlpool that attended its breakup hardly seems imaginary.[13] The fallen Phaethon is described as "still glowing" when the ship Argo passed by.[14]

[11] Velikovsky, *Worlds in Collision*, p. 382.

[12] Plato, *The Statesman*, 269D.

[13] Santillana feels that it stands for the "ecliptical world" represented by the whirling planets.

[14] Apollonius, *The Argonautica*, trans. E. V. Rieu (Baltimore, Penguin, 1959) pp. 163–164.

The powerfully visible electrical phenomena of the Saturn Configuration may not have completely disappeared immediately after its dissolution. The Bundahisn describes visible planetary trails. Gradually fading electrical phenomena would have given the ancients a very helpful orientation in the new version of the cosmos.

The ability of the ancients to measure is a central issue of historical catastrophism, but it is one that can be settled. Because geodesy requires a thorough knowledge of astronomy to produce correct data such as latitude and longitude (no matter what the attitude of the terrestrial pole), establishing the correctness of ancient geodesy would bridge the gap between ancient and modern measurement. Stecchini claims that "... by the time of the first dynasties the Egyptians had measured down to the minute the latitude and longitude of all the main points of the course of the Nile, from the equator to the Mediterranean Sea."[15] For anyone who could do these calculations, tracking equinox and solstice would be child's play.

The Platonic metaphor of the ship of the Cosmos describing the Saturn Configuration surfaces elsewhere, notably in the Mithraic Mysteries. "Mithra is the rope by which the middle of the ship is bound." In the quote heading this chapter, Julian is advised by Hermes to secure "a cable and sure anchorage,"[16] a reference to Mithraic belief. Egyptian cosmology provides the Great Mooring Post to tie up to. In Greek myth, Canopus is the steersman of the Argo. Mithras appears to be the bond that holds the cosmos together, to be equated with the forces which govern planetary precession, a bond that holds through uniform and catastrophic motion. Those who genuflect at the shrine of gravity can fairly be compared to those who worshipped Mithras, although the latter demonstrated more effective contact with reality.

We return to the icon of Mithras holding out the cosmic pole to the sun, Helios. If the sun is now in the center at the cosmic pole which stretches between the poles of the precession, it becomes possible to take an entirely different view of the solar system. The quality that is perceived at the ecliptic as gravity could simply be an expression of a balance of electromagnetic forces. As Earl Milton puts it, "... if the interaction between the planets is analyzed, they move so as to maintain the greatest possible distance between one another, passing closest at the fewest places along their orbit, in such a way that the close points are as far apart as possible ... In the literature, this commensurability has been described as the Principal of Least Interaction Action (or perhaps the least attraction). A superior description is that the planets repel one another as strongly as possible."[17] When the form of the Saturn Configuration dissociated, its member planets fell into an alternate pattern, but the same forces were still being expressed.

The attentive reader of *Worlds in Collision* is left with the impression that historical catastrophes of an interplanetary nature tend to occur in spring and fall close to the equinox.[18] Hopefully, we have brought the problem into clearer focus, showing that the equinox is a far more movable feast than we ever imagined, prone to sudden movements in time and space. That it also provides hinges for a tilt

[15] Stecchini, "Astronomical Theory," p. 167.

[16] Julian, "Caesars," trans. Wilmer Wright, *Works of the Emperor Julian* (Cambridge, Harvard, 1913) Vol. II, p. 415.

[17] Earl Milton, *Kronos* VI: 3, p. 6.

[18] The earlier catastrophes seem to have occurred in the fall: October Horse, Day of the Dead, Thesmophoria; the later ones occurred in spring. The 115-degree tilt caused by Venus exchanged the seasons.

demonstrates a certain stabilizing influence. Now when we read in Chinese history that officials of the first three dynasties—Hsia, Shang and Chou[19]—were put in charge of calendar, solstice, and equinox, we know that they were not dealing with trivia. These men were deeply involved in the science of their time.

In the same way, knowledgeable men in cultures all over the world tracked the equinox, well aware that it could be a dangerous time, given the appearance of any threat in the heavens. The different branches of the Pueblo Indians in the American Southwest have shown themselves to be skillful at tracking Earth cycles. The Sun Dagger at Chaco Canyon records solstice, equinox, and the metonic cycle of the Moon.

It is, however, left to the Zuni Indians, Ashiwy as they call themselves, to record a remarkable instance of the observation of the equinox. Their creation myth describes how they actually tested the dangerous period of the equinox in the section, "The Custom of testing the Middle in the Middle Time."[20]

"Now at last nevermore did the world rumble; yet the fathers of the people questioned in their hearts, fearing further misfortune to the children, if so be they still erred in the resting place of the sacred mysteries whatsoever. So when the sun had reached the middle between winter and summer, they devised an ordinance and custom whereby this might be tested. They brought out the things of lightning and the earthquake; even the keepers of the great navel-shell were summoned as having canny and magic skill. And as now we do in observing the custom of the Middle-arriving, all the people fasting, all the fires close kept, so then, for ten days they made ready, and on the last night the shell was laid by the sacred fire in Hein Kiwitsina of the North, and watched all the night through, by its keepers and the fathers foremost, and the Priests of the Bow. Meanwhile the incantations of dread meaning, taught of the Twain in Hanthlipink.ya, were chanted, yet the world only rumbled deeply and afar down, but it trembled not, neither did the Seven Fell Ones breathe destruction—only storms. Then, said the fathers, 'O, thanks! In peace-expecting mood may we kindle afresh the fires of our hearths for the year that is dawning.' And they sent forth new fire to all houses, causing the old to be cast out as is seen and known to us all in the custom of this day of the Middle arriving!"

The Priests of the Bow[21] are in charge of warfare while the Twain refer to the twin aspects of Mars. The Hopi, cousins to the Zuni, claim remarkable powers over the weather, deliberately living in a harsh desert environment to keep their skills (those of the pavansinom) honed. The "things of lightning and the earthquake" were not brought into play lightly.

Using ancient media and the records of the ancients, we have plotted the catastrophic motions of Earth that are associated with its precessional motion, but this tells us nothing of the shape of Earth's orbit or its distance from the sun. As Velikovsky suggested half a century ago, we must look to developing catastrophic theory and an expanded view of electromagnetism to understand how the cosmos works.

---

[19] These are analogous to the Silver Age, Age of Heroes, and Iron Age.

[20] Frank Cushing, *Zuni Creation Myth,* p. 429. This document is available in all depository libraries.

[21] The "bow" is particularly well attested in the culture referred to as Anasazi. Many of the large communal dwellings, including the well-known Pueblo Bonito, are built in the shape of a bow. Hopefully, readers will now recognize the crescent of the Saturn Configuration.

In *Earth in Upheaval*, Velikovsky examined Earth and climate changes detectable on the planet. In the following chapter, we will carry his work back to the Saturn Configuration, examining some of the most stressful periods mankind has ever experienced.

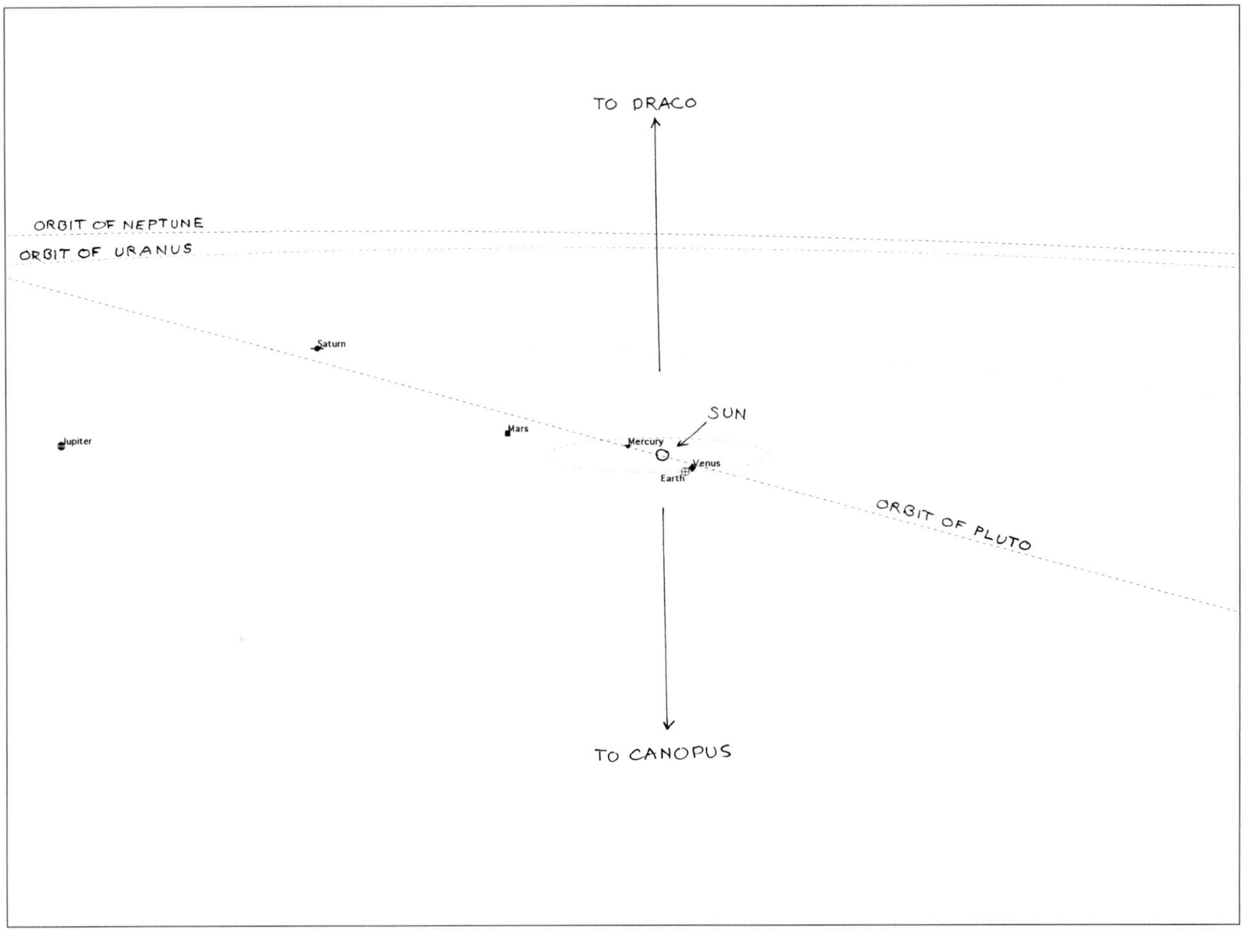

The solar system illustrates a weaker form of electromagnetic force.

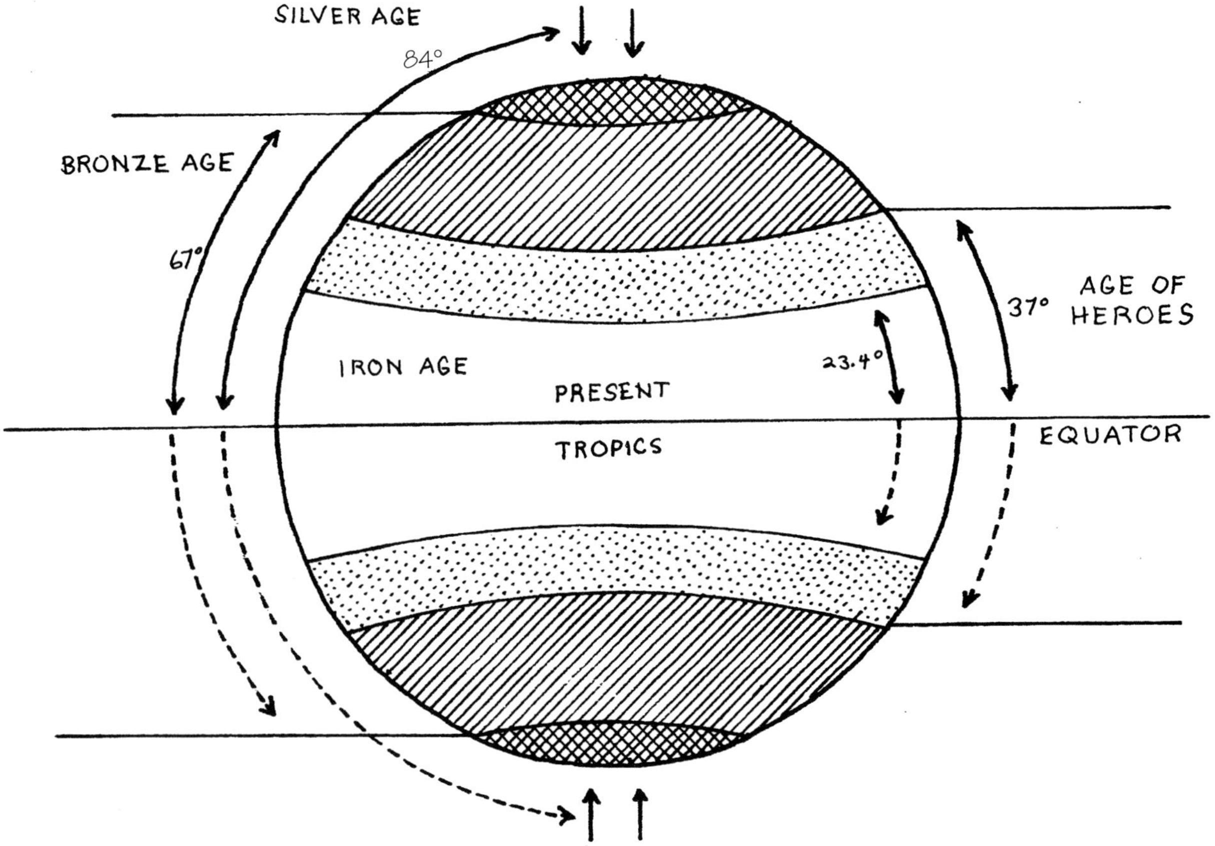

The sun travels north and south of the equator in recent world ages.

29. Climate Change: Men and Gods Go South

Hast thou commanded the morning since thy days; and caused the dayspring to know his place; That it might take hold of the ends of the earth, that the wicked might be shaken out of it.

– Job 38:12–13

When the sea rolled its fathomless billows
Across the broad plains of Nebraska;
When around the North Pole grew bananas and willows,
And mastodons fought with the great armadillos
For the pine-apples grown in Alaska.

– Anonymous

Although it was formerly a popular pursuit, Western man has long since given up the search for the terrestrial paradise described in ancient legends and holy books. Early explorers in the mainstream of Western Civilization were always on the lookout for this Garden of Eden. Efforts of this nature have been out of fashion for several centuries and are now only an embarrassment to us. But given what we have learned about the relatively recent tilts and spins of Earth, let us reexamine some knowledge that has been discarded by orthodoxy, other knowledge that is pooling up outside the paradigm.

In the second half of the 19th century, when the former existence of palm trees and corals within polar regions first broke upon the world, the news was not well received by a skeptical English-speaking public as reflected in the above rhyme. The offending flora were promptly relegated to a distant past. Sir Charles Lyell, the cautious uniformitarian, expressed himself on the subject. "In the greater part of the Miocene and the preceding Eocene epochs the flora and fauna of central Europe were subtropical, and a vegetation resembling that now seen in Northern Europe extended into the Arctic regions as far as they have yet been explored and probably reached the pole itself."[1]

An unusual perspective had been developing. Sir Joseph Hooker in 1859 saw the phenomenon thus. "When I take a comprehensive view of the vegetation of the Old World, I am struck with the appearance it presents of there having been a continuous current of vegetation, if I may so fancifully express myself, from Scandinavia to Tasmania."[2] A decade later, Oswald Heer of Munich published a work on the fossil flora of the Arctic in which he advanced the idea that "a great continuous Miocene continent within the Arctic Circle" was the "Bildungsherd" or mother-region of floral types.[3]

[1] Charles Lyell, *Principals of Geology,* 11th Ed., Vol. 1, p. 231.

[2] Joseph Hooker, *The Flora of Australia* (London, 1859) p. 103.

[3] Oswald Heer, *Flora Arctica Fossilis* (Zurich, I. Vorwort, 1868) pp. iii–iv.

Geikie, in his *Textbook of Geology*, refers to that "great invasion of Arctic plants and animals which in the beginning of the Quaternary ages came southward into Europe."[4] Wallace went further. "We have now only to notice the singular want of reciprocity in the migrations of northern and southern types. In return for the vast number of European plants which have reached Australia, not one single Australian plant has entered any part of the north temperate zone, and the same may be said of the typical southern vegetation in general, whether developed in the Antarctic lands, New Zealand, South America, or South Africa."[5]

Wallace and others took much the same view as to fauna, seeing the northern hemisphere as the birthplace of animal classes.[6] The new sciences of geology, of biology and botany, were still only scratching the surface of the Earth in the 19th century, and what they found was the most recent past. Western science's habit of pushing uncomfortable phenomena into the remote past is a matter of record.[7] Ancient media have a different tale to tell, and they suggest a very recent north polar origin of plants, animals, man, and his gods.

Under the Saturn Configuration, mankind would have chosen to live in the northern hemisphere where the proximity of Saturn, the old vegetation god, provided the ideal environment that is recorded in the Golden Age myths of all cultures. There was no ice cap within the Arctic Circle; palms and corals were growing there, the same ones which were discovered in the 19th century.

Apparently a "Miocene" polar continent was in place under the Saturn Configuration, when myth suggests the lithic bulge of the world mountain. Magnetic forces held the Saturn Configuration in line, but at the release of this force, an implosion could have produced Earth's equatorial bulge which has been the subject of much speculation. The slight pear shape of the planet would be all that was left of the influence of the previous order of the planetary system. The land mass at the pole sank, compounded by the subsequent flood that gave the rainy Hyades their name.

At the Fall, with the dissolution of the Saturn Configuration, the Earth's pole tilted almost into the ecliptic, causing the seasons to begin. As discussed earlier, these seasons were highly extreme compared to those we are familiar with today. The sun traveled not from the Tropic of Cancer to the Tropic of Capricorn, but almost from one pole to the other. Still, the pattern of climate would follow that of today, warmer at the equator, colder at the poles. As the centuries passed, alternate freezing and thawing at the poles would facilitate the build-up of ice and snow.[8] In Siberian mythology, the sky is the roof of a tent, the Pleiades "the worst hole" which let in wind and cold.[9] The Evil One entered and "by his witchcraft counter-created winter and the worst of the plagues."[10] Thus the Silver Age saw the beginning of the formation of the polar ice caps.

[4] Archibald Geikie, *Textbook of Geology* (London, 1882) p. 874.

[5] Wallace, *Island Life* (London, 1880) p. 486.

[6] Wallace, *The Geographic Distribution of Animals* (New York ed.) Vol. I, p. 173.

[7] "Large Igneous Provinces," *Scientific American*, Oct. 1993. Here the Columbia River Flood Basalts are pushed 16 million years into the past.

[8] As Velikovsky noted, cold alone cannot create large masses of ice. Heat is necessary for evaporation and precipitation. *Worlds in Collision*, pp. 22–23.

[9] Otto Dahnhardt, *Satursagen* (Leipzig & Berlin, 1907–1910) Vol. III, pp. 32–44.

[10] Darmesteter, *The Avesta* (Fargard) i.3.

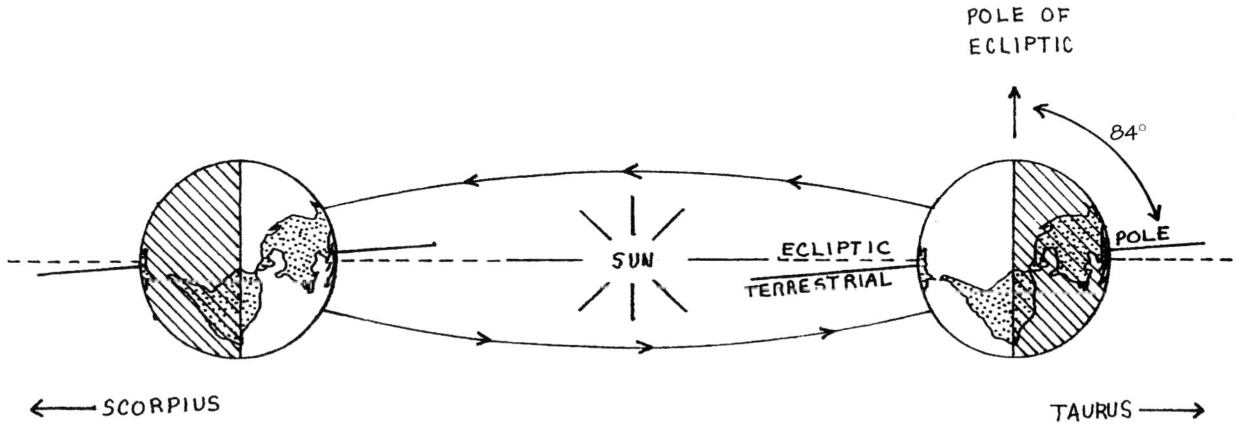

The Silver Age.
The sun travels nearly from pole to pole in the skies of Earth.

In the high latitudes of polar regions, Earth's rotation makes sun, Moon, and planets, as well as stars, appear to circle around the sky. Their path is more or less parallel to the horizon according to the distance of the observer from the pole. And so it was at the beginning of the Silver Age when such civilization as existed was high up in the northern hemisphere. The Greek astronomer Anaxagoras described the motion. "In the beginning the stars revolved in a tholiform manner," meaning in a horizontal plane, not υπο (underneath) but περι (around) the Earth.[11]

Over the course of a year, as seen from the North Pole and its vicinity, the sun would appear to circle the sky as it moved almost from one pole to the other and back again. Soon after the Fall, it would have appeared in the south at each turn of the Earth and gradually be seen to be moving east to west as it lingered longer in the sky each day, eventually remaining in the sky as it circled toward the zenith. A year would then be thought to consist of one day and one night, and the sun would have risen in the south.[12]

Traces of this cosmology still exist. "A year of mortals is a day and night of the gods." Two kinds of days are referred to in both *Iliad* and *Odyssey*: that of the gods a year's duration, that of men the one with which we are familiar.[13] In the Navaho mythology, "A year then was but as a day of our time."[14] *Hamlet's Mill* repeats the tale of Rabba bar Bar-Hara, who "left a basketful of bread at the edge of the earth on the heavenly window ledge while he prayed. He had to wait until the next day to recover it when the wheel of the Firmament turned round."[15]

At the equator day and night would be experienced during the Silver Age. Toward the solstices the sun would be low on the horizon during the day and never far below it at night. Closer to the equinoxes there would be full day and night.

[11] "Des Ecrits et de la Doctrine D'Anaxagore," in *Histoire de l'Academie des Sciences et Belles Lettres de Berlin* (Berlin 1755) Vol. ix, p. 378 ff.

[12] Frederik Klee, *Le Deluge* (Paris 1847) p. 224.

[13] See Anton Krichenbauer, *Beitrage zurhomerischen Uranographie* (Vienna 1874) pp. 1–34.

[14] "The Navajo Mythology," *American Antiquarian and Oriental Journal*, Chicago, July 1883, p. 209.

[15] Santillana, *Hamlet's Mill*, p. 35.

Our sun Helios did not always enjoy the importance he does today, not in the direct descent of the ruling gods of the Greeks, not even an Olympian but the offspring of two Titans, Theia and Hyperion, and only the nephew of Kronos.[16] The Titan couple produced two other children, the Moon Selene and the Dawn Eos. At first blush, dawn seems too fleeting a phenomenon to be equated with sun and Moon, certainly in the present order of the heavens. But each world age has its own style, and the Silver Age, the beginning of a set of "fallen" cycles, very fittingly had a magnificent dawn. Arriving sporadically in the northern hemisphere and initiating the warm season, dawn circled the southern horizon for weeks as it gradually broke upon the world, well deserving to be counted a goddess. It was during the Silver Age that Sun and Dawn would have ridden in the same chariot.[17]

When we examine early cultures, the Moon is found to have a surprising importance, far out of proportion to the cycle we presently observe in the sky. In the Silver Age, it probably moved in the ecliptic, only later being set at an angle by a planetary catastrophe. One can only speculate on the cycle of the Moon, but it could only have "come out" during the winter half of the year when the sun was below the horizon.

Lest we be concerned that mankind was plunged into long months of darkness during winter, it is important to understand that more light is available at the pole than is commonly understood. The northern lights, aurora borealis, are a regular phenomenon in the far north. In addition, light "bends" more as we approach the pole, so that the sun can often be seen when it is actually below the horizon. It appears to be a phenomenon like ocean tides, which cannot be calculated but only observed. Imperfectly understood, it is probably, like the aurora borealis, an electromagnetic phenomenon.

Following ancient media, Jupiter would have been giving more light to Earth during the long nights of the Silver Age. The long cycle of Jupiter would not have made its light available in the northern hemisphere every year/night, however.

After the Fall, then, man began to move south under the pressure of changing climate, following the sun and learning the new cycles of earth and sky. This would have been an age of hunter-gatherers, actually recorded by the Chinese. "Moreover, the dozen or so centuries following the Yellow Emperor [Saturn] was a period in which the pendulum of Chinese society was swinging back and forth between the nomadic and agrarian way of life."[18]

Early in the 20th century, the once-accepted ability of early man to cross oceans fell out of fashion, so that the remarkable migrations and voyages implied in this chapter will not be well received in some quarters.[19] But one need only read the popular literature of sailing to find accounts of cruising sailors who cross oceans for the fun of it. We are a hardy race, as much in need of adventure and challenge as of the security and stability so heavily stressed in our present culture. When we finally accept that we are as capable as our ancestors, fashionable thinking may change again.

Mankind wasted little time in setting out to explore the Earth in the Silver Age; rather quickly we began measuring and mapping it.

---

[16] Hesiod, *Theogony.*

[17] F. Guirand, *New Larousse Encyclopedia of Mythology* (London, 1972), p. 143.

[18] Kuo-Cheng Wu, *The Chinese Heritage* (Crown, 1982) p. 36.

[19] The real issue here is who owns the land. Western Civilization has generally taken over the world as if indigenous populations had few or no rights.

For some decades now, the existence of a number of remarkable ancient maps has been known to us. In *Maps of the Ancient Sea Kings* and *Path of the Pole*, Charles Hapgood apprised the public of exciting work being done in his college classroom in interpreting old maps. Among other startling finds, it was learned that the coastline and rivers of Antarctica had been accurately represented on the Oronteus Fineus map published in Europe in 1531. This is an area centered at the South Pole and buried under thousands of feet of ice and snow, impossible to map by conventional methods. Only in the middle of the 20th century would modern techniques of seismology be able to validate a map that could only have been created in the Silver Age when the poles were free of ice.

Another remarkable map produced by early explorers was one in which Greenland was depicted as three islands! It must be emphasized that these valuable maps were copies of copies of copies reaching far into the past, treasured by seamen throughout world ages. It was only in 1951 that the map of Greenland, published by one of the descendants of Antonio and Nicolo Zeno in 1558, was discovered to be correct. Again, because of rapidly spreading polar ice caps, this work of exploration and mapping must have been done early in the Silver Age, a testament to the adventurous spirit and practical science of our ancestors.

The evidence of ancient maps has been corroborated elsewhere, indicating an ice-free South Pole in recent times. During the Byrd Expedition of 1947–48, cores were taken from the bottom of the ocean off the Ross Sea of Antarctica. Radioactive dating of the sediments indicated a temperate climate in the Ross Sea as recently as 6,000 years ago, and extending back to 15,000 years ago.[20] Thus carbon-14 dating testifies to an ice-free South Pole, showing that "Antarctica's ice is less than 6,000 years old. Holmes writes: 'Algal remains dated at 6,000 B.P. (before present) have been found on the latest terminal moraines.' "[21]

Mankind's early migration from north polar regions brings the prime meridian of the Babylonians into focus, used for timekeeping and measurement as we use the one in Greenwich, England, today. The ancient prime meridian runs through the present North Pole and "the location where the gods had resided in the north," 56.35 degrees N, 52.38 degrees E, a location at some distance from the actual location of Babylon, 32.33 degrees N, 44.25 degrees E. This 8.03-degree difference in longitude is sometimes cited to explain why the Babylonians began their calculations at 8 degrees into Aries rather than at 0 degrees.

Reviewing the catastrophic Earth movements of the 8th and 7th centuries B.C., it is more than likely that the Babylonians calculated from 8 degrees Arietis because that was the point on the precessional clock, marked by the sun at the Spring Equinox, where the present age, the Age of Iron, began.[22] There are discrepancies in figures from this period that require more study. The "location where the gods had resided in the north . . ." is simply another north-pointing indicator of where civilization most recently began.

Major climate change occurred again when Earth tilted upward at the end of the Silver Age. After that, the sun would stop well short of the poles, which would grow even colder, the ice pack more extensive. This would curtail

[20] Charles Hapgood, *Path of the Pole* (Philadelphia, 1970) pp. 106–107.

[21] R. Daly, *Earth's Most Challenging Mysteries* (Nutley, N.J. 1975) pp. 227, 264.

[22] Norman Schwarz, "Old World Maps—A Response," *The Velikovskian,* Vol. II, 2. Also see Neugebauer's *Ancient Mathematical Astronomy,* pp. 368–369, 593–600.

polar exploration and voyaging and press mankind toward the equator as that region became correspondingly warmer. Each time the Earth tilted closer to the pole of precession at the end of an age, the same pattern would repeat and there would be a new surge of migration from extreme northern regions toward the equator. These terrestrial tilts are the "declination of the bodies" mentioned in Plato's *Timaeus*, the path of the ecliptic tilting in relation to the terrestrial equator.

Not every creature was able to move fast enough to save itself in the wake of the catastrophe of planetary tilt, however. The well-known case of the mammoths frozen in Siberia would have occurred at one of the rapid, climate-changing terrestrial tilts. One can imagine an ocean of polar air and water rushing down on one side of the globe, freezing everything in its path.

In *Earth in Upheaval,* Velikovsky adduced evidence in support of climate change at both the Venus and Mars series of catastrophes. Rutger Sernander's work in Scandanavia indicates strong disturbances in climate at both those times. Lake dwellings in Scandinavia, Germany, Switzerland, and Northern Italy suffered "high water" catastrophes accompanied by "very great and sudden climatic changes, Klimasturze," at close to the same times. Chronological scales based on pollen analysis indicate "radical change of life conditions" while connecting the high-water catastrophes to climatic change.[23]

As men traveled south, their gods went with them. Those of the Greeks have been particularly well noticed: big, often blue- or gray-eyed, marked by their rough, northern character, generally behaving like a crowd of unruly berserkers. The northern Hellenes were "the warriors who led the expedition against Troy, the first collective enterprise, according to Thucydides, that gave unity to Greece; their leader was Achilles."[24]

This sulking hero of the *Iliad* carried a remarkable shield into battle, often commented on because of the constellations figured on it: the Hyades and Orion,[25] the Pleiades, and Ursa

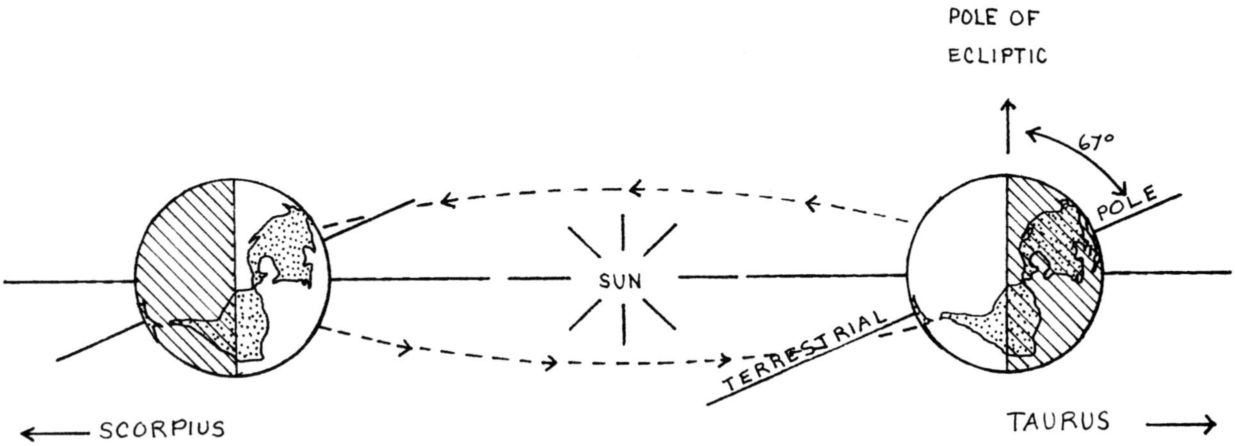

The Bronze Age.
Day and night become a more regular phenomenon over the Earth.

[23] Velikovsky, *Earth in Upheaval* (Garden City, N.Y., Doubleday, 1955) pp. 173–180.

[24] J. A. Harrison, *The Religion of Ancient Greece,* p. 27.

[25] The Egyptians often depict Orion as holding in his hand the prominent star of Aldebaran in the Hyades.

Major.[26] By now we recognize them as polar, Ursa Major actually at the pole during the Trojan War. Achilles was a northman and knew where the gods resided. He was proud of his connections and flaunted them on his shield. The Pleiades and Ursa Major are also well connected in Indic culture, the Pleiades thought to be wives of the seven Rishis, the seven stars of Ursa Major.

Efforts to sort out ancient history will be vain until we begin to listen to the ancients, beginning with Hesiod and his five world ages, separated by the Fall and the catastrophes described by Velikovsky, and implied by the *Stratigraphie Comparee* of Claude Schaeffer. Ancient media offers us a star map of Earth's movements against the background of the precessional clock, but until we grasp the vastly different world that confronted mankind in each of the five most recent world ages, we have no hope of understanding our own past.

A major work which gives us a sense of what went on in the Age of Heroes is the Daytons' monumental *Minerals, Metals, Glazing, and Man.* A valuable new tool has been added to the archeological kit, the analysis of metal and glazing, demonstrating that metallurgy and the glazing of pottery marched in step in the recent history of mankind. The book contains a valuable exposition of an Indo-European trade axis which extended from the British Isles through Europe and the Mediterranean countries to the Middle East and India. Other work would extend it to the New World across both Atlantic and Pacific Oceans.[27] Barry Fell suggests that New World copper supported Old World bronze.[28]

Although modern civilization with its hallmark of writing began in the Middle East, technology began in Europe following the Daytons, bringing metallurgy, polychrome glazes, blue paste, and glass down to the Mediterranean and Middle East (as well as rock crystal, amber, and amethyst). Although the book is dedicated to Claude Schaeffer, the Daytons are uncomfortable with catastrophe and its euphemism, hiatus. Their cure for the awkward "Dark Ages" is to move the "Grosse Wanderung,"[29] which is postulated after the tells of Central Europe were destroyed, from

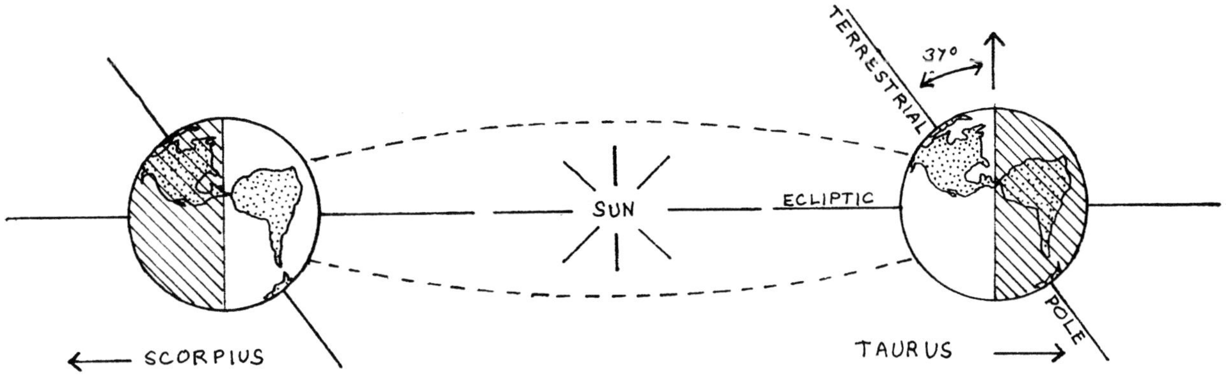

The Age of Heroes, most like our own Iron Age.

[26] Homer, *Iliad*, Book XVIII.

[27] John L. Sorenson & Martin H. Raish, *Pre-Columbian Contacts with the Americas Across the Oceans: An Annotated Bibliography* (Research Press, 1989).

[28] Barry Fell, *Bronze Age America* (Boston, Little, Brown, 1982).

[29] John and Ann Dayton, *Minerals, Metals, Glazing and Man,* p. 423.

the 13th to the 9th centuries B.C.[30] The descendants of the Urnfield Cultures would then establish in Anatolia, around the Adriatic, and in Italy. The Daytons reach for climate change to explain the obvious movements of population, and since major climate change is integral to a tilt of the pole, the pattern within these pages would serve their purpose admirably. Without doubt, there will be major problems adjusting dates, but the Daytons suggest a sensible way to start, using uncorrected carbon-14 dates for relative dating.

The catastrophes attending the Earth-Mars Wars were sufficient to break the back of the Indo-European trade axis of the Age of Heroes. The climate change caused by the tilt of Earth effectively cut off the New from the Old World. An easy summer sea route over the North Atlantic by way of Iceland and Greenland, etc., would become too rigorous for all but hardy Vikings. The visits to the New World would be limited to the seasonal use of Atlantic Ocean winds and currents. These longer, open-sea voyages, requiring navigational skills to make a desired landfall, gradually became secret and were finally forgotten.

During the Age of Heroes, a far-flung civilization bloomed over the Earth. In the historical tradition of Western Civilization, this was obscured by the rise of Classical Greece and Rome, and eventually Christianity. Earlier knowledge was lost or deliberately overlooked. Over the next millennium the dualistic thought of the Judeo-Christian tradition would be honed to a fine edge. When men of the Old World next confronted their brothers at the "discovery" of the Americas, they would find them alien indeed.

[30] The exposition here would suggest 717 B.C.

Tet from the papyrus of Hunefer.

30. Mankind in Amnesia

Coming as he has out of the psychoanalytic tradition, Immanuel Velikovsky's work represents a major attempt to reintegrate man's forgotten past with his rationalized awareness of himself and his history.

– Thomas A. Parry, "The New Science of Immanuel Velikovsky"

The task I had to accomplish was not unlike that faced by a psychoanalyst who, out of dissociated memories and dreams, reconstructs a forgotten experience in the early life of an individual.

– Immanuel Velikovsky, *Worlds in Collision*

The first rule of the game is that the game must continue.

– Alan Watts, *Psychotherapy East and West*

It is difficult for the man on the street to understand why we do not, collectively, recall the great historical catastrophes described by Velikovsky in *Worlds in Collision.* We flatter ourselves that we are paying attention, are lucid, rational, in control. If these terrible things really happened, we would remember them, right? To attribute this loss of memory to unconscious repression is hard for most of us to accept. Most men are not comfortable with the concept of the unconscious and would prefer to laugh with H. L. Mencken who says somewhere that psychiatry is the disease it is trying to cure. It is a good joke, but as long as mankind is unable to behave logically, rationally, we cannot laugh too hard.

The modern myth and dream factory, Hollywood, has occasionally taken up the cudgels for psychology, using the theme of Freud's repetition-compulsion. The mechanics of it are fairly easy to understand. A child is traumatized by a bad experience, represses and forgets it, but later is under compulsion to repeat the experience in some form until the original experience is brought to conscious attention and healed. The pattern is a helpful one, the repression a device to protect the immature self from an experience it cannot deal with or make sense of. Child abuse is a common example. The repressed material tends to rise into consciousness when the individual is more mature, capable of resolving it.

In following Freud's repetition-compulsion as well as Jung's conception of the collective unconscious, Velikovsky developed the theory set out in *Mankind in Amnesia.* Early man was traumatized by historical catastrophe, and is repressing memories so terrible that he has not yet been able to admit them to consciousness. Consequently we are in amnesia, in a state of compulsion to repeat similar widespread destruction on Earth. Growing as it does out Freud's work with individuals, the idea is persuasive, an explanation for man's destructive, irrational behavior. War and the destruction of the planet leap to mind, but even smoking could be a compulsive repetition of inhaling the noxious fumes produced by the vulcanism of interplanetary catastrophe. Deliberately inhaling smoke into one's lungs is hardly a rational act.

The establishment has always had a hard time winnowing the wheat of Freud's recognition of unconscious forces at work in man's psyche from the chaff of his skewed vision of the importance of sex. His work is not generally accepted as science. That the unconscious does not impinge on the world in a way that is easily measurable by the scientific method in no way lessens its importance.[1] But science, the flagship of Western Civilization, sets the style for how the rest of us think, and this is reflected in the media. Stecchini refers to the great mass of natural scientists as "naive realists."[2]

Yes, war is hell, violence a bad idea, but we always find reason for them, logical necessity. It is at this point that logic and reason fail us. Unsupported by intuition and spirituality, they lead us into places we do not want to be, behavior like burning books and witches, bombing innocent people, destroying the environment. We must call a halt to this behavior; it is unreasonable, illogical. But somehow we cannot. And this is why we study Freud, Jung, and Velikovsky, for mankind's continuing experiment with war, violence, and degradation of the environment has all the flavor of compulsive repetition.

If we are consciously controlling our lives, we are making a bad job of it. Parry puts it well. "Psychoanalysis represents the first realization in Western culture that the neglect of the receptive dimensions of man's nature has serious adverse consequences on his behavior, both personally and collectively."[3]

Velikovsky's hypothesis is a ray of light in a dark world, for most of us believe that man's destructive behavior is embedded in his nature, the way we are, a warlike race. If, instead, we are merely in a state of compulsion to repeat, then recalling historical catastrophe would clear the air and allow us to approach our ideals of peace and brotherhood. As we have already learned, some cultures do remember. It is Western Civilization and its heirs, those whose thought we have undermined, who are in amnesia.

For us, historical catastrophe is embedded in myth and ritual, in art and legend, in ancient records which we do not consider to be true in the same sense as the latest pronouncements of science. But now we have seen a previous celestial equator described in the Labors of Hercules, the Fall depicted on the ceiling of an Egyptian tomb, and the end of several world ages in the tauroctony of the Mithraic Mysteries. Hopefully, the reader's eyes have been opened to the storehouse of knowledge contained in material which appears on the surface to be fanciful. Ancient media can handily contain what we think of as science, and impose order on chaos at least as well.

In 1950, before the publication of *Worlds in Collision*, virtually all of us were in a state of amnesia regarding planetary catastrophe in historical time. In retrospect, we see that a vocal, controlling part of the scientific establishment was determined to remain in amnesia at any cost.[4] Exactly what went on in the minds of the general public who read *Worlds*

[1] But see Sheldrake, *Seven Experiments That Could Change the World* (New York, Riverhead Books, 1995).

[2] Stecchini, "The Inconstant Heavens," *The Velikovsky Affair*, ed. De Grazia (New Hyde Park, N.Y., Univ. Books, 1966) p. 115. ". . . the great mass of natural scientists . . . clings to scientism, the crude mechanical determinism of the eighteenth century, with insufficient awareness of all the knowledge that has been accumulated in two hundred years on the problem of human perception."

[3] Thomas A. Parry, "The New Science of Immanuel Velikovsky," *Kronos* I, 1, p. 3.

[4] The lamentable, often neurotic behavior of orthodox scientists in response to *Worlds in Collision* is well recorded. See A. De Grazia, *The Velikovsky Affair* (New Hyde Park, N.Y., Univ. Books 1966), particularly Stecchini, "The Inconstant Heavens," p. 112. "When one examines the reviews of *Worlds in Collision* written by some one hundred luminaries of our age, he observes that the civil liberty aspects of the affair (the effort to prevent the printing, the academic pressure exercised to keep reviewers in line, and the refusals to publish corrections of misstatements) recede in the face of the frightening realization that the experts to whom is entrusted the human inheritance of scientific thought, our most precious possession, can be the victims of collective hysteria."

in Collision, making it into a bestseller, has never been considered. Presumably they were free to work their way through the extensive descriptions of cataclysm in the book and thus relieve compulsion to repeat. In fact, *Worlds in Collision* was a touchstone, separating those who were comfortable with historical catastrophism from those who weren't. It is a matter of record that many thoughtful people, including some scientists, not only accepted Velikovsky's thesis, but did their best to carry it forward. Thus an undetermined part of the population is free from amnesia, no longer repressing planetary catastrophe nor compelled to repeat its violent character in some form.

Velikovsky was first of all a physician and healer. That he is the natural successor of Freud and Jung is not grasped by orthodoxy. The present vision of extending knowledge by specialization makes it difficult if not impossible for scientists to understand a polymath like Velikovsky. As he uncovered the horrendous experiences of mankind in researching *Worlds in Collisio*n, he began to see patterns in collective consciousness that mirrored those that Freud saw in the individual. After confronting the patient, mankind, with *Worlds in Collisio*n, he wrote *Mankind in Amnesia* many years later to explain his theory. If Velikovsky is correct in drawing the analogy between individual and collective consciousness, then the handwriting is on the wall, and it is only a matter time before historical catastrophism is accepted. That some undetermined percentage of the public already has, is significant.

The compulsion to repeat that Velikovsky feared was the use of nuclear and hydrogen bombs, considered to be a serious threat for many years after the bombing of Hiroshima by the United States at the close of World War II. The present style of destroying the Earth is pollution and degradation of the environment, which is not to say that a nuclear bomb will not be used. However, it has not happened yet. Whether the publication of *Worlds in Collision* defused to some degree the compulsion to repeat is an interesting question, something to keep in mind as future events unfold.

The concept of Western man's being in amnesia with regard to ancient cataclysms fits the facts of the case neatly. Exactly how the amnesia developed is a point on which I differ from Velikovsky. Catastrophes styled by insurance companies as acts of God: tidal waves, erupting volcanoes, devastating storms and earthquakes, are not alien to the Earth. The difference between them and the devastating Earth changes described in *Worlds in Collision* is only one of degree. The horror that Velikovsky and others perceive in interplanetary catastrophe is impersonal, not to be compared with that of one human being deliberately trying to injure another.

My own belief is that historical catastrophes were repressed, but not because they represented an ultimate horror. The ancient cyclic view of the world took account of endings and beginnings. If Earth were destroyed by devastating fire, the ancients knew the grass would grow again; a flood would recede and the sun come out and engender new life. If the climate changed, they would move to a better one. Their connection to the Earth, their world view, was different.

The vision of historical catastrophe as ultimate horror is the present world view of Western Civilization projected into the past. With our safety nets of allopathic medicine, insurance and technology, theoretically protecting us from everything from germs to acts of God, we simply cannot envision being naked before the forces of nature. The ultimate horror for us, then, is not Earth burned to a cinder, drowned or poisoned, human and animal populations destroyed. These are effects that we tolerate every day as long as they go on slowly and at a comfortable personal distance. The ongoing rape of the Earth (chinga tu madre) is merely our style, our hero journey.

For Western Civilization, the ultimate horror is loss of control.

Catastrophic events are repressed because they are patently uncontrollable, and in the 7th century B.C., mankind was embarking on a new hero journey of domination and control. We must be cautious about pushing our linear perception of cause and effect too far back in time. It was just coming into its own after the Earth-Mars Wars. Following the historian of religion, Mircea Eliade, "Judaism represents an innovation of the first importance. For Judaism, time has a beginning and will have an end. The idea of cyclic time is left behind. Yahweh no longer manifests himself in cosmic time (like the gods of other religions) but in historic time, which is irreversible . . . his gestures are personal interventions in history."[5]

What appears to have happened is that we embarked on our new hero journey and only gradually realized that the great planetary catastrophes would not fit into the picture. Thus, repressing the memory of them was a fairly conscious process which went on for a long time. The curtain of forgetfulness that rings down between ages was never so complete as after the Earth-Mars Wars, for then we deliberately cut ourselves off from our past. Cyclic time came to an end, and Western Civilization was born in linear time.

Cyclic time occurs in the relaxed, expansive style of myth and dream. That the Hopi and Eskimo still reside in part in cyclic time is reflected in their using only the present tense in their language. Linear Western time sees one event after another in a cause-and-effect perception of the world. It is perfectly reflected in our languages with their many time-related verbal tenses, which unfortunately makes it difficult to discuss cyclic time. We will return to these concepts presently.

It would appear that Western Civilization owes its linear style of time to the Jewish prophets who perceived the Earth-Mars Wars as Yahweh's "personal interventions in history." It was the beginning of history and of our compulsion to write and rewrite it. It was also the last time the Jews would consciously bring Heaven to Earth. Isaiah was greatly troubled by Venus, Queen of Heaven, for whom Jewish women were still baking cakes. Under his influence, the Jews would eventually give up their visible sky-goddess and devote themselves exclusively to the invisible Yahweh, opting against any representation of him at all.

Western Civilization has always been understood to be founded on Greek and Judeo-Christian traditions. It is in perceiving the catastrophic background against which it was forged that we can begin to see the total picture. Developing ideas of free will in the Greek world gave us leave, but it was the rigid, punishing, controlling Yahweh, as perceived by the Jewish prophets, who provided the pattern for the new Age of Iron. The goddess Venus, formerly one of his consorts, Anatha Baetl, was on her way out and a patriarchal takeover under way.

Yahweh would rule mankind with a heavy hand and from a distance, with all the power necessary to protect the Jews from their enemies. There would be no more comfortingly visible goddess. Unfortunately, this was a pattern which would devolve into other areas. The pattern of kingship, ritual regicide, was well established, having come down from the sky, but now it would change, escape the power of the goddess, become stronger as the light of Venus declined.

Yahweh would rule the Jews as kings would rule their people, as mankind would rule the Earth, as man would rule woman,

[5] Eliade, *The Sacred and the Profane*, trans. Willard Trask (N.Y., Harcourt, Brace, 1959) p. 110.

and as the developing ego would rule the psyche, in an alienated style, submerging the intuitive feminine. Contact with nature would be severed, both from earth and sky. Our Mother Earth would become alien, "nature red in tooth and claw," an objective reality to be dealt with using the reason and logic developing in the Greek world.

Jumping off into Western Civilization can be viewed in a number of ways: as a game to be played with new rules and a tight focus, as a hero journey as described in the Bearson legend, or as a necessary development of the ego.[6] And it is more than likely that there are lessons involved that we cannot now perceive. Viewed as a game, it might be called "Control the Earth." As we gradually settled into the new, alienated mode and began to play, memories of historical catastrophe threatened to spoil the game. Previous planetary cataclysms were obviously uncontrollable. The game occupied all our attention, so we decided to ignore, make a secret of, then forget catastrophe. After a while, we became angry when anyone brought it up. The new game, playing with the rich bounty of the Earth as objects, was consuming. It is no wonder we have played it through so far.

In the Great Forgetting that we consciously consented to, that of historical catastrophe, much associated information was lost as well: ancient science, the New World, and our brothers scattered over the Earth. We would be put to the trouble of reinventing much science, "discovering" what was old stuff to the ancients.[7] Technology is all our own, clever trinkets fashioned by Hephaestus the smith, love gifts to the goddess of wisdom, Athena, who scorned them. Unlike her, we are still under their spell.

In the following chapter, we will examine the period of forgetting subsequent to the Earth-Mars Wars.

6 Jung always maintained that the egocentric mode of consciousness is a universal and necessary step in the development of mankind.

7 Velikovsky, "Precursors" *Kronos* VII, 1, pp. 48–55.

31. Keeping Secrets

To live today in the awareness of a mystical, transcendent order of being and of unknown dimensions of time and human experience opens up fearful and undreamed of worlds. These realities are frightening, for they awaken us from sleep we had not known as sleep. It is much easier to pull the covers of culture over our heads and sleep a little longer before awakening.

– Gregor Goethals, *The TV Ritual, Worship at the Video Altar*

Wherefore be of good cheer and try to discover by recollection what you do not know, or rather what you do not remember.

– Plato, *Meno*

Memory is inherent in nature.

– Rupert Sheldrake, *Presence of the Past*

It is a paradox in the thought of Western Man that he perceives value in both openness and secrecy. On the one hand we have public libraries, a press free in theory, an ideal of full disclosure; on the other we have secret negotiations, secret societies, media control, and often a less than candid interface with our fellow man. We want knowledge to be accessible, but the desire to know is balanced by the desire not to know, some knowledge even considered to be dangerous. The mythology of secrecy is a powerful one.

The myth of Pandora's box warns us not to open it. Bluebeard gives his young wife the keys to the castle, but tells her not to open one door. When the ark is finally opened in the film, *Raiders of the Lost Ark,* hero and heroine are saved by not looking, by closing their eyes. And in Christian mythology, God himself warned Eve against eating the fruit of one particular tree in Paradise, that of the knowledge of good and evil. It is frightening to raise the lid on the box, open the secret door, eat the apple. The inhibitions against it are powerful, and we must be prepared to accept the consequences.

The notion of keeping secrets is embedded in Western thought. The Romans celebrated the shortened days of the Angeronalia on December 21st. The goddess Angerona was portrayed "with her mouth bandaged and sealed, and according to one witness (Solin 1.6) with one finger on her lips in the gesture enjoining silence."[1] Giving secrets away can be dangerous. Philoctetes gave away the secret of where Hercules' ashes could be found by stamping on the ground, and was subsequently wounded in the foot by one of Hercules' own arrows which flew from his quiver.[2] Many Sufi tales suggest the difficulties that await one who gives away secrets to those who are not ready for them.[3] Were he still with us, Velikovsky could testify to the wisdom of those stories. As it happens, however, homo sapiens is a

[1] Dumezil, *Archaic Roman Religion,* trans. Philip Krapp (Univ. of Chicago, 1970) pp. 336-337.

[2] Graves, *Greek Myths,* 161.L.

[3] Idres Shah, *Tales of the Dervishes* (New York, Dutton, 1970).

sophisticated creature who often keeps secrets from himself, loves games and mystery stories, and waits for Christmas to open the packages.

The mystery religions, those of Mithras, Isis, and Eleusis, as well as a host of lesser cults and oracles, are among the best kept secrets of the ancients. Only recently has the tauroctony of the Mithraic Mysteries even been connected to the sky. But they are mysteries only to us. To our ancestors in cyclic time, they were an open book written in the heavens. The "many changes in the direction of the foundations at Eleusis . . ."[4] were made consciously as the terrestrial axis tilted again and again. The same is true in the mysteries of Isis, as the ancient Egyptians sought the help of the goddess Sesheta to orient their temples.

The Mithraic Mysteries are something else again, secret from their inception. Ulansey may not convince us that they grew out of the discovery of the Precession of the Equinoxes in 128 by Hipparchus, but they have the appearance of having been devised at a particular point in time. Catastrophic elements that came down from the sky in cyclic time were incorporated into ritual and dogma in both Mithraism and Christianity, but well after the fact. This is not the same as the Mysteries of Eleusis which grew out of original time. Mithraism appeared in the 1st century B.C., then grew along with Christianity until it succumbed to the latter in the 4th century A.D.[5]

In order to understand the rise of Western Civilization and the Great Forgetting of previous cyclic time, we will consider the phenomenon of memory and the repetition that has always been assumed to assist its operation. Those of us with gray hair may remember a time when multiplication tables, rules of spelling and grammar, poetry, and the elements of the periodic table were committed to memory by repetition, often with mnemonic rhymes such as "i before e, except after c, . . ." Repetition is out of fashion now for human beings, discarded in educational theory,[6] but still considered useful in dealing with animals, notably laboratory rats.

Many readers will already be familiar with a remarkable series of experiments in which a number of rats and their progeny over many generations learned to solve a particular maze. As the years of the first experiments passed, succeeding generations of rats learned to solve the maze more and more easily. These results were unacceptable to current scientific theory, so the experiment was repeated with new rats. Inexplicably, the new rats appeared to be taking up where the first rats left off, some solving the maze the first time. In another experiment, control rats unrelated to the regulars were occasionally tested, but these, too, learned faster than was expected.

"This effect has never been satisfactorily explained. But it is just what would be expected on the basis of morphic resonance."[7] In Rupert Sheldrake's theory of formative causation and morphic resonance, patterns of memory are laid down in morphic fields similar to those of physics, strengthened by repetition, and then accessed when the same activity is repeated.

---

4 Lockyer, *Dawn of Astronomy* (Cambridge M.I.T., 1964) author's preface.

5 Ulansey, *Mithraic Mysteries*, p. 4.

6 This is true until we get to graduate school, where students are not such fools as to overlook memory hooks that help them to remember important information. The initial letters of the sentence "Fine little ladies make dancing by night pleasurable," help students remember the order of the geological layers of the Great Plains as they tilt up to form the Rocky Mountains. Examples can be quoted from many disciplines.

7 Rupert Sheldrake, *Presence of the Past*, p. 175.

The concept is proving to be a useful working hypothesis since it explains much problematical phenomena: the difficulty in crystallizing new compounds, the folding of proteins, the inheritance of acquired characteristics, dominant and recessive characteristics, the behavior of mutations, instinctive behavior, behavior of insect societies, parallel forms in marsupials and placentals. Astonishingly, the hypothesis has escaped formal science. Contests have been held to design experiments to prove its validity.[8]

Both myth and ritual are treated in Sheldrake's *Presence of the Past*, myth referring back to models in the past while modern theories of progress refer to future events. As to ritual, it is well known that its effectiveness depends on being performed in exact accordance with an ideal model, an idea strangely reflective of scientific repeatability. In explaining why this is in keeping with his theory of morphic resonance, Sheldrake confirms both ancient wisdom and Mircea Eliade.[9] "Through morphic resonance, rituals really can bring the past into the present."[10]

Now these are strong words from a respectable biologist, but we will examine them. In the light shed by the ideas of these two men, we will consider the patterns of primeval celestial phenomena and catastrophe: their repetition in cyclical time, their repression and repetition in historical time.

Exactly how the ancients were able to keep coherent records over cyclical time, when they were regularly interrupted by devastating destruction, is puzzling. There is no question but that planetary catastrophe all but wiped out civilization in various parts of the world periodically. The most ancient of civilizations who kept written records, Egypt and Mesopotamia, could hardly recall their own origins.

As Santillana remarks, "Now that documents of the earliest ages of writing are available, one is struck by a wholly unexpected feature. Those first predecessors of ours, instead of indulging their whims with childlike freedom, behave like worried and doubting commentators; they always try for an exegesis of a dimly understood tradition. They move among technical terms whose meaning is half lost to them, they deal with words which appear on this earliest horizon already 'tottering with age' as J. H. Breasted says, words soon to disappear from our ken."[11] Ashurbanipal was beset with the same problem. "I study inscriptions from before the flood, which are obtuse, obscure and confused."[12] Here we recall that Plato had hard words for writing. "For this invention will produce forgetfulness in the minds of those who use it because they will not practice their memory."[13]

We, of course, count writing as a hallmark of civilization and give scribal utterances great weight while overlooking the major burden of myth, legend, and the rest of ancient media. As it happens, the Code of Hammurabi was engraved on a conical stone which represented God. One could say, then as now, that the medium was the message.

Realizing that celestial phenomena is repeated in all ancient media is fundamental to understanding the thought of the ancients. To begin with, this repetition preserved contact between Heaven and Earth. It was

[8] Sheldrake, *A New Science of Life* (Rochester, Vt., Park St. Press, 1981) Appendix, pp. 247–249.

[9] Eliade, *The Eternal Return*.

[10] Sheldrake, *Presence of the Past*, p. 260.

[11] Santillana, *Hamlet's Mill,* p. 119.

[12] Vorderasiatische Schiftdenkmaler, V.A.B., VII, 256, 18–19=f.

[13] Plato, *Phaedrus,* LCL, Fowler, 274D–275B.

also the old way of imposing order on chaos, making sense of the world much as science does today. What also appears to be true is that it was the ancients' way of making a historical record so they could make the "eternal return."

As Eliade makes clear, archaic man counted on being able to access original time. For him it was not in a forever finished past, but in a time conveniently adjacent. "It is further of interest to note that the New Year scenarios in which the Creation is repeated are particularly explicit among the historical peoples, those with whom history, properly speaking, begins—that is, the Babylonians, Egyptians, Hebrews, Iranians. It almost seems that these peoples, conscious that they were the first to build 'history,' recorded their own acts for the use of their successors . . ."[14] One could say that these peoples demonstrated the old style of recording events in ancient media in the new style of written media.

What ancient media is all about is the creation of patterns, strengthened by repetition, which can later be accessed by similar activity. It was tantamount to making a historical record and was done deliberately and with the consciousness then available. The patterns of celestial phenomena over cyclic time exist: like those that helped the rats solve the maze, like those for folding proteins, crystallizing compounds, or making baby rabbits grow up to be big rabbits instead of owls or toads. These patterns "inform" nature, and they apparently informed the ancients who were thereby able to keep running records of a historical type during a difficult cyclic time.

The computer model of memory (a process of coding, storage, and retrieval) is so popular and highly thought of, that some scientists confidently await a time when machines develop consciousness! This remains to be seen. But given the coding implicit in ritual, art, and myth, the storage and strengthening of patterns in morphic fields, access and retrieval becomes eminently possible. In fact, it has been accomplished within these pages. With all due respect to technology, we note that the pattern for the computer is a very old one.

So much for cyclic time and its record-keeping. We move now through the 60 years of the Earth-Mars Wars which changed our reality, into linear time and the Great Forgetting of what had gone before. It would take over 1,000 years, from −687 to approximately the 4th century A.D., but the memory of of cyclic time and historical catastrophe would finally be repressed. What seems to have happened is that myth, ritual, and other primeval media fell out of fashion, retreated to their own small sphere. Unitary knowledge shattered like a mirror, and each shard would now throw back only a small part of the truth. Conscious knowledge of historical catastrophe was focused in the mystery religions where they became secret, and were ignored and finally forgotten by the general populace.

Homer and Hesiod made records of the Earth-Mars Wars soon after the event, and others continued the work, but in the old mythical style. Plato disparaged myth[15] and one could easily imagine the retreat of cults and oracles in the face of the rigorous logic and reason associated with Greek thought at that time. Eleusis went underground as well, no longer a treasury of ancient knowledge but only the myth of Persephone. On the one hand they decided to keep secrets, but on the other, fewer people wanted to know them.

The Bearson was off on his hero journey as the first great expansions of Western Civilization took place. Philip of Macedon and Alexander the Great extended the Greek

[14] Eliade, *Eternal Return*, p. 74.

[15] Plato, *Phaedrus.* See discussion in Graves, *White Goddess*, pp. 10–12.

World in a brief explosion that reached to India. The Empire of Rome would prove larger and more lasting. This is the story that we know, the adventure of Western Civilization that can be found in our libraries. As Western man reached for more control, he became more and more ill at ease with the memory of uncontrollable catastrophe. In the old cyclic style of time, man had rituals to access past information. The new linear style had nothing of the kind. That it has little to say about the Earth-Mars Wars is in part due to history as we know it having only recently taken form.

In the 1st century B.C., Mithraism became visible, an effort to find God in the forces that govern the Precession of the Equinoxes and the cyclic catastrophe that takes place against its background. In his belief that the Mithraic Mysteries were founded in the discovery of the precession, Ulansey stays within the present paradigm. Traditions of catastrophism are strong in Iran, the birthplace of Mithraism, and Cumont was correct in pushing back to the Bundahisn. But when we confront the elements of the Mithraic Mysteries—the series of seven initiations, the cave-like mithraea in which they were held, the carefully thought-out tauroctony—we see a construction of human devising which resurrects ancient knowledge well after the last historical catastrophe in −687.

Mithraism arose in the 1st century B.C., reached its peak in the 3rd century A.D., and succumbed to Christianity at the end of the 4th century.[16] The characteristic mark of Mithraism was the tauroctony, present in every mithraeum, and these were found, over 500 of them, throughout the Roman Empire. Ulansey traces the religion back to Stoic thought, astral religion, and interest in world ages.[17] Though it was a vehicle for knowledge of catastrophism, Mithraism was conceived in a logical, linear style. In practice it was even more patriarchal than Judaism or Christianity, for women were actually excluded from membership.

Whatever its background, Mithraism was a Roman religion, its seven grades appealing to the legionaries who constituted a large part of its membership, which also included bureaucrats, merchants, and slaves. At its highest levels, not only the catastrophic history of the Earth would have been known, but possibly something of the Saturn Configuration. The Saturnalia, when the god was released from his "bonds," and the world turned upside down as masters became slaves and vice-versa, would have had a deeper meaning for the Mithraic part of the population.

We do not think of the Romans as astronomers. They left no tables of star risings like the Babylonians and "there is no public festival marking the equinoxes or Summer Solstice, and no divinity patronizes them,"[18] though important feasts cluster at these points. They had a goddess of hinges, Cardea, and an astronomical term, vorta, which would cover major Earth movements. It is possible that these supposedly simple Roman soldiers had some interesting thoughts while on campaign, sleeping out under the stars.

Secret doctrines cannot be spoken of, and those of Mithraism were respected, for little has been uncovered to cast light on them. What we can see now is that the very secrecy in which they were shrouded, the determination not to reveal historical catastrophe, allies with conscious repression of the events. The great planetary cataclysms which visited destruction on Earth became secret, were repressed and ignored, then forgotten.

[16] Ulansey, *Origin of the Mithraic Mysteries,* p. 4.

[17] Ibid., pp. 75–76.

[18] Dumezil, *Archaic Roman Religion,* p. 335. The exception is the Angeronalia at the Winter Solstice.

What we also forgot about was the ancients' understanding of the system of coding, storage, and retrieval, still in place but unused. All the patterns of celestial phenomena and catastrophe still exist in morphic fields, coded in primeval media such as myth and art, and retrievable during similar activity. Once knowledge of primeval celestial phenomena and catastrophe was below our conscious attention, systematically pushed down whenever it surfaced, we no longer had easy access or any control over them. Problems began as the uninvited phenomena of repeating catastrophe continued to surface. It is as if repression of these particular events inclines them to repeat. And repeat they do, notably in warfare, and against our conscious volition. Thus Freud's repetition-compulsion in the individual, Velikovsky's in the collective.

Christians struggle to celebrate the birthday of Christ, but their efforts are swamped by the phenomena of the Golden Age, still accessible in morphic fields. We install a new god at the North Pole, Santa Claus, set up a (world) tree with a (sun) star on top without a clue that we are repeating images of the Saturn Configuration. The polar chariot and horses of the Silver Age become Santa's sleigh and reindeer. Christmas is frequently spelled Xmas, the initial letter of Saturn in Greek, Χπovos. Later in the year at Easter, a movable feast still connected with full Moon and equinox, we celebrate the World Egg. It is hard to quarrel with the good feelings present at Christmas, but these have always been associated with the Golden Age as well as the Saturnalia of the Romans at the Winter Solstice. Other archetypal patterns are not so pleasant.

It is chilling to view the old documentaries in which tens of thousands of Hitler Youth salute Der Fuhrer, raising their hands and shouting "Seig Heil." It is an archetypal gesture made by Roman soldiers to Mars, god of war, or whatever emperor or general stood in his place. The spectacle, even on film, is one of raw force, enough to raise the hair on one's spine. It is all the old wolf energy associated with the Earth-Mars Wars. Hitler, we recall, knew well the Teutonic mythos and was skillful at accessing its power. When it comes time to join the battle, engaging the force of Mars is what war is all about.

Velikovsky has dealt at length with the nuclear or hydrogen bomb repeating the destruction of planetary catastrophe. The bullfight is still reenacted on several continents. The Pentagon in Washington, D.C., echoes the five episodes of the Earth-Mars Wars. Lesser models such as the eye atop the pyramid on the dollar bill are thick around us. Many have been called to our attention within this book, but there is no room to list all our pale copies of ancient and catastrophic phenomena, nor do I wish to spoil the interested reader's pleasure in finding more.[19]

To sum up: in cyclic time, man laid down the patterns of primeval and catastrophic phenomena in morphic fields, strengthened them by repetition, and accessed them consciously. In historic time, man represses these patterns and repeats them unconsciously. Hopefully, this will place an important part of Velikovsky's work in perspective. As a psychiatrist, he was concerned to help mankind raise his past into consciousness so that necessary healing could begin.

It remains to be seen if mankind is quite ready to confront his past. There may be lessons to be learned that we have not yet assimilated. In the following chapter we will address one that we hardly know we are working on.

[19] See the images in Third World art, money around the world, but most particularly those which introduce our modern myth, the Hollywood film.

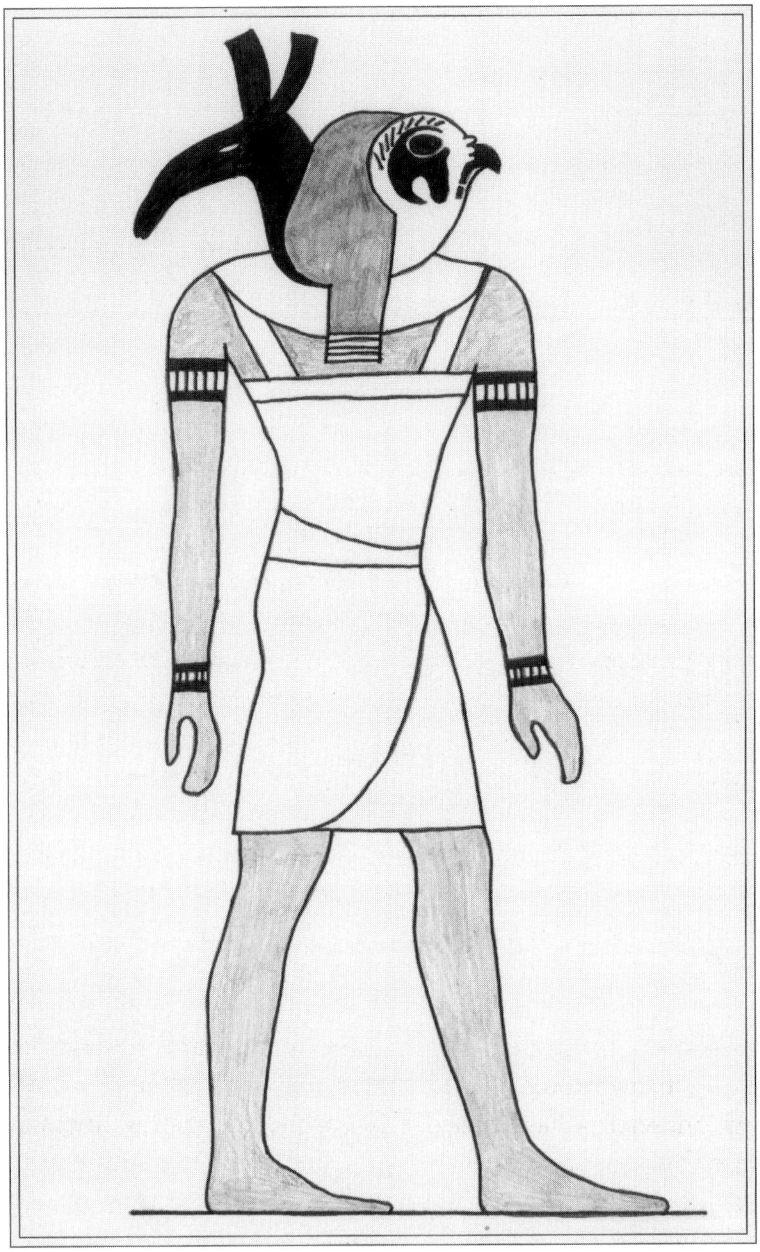

The Egyptian twins of duality: Horus and Set.

32. Dualism: The Hidden Agenda

By and large Western culture is a celebration of the illusion that good may exist without evil, light without darkness, and pleasure without pain . . .

– Alan Watts, Introduction, *The Two Hands of God*

Now in the Egyptian language, this sole relic of a primitive world, there are a fair number of words with two meanings, one of which is the exact opposite of the other.

– Karl Abel, *Sprachwissenschaftliche Abhandlingen*

You must be free of the pairs of opposites.

– Bhagavad Gita

Many years ago while visiting with a friend, I expressed a desire to learn more about my family history. As it happened, she was familiar with the methods of research and offered to help me get started. "But make up your mind, Georgia," she assured me. "You will find out things you'd rather not know." I never did pursue the family history, side-tracked by that of Earth and cosmos, but even in that larger context, her words were prophetic.

Science, the flagship of the culture, must be discomfited to realize that the present order of the solar system can only be projected back to the 7th century B.C., while an entirely different configuration of planets was in place something like 5,350 years ago (assuming the present rate of precession). Rejecting its own roots in the Judeo-Christian tradition, science has devised its own mythology and this must now lose credence. If it wishes to deal with reality, it must attend to the same ancient media in which much of religious dogma is rooted, that dogma which has long been considered to have no basis in reality. Surely this will be a bitter pill to swallow.

Religion, the most conservative force in the culture, must watch as its dogma transforms into a mix of astronomy and historical catastrophe. We are amused by the myth of the crafty Sisyphus and his three-day visit to Hell, bargaining with Persephone to return to the upper world. His punishment of having to roll a stone uphill is such a clever way to describe the movement of the pole from Ursa Major to Ursa Minor. But when we recall that the Council of Nicea determined that Jesus Christ spent three days in Hell, and a rock was rolled away from his tomb in Bible history, it hits close to home, hard enough to make a dogmatic Christian squirm. Thus reclaiming our catastrophic past is a prickly, mixed bag.

Many ancient religions grew out of the background of cyclical catastrophe. Judaism was new in that it took the idea of a single god (not new but as old as Saturn) and embedded him in linear time from which he regularly intervened in the affairs of men. The fiction of the Egyptian Akhneton establishing the first monotheism is a misreading of events. He, too, was harking back to the Saturn Configuration. Mithraism relates to the precessional forces that combine with the cosmic circular force to hold both the Saturn Configuration and the present solar system together.

In retrospect, it can be seen that the birth of Christ and introduction of the Christian ethic in his teaching parables was a welcome amelioration of the harsh model in which Western Civilization was cast. We would indeed need faith, hope, and charity to get through the Iron Age. At a time when the entire world was beginning to be seen as other, Christ reminded us to try to see him in others. He brought us the simple message of the Golden Rule which is expressed in many of the great religions. Not surprisingly, the Buddha and Krishna appeared even earlier in the period following the Earth-Mars Wars, bringing personal revelations which initiated religions that endure to this day.

That many of the messages in primeval media devolved on the Christian Church is a matter of record. Consciously or unconsciously, it was saving the phenomena of the cyclic past as were the more lucid Mithraic Mysteries. Catastrophic elements already sacralized in previous cultures became dogma: the three-day descent to Hell, ritual murder and eating of the king, the "sons of thunder" twins James and John, Peter the Rock, death on the cross, the previously feminine Trinity. The new patriarchal vision tried to exclude the goddess, but she was too powerful to be suppressed, reentering the scene during the period of the Crusades and celebrated in the medieval cathedrals to the Virgin.

What appears to have happened is that the personal revelation of Christ was swamped by primeval media. Interestingly, the same pattern can be seen in the case of the Buddha, who immediately attached the mythology of his culture. Religion, after all, must reflect as well as influence the society it grows within, and it is not uncommon for historical figures to be mythologized.[1]

When all were closer to their catastrophic origins, the interface between cultures may well have been easier. Exactly how catastrophic elements entered the Christian Church will make an interesting study for exegetes, but it clearly provided common ground with indigenous religions all over the world, facilitating the spread of the faith. People as far apart as the Danes and the Peruvians knew who the twins were, James and John, the "sons of thunder."[2] Graves has noted how easily Christianity meshed with the native religion of the British Isles.

Some of the malaise we experience on confronting ancient knowledge is due to its unitary character. All subjects were considered to be connected, whether science, art, or humanities. In contrast, modern knowledge is divided into many separate disciplines, making it exceedingly difficult to see the whole. It was only by crossing the lines between disciplines that Velikovsky was able to put together his historical reconstruction. In 1950, he was vilified for this, but times have changed as we have become aware of problems inherent in the fragmentation of knowledge. Now the concept of interdisciplinary studies is highly regarded as an effort to bring knowledge into unity and balance.

We have already noted how the critical issues of our time have grown out of the examination of their contraries, such as feminism out of the patriarchal vision. Holism is another like issue, but a deeper one, harking back to the Saturn Configuration. It has become almost a household world. All children are familiar with holograms to a greater or lesser extent, and even adults are learning they cannot tug at what appear to be loose ends of reality without affecting everything else. The world is indeed complicated, braided together, all of a piece. In a remarkable turnaround, multiplicity is becoming unity.

In order to understand how the original unitary vision of the ancients became first dual, then multiple, we must go back to the

[1] Eliade, *Myth of the Eternal Return*, pp. 39–43.

[2] Harris, *Boanerges*, pp. 9–12. Discussed in Santillana, *Hamlet's Mill*, p. 226.

Yin yang symbol.

Saturn Configuration when the one perceived God loomed overhead, connected to Earth by the axis column. The illuminated crescent of the Saturnian band in its different positions, contrasted with the darker side of the band, produced the concept of the twins, Isis and Nephthys. It is, however, far more likely that the changeful twin, Mars, propelled us faster along the road to dualistic thinking.

Mars had a single identity, but a dual character as described in the chapter "Mars, The Magic Twin." Watching the planet move up and down the axis column between Earth and Saturn must have put early man in the habit of seeing the one in two different aspects: dwarf and giant. As we have seen, Mars became the model for twins in many cultures, and by the time of the Earth-Mars Wars, Mars and his alter ego, Hercules, had attracted opposite qualities. When the "evil" Mars attacked Earth in the 8th and 7th centuries B.C., the "good" Hercules was assumed to be defending it. Several world ages passed in the course of sorting out opposite qualities, but intimations of this dualistic vision of the world were expressed early on in the Egyptian language.

As remarked by the philologist, Karl Abel, ". . . in the Egyptian language, . . . there are a fair number of words with two meanings, one of which is the exact opposite of the other . . ." He noted other combinations of contraries as well. "Of all the eccentricities of the Egyptian vocabulary, perhaps the most extraordinary feature is that, quite apart from the words that combine antithetical meanings, it possesses other compound words in which two vocables of antithetical meaning are united so as to form a compound which bears the meaning of only one of its constituents . . . a part which would have had the same meaning by itself." The compound words would eventually divide into two words of opposite meaning.

Abel solves the problem to his own satisfaction. "Since every concept is in this way the twin of its contrary, how could it be first thought of and how could it be communicated to other people who were trying to conceive it, other than being measured against its contrary . . . ?" ". . . Since the concept of strength could not be formed except as a contrary to weakness, the word denoting 'strong' contained a simultaneous recollection of 'weak' as the means by which it first came into existence. In reality this word denoted neither 'strong' nor 'weak,' but the relation and difference between the two, which created both of them equally . . ." Abel goes on to notice the phenomenon in other languages.[3]

It is not surprising that the civilization that raised the pyramids took the trouble to differentiate between strong and weak.[4] And Abel has definitely given us something to think about, though cast in the framework of a dualistic language. The Egyptians were obviously struggling with the concept of expressing opposites, something that seems terribly obvious to Westerners. A word that expressed "the relation and difference" between two opposites would be alien to us indeed, but a handful of such words are still with us.[5]

[3] Karl Abel, *Aegyptische-semitisch-indoeuropalisches* (Leipzig, W. Friedrich, 1886) (Wiesbaden, M. Sandig, 1969) quoted in Freud "On the Antithetical Meaning of Primal Words."

[4] The unitary way of looking at this duality would simply be seeing degrees of strength.

[5] See Freud, "On the Antithetical Meaning of Primal Words" (see concordance).

Rather than believe that such changes in language "just happen," I would submit that the developments in which vocables of opposite meaning were connected and eventually divided was influenced by the behavior of the changeful twin, Mars. What we have here is the division of a former unity: Mars becoming a twin reflected in the language by the conception, then division, of opposites. This would be the beginning of the concept of duality, but only a beginning. More was to come at the dissolution of the Saturn Configuration.

Freud was much struck with the work of Abel because it agreed with phenomena he had noticed in dreams, that they "show a particular preference for combining contraries into a unity or for representing them as one and the same thing." He felt that this particular correspondence illustrated "the regressive, archaic character of the expression of thoughts in dreams." Archaic it is, but since dreams always come in the service of wholeness,[6] they are probably throwing up a message not to forget an original unity. The tendency of figures in dreams to turn into their opposite is called enantiodroma. "The dream interpreters of antiquity seem to have made the most extensive use of the notion that a dream can mean its opposite."[7]

The Egyptians also clarify different stages in their thinking in their art and iconography. We find representations of Horus with the dwarf Bes figured above his head: Mars as the giant close to Earth, the dwarf when far away. But we also see a single figure with two heads: the falcon-headed Horus on the left, the dark, ass-eared Set on the right. Here the separation of good and evil has begun, the two heads prefiguring the division to come, for Set will become the leader of the forces of evil.

In retrospect, it is natural and even obvious that the concept of polarity should have first been observed at the pole; it would have been a perfect way to observe its operation. Mars moving back and forth between Earth and Saturn would have been gradually acquiring, then dissipating, an electric charge. We know that the polarity of a magnet cannot be divided. Cutting it in two produces two magnets. But now comes the confusion between polarity and duality.

The Egyptians were embarking on an adventure in consciousness, seeing the world in terms of opposites, the beginning of a dual vision. Now extending the concept of opposites to language causes us to stumble into a problem, for it is possible to separate concepts in language that will not come apart in reality. As children, we study lists of homonyms and antonyms. It seems very obvious to us, brought up in a system of dualistic thought, that "strong" and "weak" are opposite qualities. In fact, they may be seen as points along a progressive scale. "In order to facilitate languaging, the mind chooses some point along a progressive scale from which to segregate all measurements or qualities into two contrasting groups. They are then perceived dualistically as 'opposites' and become a source of conflict."[8]

Black and white duality symbol.

[6] Taylor, *Dream Work* (New York, Paulist Press, 1983) p. 138.

[7] Freud, "On the Antithetical Meaning of Primal Words."

[8] Hawkins, David R., *I, Reality and Subjectivity* (Veritas Press, 2003) p. 169.

There are advantages to seeing the world in terms of opposites. The style lends itself to rigorous examination, but it is fraught with a certain danger. Eventually one of each set of polar qualities would be chosen over the other as better or more desirable, and we would have our present dualistic vision of the world, for duality sees the world in polarized sets of opposites and chooses one over the other. It is the screen through which we have viewed the world for the last four world ages, a screen which eventually became invisible to us. It now appears to be the hidden agenda of the current series of world ages.

The problem of good versus evil is a vexing one, but we have seen how it has been handled by religion, and the results are far from satisfactory. The extremes of evil have been proscribed and outlawed through the ages while concepts of good have remained relatively static. Most of the energy has been focused on evil, making the "devil" very interesting, and God rather dull, not at all what we had in mind. Since the polarity of good/evil is intimately connected with planetary catastrophe, we will consider it here.

Following Judeo-Christian myth, the effort to separate good and evil began when Adam and Eve ate the apple and were put out of the Garden of Eden. The serpent, hanging from the wall (Venus spinning off the Saturn Configuration), was seen as evil just as Ahriman in the Bundahisn or Set of the Egyptians. At the Fall, the dissolution of the Saturn Configuration, order was dissolved into chaos as Earth and Saturn (the visible God) were cut off from each other and the northern hemisphere of Earth was plunged into darkness. That we subsequently entered an age where alternate periods of light and dark lasted long months invested that particular polarity with considerable force. Cut off from his former visible God who had loomed above the Earth, mankind was left with the strong impression that above was better than below.

Primary dualities that grew out of the Fall would have been: good/evil, order/chaos, above/below, and light/dark. But as in the Christian myth of Paradise and the Fall, all were inextricably linked together. Man "fell" into a dark, evil, chaotic condition under catastrophic circumstances. In that happening, unity and polarity were ruptured, and duality was born as man chose: above, light, order, and good over below, dark, chaos, and evil.

It was a decision that would color all subsequent ones. Polarity had become duality. Extremes were noted, and one seen to be "better" than the other. Anything that smacked of darkness, chaos, or the below was associated with evil and immediately suspect. It was a new, exhaustive way of looking at the world; we would learn much about it. But eventually we would forget the screen and make bad decisions.

It would become obvious and "true" that some plants are useful and others are not, when in fact we had simply not found a use for the weeds yet. Weeding a garden is a sensible enterprise, allowing desirable vegetables and flowers to flourish. It is when we decide that weeds are "bad" and poison them that balance is destroyed. What appears to be a solution simply raises another problem as we gradually poison the Earth and ourselves.

In another instance, it seems very obvious to choose knowledge over ignorance, forgetting that they exist in a balance. Ignorance covers material we do not know yet, like historical catastrophe. On the other hand, what we consider to be knowledge can cover incorrect assumptions, like retrocalculation into the remote past. This particular balance is complicated both by premature closure and by man's regrettable habit of hanging on to his ideas like grim death, almost as if they were a part of his identity.

The current rise of interest in holism suggests that dualistic thinking may be running out its string in Western Civilization. Examination of good and evil has been creative and thorough, going on throughout society rather than just under the aegis of religion. Femininity has certainly been seen in its extremes: on the one hand a woman so pure that she can conceive and give birth without sacrificing virginity, on the other, witches so evil that they must be burnt. The true polarity of masculine and feminine was clarified by Jung who pointed out the anima (feminine principle) in man, the animus (masculine principle) in woman. Many women now feel free to claim logic and reason, while Velikovsky is an outstanding example of an intuitive male.

In this cursory discussion of polarity and duality, we can see that the issues under discussion are interconnected. Clearing and cleansing in the service of one will help all. A particularly bright spot is the holistic movement in health care, indicating a turn toward balance and underlying unity. This will properly spread to other areas.

The good news is that man does not have a fallen nature, and by balancing reason with intuition and spirituality, may yet exercise his free will. However, our brave experiment of going it alone in the universe at a distance from immanent, transcendent power has fizzled in one respect. We are still attached to our visible gods; Saturn and other planets and stars, heroes, and saints with their Saturnian nimbus are now replaced by gurus and "stars" of film and sports. Even Christianity awaits a second coming.

It is from science that we learn of fields that influence us, of enfolded and unfolded orders of events. There is something out there connected to something in here. Whether we choose to think of it as the Einstein-Podalsky-Rosen effect, synchronicity, or the grace of God, it is not distant but intimate. In the spread of holistic thinking, the awakening of intuition and spirituality, we may eventually feel free to explore our own inner connections to the universe.[9]

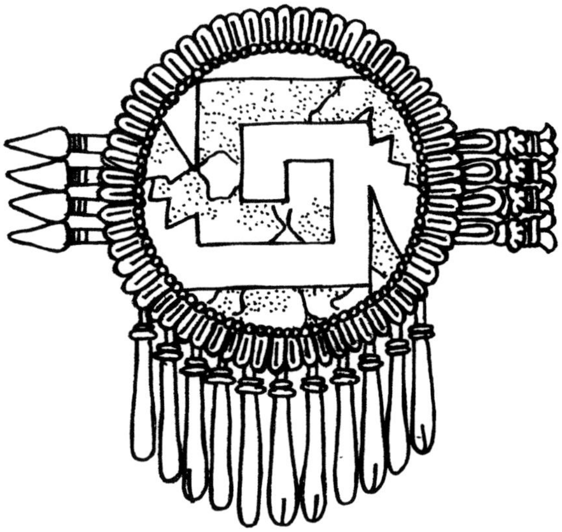

Mixtec duality symbol.

[9] See *Power vs. Force,* Davis R. Hawkins, M.D., Veritas Press, 1995, or veritaspub.com.

33. Through the Looking Glass

The fates lead him who will; him who won't, they drag.

– Seneca

Indeed this new landscape is unfamiliar, but with one more turn of the screw on the microscope, we shall again see ourselves.

– Alan Watts, *Psychotherapy, East and West*

Everything looks different on the other side.

– Lewis Carroll, *Through the Looking Glass*

It is well known among creative artists that ideas and inspiration that are not acted on, brought into reality, can slip back below consciousness. The same is true in dream work; insights must be used, acted upon, or they will be lost. Velikovsky brought up a remarkable body of cohesive knowledge in *Worlds in Collision,* but much of it has slipped away in the interim since publication, or at least that is the impression one receives from perusal of the journals devoted to his work. Interest in the Earth-Mars Wars waned early, long before space probes brought us pictures of what is agreed upon to be a planet devastated by catastrophe. Characteristically, the events are pushed into the remote past by orthodox astronomy.

Brisk tap-dancing around the unexpected heat of Venus leaves us with an unsatisfactory "greenhouse effect" to explain a planet literally covered with volcanoes. Venus, of course, will continue to cool over the centuries, but for the present it is swept under the rug. George Talbott developed a heat-transfer model to account for Venus, incandescent in 1500 B.C., having the high temperature it does today, but it has been little noticed.[1] As it happens, the cometary Venus continues to be the most common phenomenon associated with the name of Velikovsky.

Other important work has been accomplished outside the paradigm. Rose and Vaughn developed a computer model of planetary orbits in which the historical contacts between Venus, Earth, and Mars could have occurred, in which Mars was once an inner planet.[2] Velikovsky's vision of electromagnetism, broader than that allowed by orthodoxy, has been pursued by others. Hannes Alfven, the Nobel prize winner, has left us with a vision of a plasma universe at odds with orthodox theory.

That so many of Velikovsky's predictions have come to pass has kept his work alive to science in a sense, while other disciplines have been looking over his shoulder since the beginning. For long-term focus of interest in historical catastrophe, however, nothing equals the examination of mythology which has been going forward over the last few decades. Following Velikovsky, many have tried to extract visions of the past from what have been considered fanciful stories. That most of the work has been centered on Saturn and the

[1] George Robert Talbott, "The Cabots, the Lowells, and the Temperature of Venus," *Kronos* IV, 2, pp. 3–25.

[2] Lynn Rose, "Could Mars Have Been an Inner Planet?" *Pensee,* Vol. 2, No. 2, p. 42.

Saturn Configuration should not be surprising since it represents the seminal mythos of the Golden Age and is recorded plentifully in all primeval media. It is the critical Earth-Mars Wars that have slipped below the conscious attention of even historical catastrophists.

Scientific orthodoxy was not ready for *Worlds in Collision* in 1950, nor for *The Saturn Myth* in 1980. The present work makes room for both within the orderly matrix of a new vision of the forces behind the Precession of the Equinoxes. How much time must pass before these ideas are accepted is problematical, but the outcome is a foregone conclusion. Our cyclical past as recorded in ancient media is like water or love: it will find a way. Pushed down below consciousness, it will surface again and again until it is admitted.

Velikovsky's work and its continuation within these pages is not just a set of ideas difficult to integrate into Western science; it raises the issues of our cyclical past and consciousness itself. It is Western Civilization that must fit into the larger scheme as the Iron Age, a degenerate one in Greek and Indic thought. Now that we can see the break between cyclical and linear time at the Earth-Mars Wars, it becomes clear that Western Civilization reflects a style of consciousness rather than the way man really is. Mankind and his environment are both larger and more complicated than we suspected. We chose a path then and will choose again, but hopefully not before dealing with our past.

We are not really involved in a battle of ideas: uniformitarianism versus catastrophism. Both are correct and obtain in succession. That we tend to see them as juxtaposed, one preferable to the other, is just another example of our dualistic perception of the world. The ideas merely need airing. Unfortunately, in this degenerate Age of Iron, we tend to do this in a quarrelsome way with name-calling and unpleasant tone of voice. Until we can better access primeval media, which still works like a charm, we are dependent on the present linear, technological style of communication in which confrontation and humor are often necessary to move the information along.

Worlds in Collision breathes with an expectation of dialog, but it was too soon, Western man not ready. Dialog will begin when a critical mass of humanity tires of the game of "Control the World." Velikovsky's narrative hypothesis has proved to be a successfully predictive one again and again, but either Western man is ready to move outside the paradigm, or he is not. Thus far, he is not, but the event grows closer.

The notion that we are approaching the end of the current age is hardly novel; that a new one will begin is more so. There have been a host of predictions of calamity concerning the present period, from those of Edgar Cayce (which we appear to have outrun) to whatever events we may expect at the end of the current Mayan cycle at 2012, or at the appearance of the Hopi Blue Star Kachina which is believed to signal the end of the present world.

The prospect of a significant paradigm change has been treated most thoughtfully in *The Aquarian Conspiracy*,[3] conspiracy used in the Latin sense of breathing together. Oddly enough, the Silver Age began with the Spring Equinox in the sign of Aquarius, but we do not look for that event to repeat for about 150 years. In her book, Marilyn Ferguson examines the efforts of some people to bring intuitive balance to our very left-brained world. It is happening, is exactly what is needed, and we can only hope that the Earth will survive until the benefits of balanced human thinking begin to appear.

Exactly how the paradigm breaks or shifts is a mystery, but at least one group, the Institute

[3] Marilyn Ferguson, *The Aquarian Conspiracy* (Los Angeles, Tarcher, 1980).

of Noetic Science,[4] is dedicated to observe and possibly midwife the event. They use the Copernican revolution as a model for paradigm change. Before "De Revolutionibus," Earth was thought to be at the center of the universe; afterward we began to realize that the sun was the center of the solar system. Thus there occurred a changed perception of the world: not a change in the heavens as occurs at the beginning and end of a world age, but a change in the perception of them.[5]

The large issue looming over us is fitting Western Civilization into the big picture. That we have been at the top of the technological heap for some time should not be allowed to muddy the waters. This will pass in the natural order of events. We have terrible faults and wonderful virtues, all of which we already know. When we are ready to understand ourselves as the alienated and destructive but ebullient and lovable Bearson, then there are things to be done: give over our determination to control and plunder the Earth, cultivate humility vis-a-vis other cultures in both present and past, rewrite history following the lead of Velikovsky.

Other corollary issues have already been under discussion for some time, those that were slated for development during the Iron Age and must now be brought back into balance. We know them by their opposites. The patriarchal adventure has raised the issue of feminism. The rule of logic and reason forces an examination of intuition and spirituality. Our present predilection for war makes us cry for peace. Despoiling nature has turned her against us, and we cannot eat food or drink water without concern. The divine right of kings gave way to representative government, but increase in personal freedom proves illusory as we chafe under the technological apparatus of control.

These problems are interconnected. If they could have been solved by logic and reason, they would have been. Feminism, for instance, has women mistakenly reaching for the apparatus of control. But only a new world view, a transformation of consciousness, will allow us to set our world in order. The hardest nut we have to crack, that which reflects all we see on the outside, is within. It is our own muscle-bound ego, the "I" with which each of us identifies, determined that it is in control, sure that it can fix everything, nervous and fearful about allowing right-brain input, hanging on by its fingernails to keep its skewed vision in place. The ego fears dreams and intuition as science fears religion and mythology.

But there is nothing to be gained by putting down the left-brained ego, especially since its tight focus helps us to solve algebraic equations, and cross the street in heavy traffic. It must be teased and cajoled, played with, using exactly the consciousness-raising techniques which are already in place, treated in *Aquarian Conspiracy*. One's best teacher is always within. Intuition must be checked against logic and reason, our most familiar tools for dealing with reality, until they can be perceived to complement each other.

We grow justifiably angry with our scientific establishment and infrastructure. Their ongoing experiments now appear questionable in some instances. We begin to suspect them of doing harm as well as good. But again, "them" is "us," deeply into the game, charmed with the latest gadgets and toys, and anxious for the many safety nets that science provides.

Science cheerfully cuts itself off from its past at every major breakthrough as described by Kuhn, while reciting the names of Galileo, Copernicus, Newton, and Descartes as a litany

[4] IONS, 475 Gate Fives Road, Suite #300, Sausalito CA 94965, noetic.org.

[5] Copernicus was well aware that this was not a fresh idea. See Velikovsky's "Precursors" *Kronos* VII, 1, pp. 48–55.

that is somehow going to save us. The golden age of science described in the Sunday supplements recedes into the future as new degenerative diseases appear before old ones are cured. We stand on a narrow ledge of rational certainty, sorely in need of real roots in the past and a path to the future.

Velikovsky hacked out a path into the past, overgrown and difficult to see; he rewrote the history of the cosmos which has been continued within these pages. But the path leads into the future as well, a reality in which the same forces activate both the atom and the celestial realm, both microcosm and macrocosm, a reality in which both reason and intuition are brought to bear on our problems. There is no reason to fear narrative hypothesis; we can subject it to the same testing as a mathematical model, whereupon it either proves useful or it doesn't.

Orthodox knowledge in general is in an unenviable position, up against the wall of the present paradigm while anomalous material continues to pool up outside. Independent scholars rail against orthodox dogma while graduate students experience frustration at not being able to work on the real edge of knowledge. Looking toward the past, no transoceanic voyaging is allowed before Columbus. Mesoamerican studies begin to perceive the end of the Jaguar Age without being able to define it. Ancient history and archaeology are in a snarl attempting to deal with dark ages and hiatus. The history of art has been described by one scholar as having the appearance of "an improperly closed zipper."

The orthodox path leading into the future fails to reassure. Science pays lip service to electromagnetism while hanging on to gravity as if failure to believe would cause us all to fly off the planet. Gravitational theory is in trouble, forcing outrageous speculation to fill the holes in dogma. The various ad hoc solutions seem overly creative, having the flavor of "phlogiston" and "miasma" of earlier scientific paradigms.

The cracks in Earth's crust caused by historical catastrophe have become the ground for theories ranging from plate tectonics to the complementary sea floor spreading. Magnetic anomalies have produced a creative calendar for geomagnetic reversals reaching back millions of years, swallowed without gagging by the scientific establishment, parroted by the media, and spoon-fed to tuition-paying students who deserve better from their teachers. The cadre of the faithful must be ever alert to hierarchical signals so they know which way to jump. Truly we are in Cloud-cuckooland, not thinking for ourselves.

The sad truth is that in accepting our rather recent cyclic past, we must give over the bulk of our high-flown retrocalculations involving millions and billions of years, and drop back for a while to thousands. It is the penalty we must pay for stuffing catastrophe into the remote past. The grandiose visions of science prove ephemeral, because we have all learned so many things in school that aren't so. Sandbagging efforts to hold the dike in place are doomed; the only question is when the structure will give way. Activity in the orthodox infrastructure reflects critical mass in the population. At some point the balance will tip and the paradigm will shift.

The great adventure of Western Civilization is almost finished. We have gone as far as we can go and are in danger of destroying the Earth in our alienated mode. Breakthroughs in knowledge, the growing amount of phenomena that doesn't fit in, have the effect of stressing the present paradigm, and there are signs that it is ready to change.[6] Field theory in physics is

[6] The American Association for the Advancement of Science (AAAS) suffered a split in 1994 in which several hundred members left to form their own group, the Natural Philosophy Alliance.

a crack in the wall that is widening, and now we have the morphic fields of biology which immediately widen out to be interdisciplinary. Human perception is acknowledged to influence experiment.

Once we grasp the reality of the present Iron Age, the setting out on a new hero journey, the burying of Earth's history in the mystery religions, the forgetting of our past, we can better see the pattern in our history. It is Western Civilization out to dominate the Earth, but punctuated by intuitive bleed-throughs in which we remember the past: catastrophic events, past world ages, previous scientific knowledge. We remember briefly, then sink back into the action going forward, playing out the game of Control the Earth.

It is embarrassing to notice how we have struggled to free ourselves from religion, then childishly made a religion of other things, notably science. Our Bible must be reexamined as well as the myths and legends of other cultures. Different ourselves, we will relate to them differently. It is only going back and looking at "myself when young." There was no terrible mistake, no fall from grace, but a voluntary choosing of a path, man alienated from nature and seeing what he could do with it.

And now we know and can move on, picking up other previous knowledge and putting it in balance. Eating a little humble pie is a small price to pay for reconnecting with our past. Happily we can be other than we are; we do not have to follow old patterns, we can change them. But first they have to be brought into consciousness and examined.

This is a work of synthesis, reflected by the many quotations and references in the text. Reading ancient media and perceiving Earth movements that culminate in a new view of the Precession of the Equinoxes is simply another intuitive breakthrough, grown out of work on the real edge of knowledge, the historical reconstruction of Immanuel Velikovsky.

What are new ideas to us were old stuff to the ancients, but that is as it happens when we deliberately choose to forget our past. A time of coming to has already begun, waking up to the game we have been playing. We have closed the books on reality prematurely, but now they must be opened and stay open, at least until we choose a new game.

It remains to be seen if Western Civilization can move into a new age with any elegance. In the past we have been rowdy boys, prone to forget our ideal of the Christian ethic and inclined to the rough and tumble way. It is not necessary that everyone grasp the ideas underlying Plato's *Timaeus*; not all the rats must solve the maze. Others will pick up necessary information and use it to their advantage in the strange way that ideas move.

The insights herein are offered in the spirit of the ongoing human adventure. If no one is ready to move on, we will simply drop back to less enlightened games until we are finished with them. It will not be time yet as it was not time in 1950. Pulling the covers of culture over our heads is a distinct option, going back to sleep, dreaming the destructive dream onward.

But once awake, breathing the chilly air under a night sky that we know can change drastically, it is exciting to imagine seizing hold of our destiny again. We can be other than we are, see the world with new eyes, heal old wounds, and plan a new adventure. Each world age grows out of the one before, so it would behoove us to make what preparation we can. We may not need a tilt of Earth to shake us out of our world view. There is still time left, for the Hopi Blue Star Kachina has not yet appeared in the sky.

As we enter the third millenium, the Spring Equinox is in Pisces, moving closer to Aquarius. More to the point, the Summer Solstice will move into the sign of the bull Taurus, where it has not been since the Silver and Bronze Age. Thus, without ever having made the full 26,000-year precessional tour, and thanks to catastrophic interaction in our solar system, the summer solstitial colure will soon be back where it was at the time of the Fall.

Our ancestors lived in the knowledge that planetary catastrophe would recur at intervals, but we have yet to accept such information. Exactly how our past is admitted to consciousness and dealt with should prove an interesting spectacle. Those who are comfortable in the rarefied air of historical catastrophe should ask themselves as Seneca and Velikosky did, "Which part of the work belongs to us?"

THE END

Index of Subjects

Index of Names